Symbolic Logic

Symbolic Logic

Syntax, Semantics, and Proof

David W. Agler

ROWMAN & LITTLEFIELD PUBLISHERS, INC.
Lanham • Boulder • New York • Toronto • Plymouth, UK

Published by Rowman & Littlefield Publishers, Inc.
A wholly owned subsidiary of The Rowman & Littlefield Publishing Group, Inc.
4501 Forbes Boulevard, Suite 200, Lanham, Maryland 20706
www.rowman.com

10 Thornbury Road, Plymouth PL6 7PP, United Kingdom

British Library Cataloguing in Publication Information Available

Library of Congress Cataloging-in-Publication Data

Agler, David W., 1982–
  Symbolic logic : syntax, semantics, and proof / David W. Agler.
    p. cm.
  Includes bibliographical references and index.
  ISBN 978-1-4422-1741-6 (cloth : alk. paper) — ISBN 978-1-4422-1742-3 (pbk. : alk. paper) — ISBN 978-1-4422-1743-0 (electronic) 1. Logic, Modern. 2. Logic, Symbolic and mathematical.
I. Title.

 BC38.A35 2013
  511.3—dc23   2012026304

∞™ The paper used in this publication meets the minimum requirements of American National Standard for Information Sciences—Permanence of Paper for Printed Library Materials, ANSI/NISO Z39.48-1992.

Printed in the United States of America

This book is dedicated to my parents,
Thomas Agler and Monica Agler.

Contents

Extended Table of Contents

It is pretty generally admitted that logic is a normative science, that is to say, it not only lays down rules which ought to be, but need not be followed; but it is the analysis of the conditions of attainment of something of which purpose is an essential ingredient. It is, therefore, closely related to an art; from which, however, it differs markedly in that its primary interest lies in understanding those conditions, and only secondarily in aiding the accomplishment of the purpose. Its business is analysis, or, as some writers prefer to say, definition.

—Charles S. Peirce, *Collected Papers 4.575*

Acknowledgments

In writing this book, I am very grateful to have had the support of my colleagues and friends. At the Pennsylvania State University, I received comments and encouragement about the text from a variety of graduate students and faculty: Kate Miffitt, Stuart Selber, Ayesha Abdullah, Elif Yavnik, Ryan Pollock, David Selzer, Lindsey Stewart, Deniz Durmus, Elizabeth Troisi, Cameron O'Mara, and Ronke Onke. I owe special thanks to Mark Fisher and Emily Grosholz, without whom this book would not be possible.

In addition, I owe thanks to my students for catching a number of typos, for the style in which this book is written, and for their feedback on various sections. These pages began as my student notes, which transformed into lecture notes, then into handouts distributed in summer 2009, then into a course packet used in the fall of 2009; finally I used them as a textbook in the summer and fall 2010 and spring 2011. Later, a number of my colleagues used this book in the spring, summer, and fall 2011 and spring 2012 semesters. Some notable students include Isaac Bishop, Kristin Nuss, Karintha Parker, Sarah Mack, Amanda Wise, Meghan Barnett, Alexander McCormack, and Kevin Bogle. I owe special thanks to Courtney Pruitt, who provided me with a number of her solutions for exercises, and Robert Early, who pointed out a number of typos in a draft of the book.

Introduction

I.1 WHAT IS SYMBOLIC LOGIC?

What is logic? *Logic* is a science that aims to identify principles for good and bad reasoning. As such, logic is a *prescriptive* study in that one of its goals is to tell you how you ought to reason. It is not a *descriptive* science in that it does not investigate the physical and psychological mechanisms of reasoning. That is, it does not describe how you do, in fact, reason. Thus, logic lays down certain ideals for how reasoning should occur.

Symbolic logic is a branch of logic that represents how we ought to reason by using a formal language consisting of abstract symbols. These abstract symbols, their method of combination, and their meanings (interpretations) provide a precise, widely applicable, and highly efficient language in which to reason.

I.2 WHY STUDY LOGIC?

This is a good question and depends a lot upon who you are and what you want to do with your life. Here are a variety of answers to that question, not all of which may be relevant to you:

(1) *Circuit design:* Logical languages are helpful in the simplification of electrical switch/circuit design. This means that you can manipulate the formal language of logic to simplify the wiring in your home, a computer, or an automobile.
(2) *Other formal languages:* If you are interested in other formal or computer languages (e.g., HTML, Python), some components of symbolic logic overlap with these languages. This means that learning symbolic logic can supplement your existing knowledge or serve as a stepping-stone to these other languages.
(3) *Rule-following behavior:* Some of the procedures in learning symbolic logic require that you pay close attention to precise rules. Learning how to use these rules is a transferable skill. This means that if you can master the many precise rules of

symbolic logic, then you should be able to take this skill and transfer it to something else you want to do in life (e.g., law, law enforcement, journalism, etc.).

(4) *Problem-solving and analytic skills:* To some extent, this text will teach you how to argue better by giving you a method for drawing out what is entailed by certain propositions. Learning various rules of inference maps onto uncontroversial ways of arguing, and this has application for problem solving in general. If you are planning on taking an entrance exam for college, law school, graduate school, and so on, many students claim that logic offers one way, among a variety of others, to solve problems on these exams.

(5) *Study in philosophy:* Many introductory courses in logic are taught in the philosophy department, and so it is important to emphasize the importance and use of logic in philosophy. There are a variety of ways to show why logic matters to philosophy, but here are two simple ones. First, the precise syntax and semantics of logic can be used as a philosophical tool for clarifying a controversial or unclear claim, and the precise set of inference rules allows for a canonical way of arguing about certain philosophical topics (e.g., God's existence). Second, some philosophers have argued that the syntax and semantics of natural language (e.g., English) are sloppy and ought to be revised to meet the standards of precision of our logical system. This has certain implications for what we count as meaningful and rational discourse.

I.3 HOW DO I STUDY LOGIC?

In order to gain competency in logic, you will need to read the textbook, do a variety of exercises (of varying difficulty) throughout the book, and check your answers against the solutions that are provided. Learning symbolic logic is much like learning a new language, becoming skilled at a particular athletic sport, or learning to play a musical instrument: competency requires regular and engaged practice. Insofar as competency requires regular practice, you will want to read the text and do the exercises in short, spaced-out increments rather than one, long block. Just as weightlifters, tennis players, or saxophone players don't become great by practicing fifteen hours straight for one day, you won't retain much of what you learn by solving a hundred proofs in one sitting. In addition, insofar as competency requires engaged practice, it is important that you use a pencil and paper (or Word document) to actually write down your solutions to various exercises. Just listening to great singers doesn't make you a great singer. Being good at various logical operations requires hard work.

I.4 HOW IS THE BOOK STRUCTURED?

The book is structured in three main parts. Part I begins with a discussion of a number of central concepts (e.g., argument, validity) in logic as they are used in everyday English. Part II articulates a system of symbolic logic known as *propositional logic*. The formal structure, semantics, proof system, and various ways of testing for logical properties are articulated. Part III investigates a second, more powerful system of logic known as *predicate logic*. The formal structure, semantics, proof system, and testing procedures are also investigated.

As there are many ways to learn symbolic logic (or teach a symbolic logic course) without reading every chapter, three different routes are provided in table I.1.

Table I.1.   Suggested Course Design or Reading

	Complete	Shortened	Shortest
Propositional logic	1 (elements)	1 (elements)	1 (elements)
	2 (syntax, semantics)	2 (syntax, semantics)	2 (syntax, semantics)
	3 (truth tables)	3 (truth tables)	
	4 (truth trees)		
	5 (derivations)	5 (derivations)	5 (derivations)
Predicate logic	6 (syntax, semantics)	6 (syntax, semantics)	6 (syntax, semantics)
	7 (truth trees)		
	8 (derivations)	8 (derivations)	8 (derivations)

Chapter One

Propositions, Arguments, and Logical Properties

What is logic? *Logic* is a science that aims to identify principles for good and bad reasoning. As such, logic is a *prescriptive* study in that one of its goals is to tell you how you ought to reason. It is not a *descriptive* science in that it does not investigate the physical and psychological mechanisms of reasoning. That is, it does not describe how you do, in fact, reason. Two concepts central to logic are *propositions* and *arguments*. Propositions are the bearers of truth and falsity, while an argument is a series of propositions separated by those that are premises (or assumptions) and those that are conclusions.

There are three major goals for this chapter:

(1) Define the concept of a *proposition*.
(2) Define the concept of an *argument* and develop the ability to distinguish arguments from nonarguments.
(3) Introduce an informal definition of *validity*.

1.1 PROPOSITIONS

Typically, in order to say something true or false, you must utter a sentence. Thus, if Victor says, 'John is tall,' then Victor has said something true if and only if (iff) *John is tall*, and Victor has said something false if and only if *John is not tall*. Other examples include the following:

Corinne is playing guitar.
Water consists of hydrogen and oxygen.
Tomorrow the sun will rise.
Toronto is located in Canada.
Either John is guilty, or he's not.
If John is guilty, then there should be evidence at the crime scene.

Each of the above sentences is either true or false. However, not all sentences are either true or false. For example, *questions* like 'How tall are you?' cannot be true or false, *commands* like 'Close the door!' cannot be true or false, and *exclamations* like 'Oh my!' cannot be true or false. Typically, although not always, it is only *declarative* sentences that express something that can be true or false.

The branch of logic that will be considered here is concerned with sentences that express contents that can be true or false. The content expressed by sentences that can be true or false is called a *proposition*.

> Proposition   A proposition is a sentence (or something expressed by a sentence) that is capable of being true or false.

Propositions can be understood *concretely* or *abstractly*. Under a concrete interpretation, propositions are just the sentences (the physical ink or sound) that can be true or false. If you understand propositions in this way, it is a certain set of sentences that are true or false. Under an abstract interpretation, a proposition is just the abstract meaning or thing expressed by sentences. If you understand propositions in this way, it is what the sentences express that is true or false. Since it is not the purpose of this textbook to decide how best to understand propositions, this book will treat the issue indifferently.

Many of the propositions above give the impression that propositions are sentences whose content is completely articulated by the words in the sentence. However, in normal conversation, speakers often leave much of what they are trying to express unsaid or unarticulated. In these cases, the *context* in which a sentence is uttered 'fills in' or 'enriches' the sentence so that it is capable of expressing something that is either true or false. Here is an example:

(1) Yes.

In looking at (1), we don't know what 'yes' is a response to, we don't know what question it is an answer to, and without this information, we cannot really say that (1) expresses a proposition. However, given a broader context, for example, a conversation, we might say that (1) does express a proposition. For instance, consider the conversation below:

John: Did you eat lunch?

Victor: Yes.

In the above example, notice that we understand Victor to express the proposition *I ate lunch* even though he does not fully articulate this proposition by saying, 'I ate lunch.' Instead, we understand Victor's one-word utterance 'yes' to express a complete proposition, but only because a certain amount of missing information is filled in by the context. Such a proposition would be true if and only if *Victor ate lunch*, and it would be false if and only if *Victor did not eat lunch*. In sum, not every language expression will express a proposition on its own; sometimes we need to look at the context in which the sentence is uttered to determine what proposition is being expressed.

Finally, a single sentence uttered in conversation or in an argument can express different propositions depending upon whether we analyze what a sentence means or what we think a speaker means in uttering a particular sentence. That is, our focus can be upon what a sentence says (*literal meaning*) or on what a speaker means in uttering the sentence (*speaker meaning*). Consider the following example:

> Corinne: Is John a good dancer?
>
> Victor (in an ironic tone): He definitely attracts attention.

In the above example, what Victor literally says is that *John definitely attracts attention*, but what Victor means is *John is not a good dancer; he is wild, and this causes people to notice*. Literal meaning and speaker meaning are two different types of meaning. To put the difference somewhat roughly, the *literal meaning* focuses on what the individual words of the sentence mean and how these words are put together to form larger units of meaning. You can think of the literal meaning as the meaning of a sentence in isolation from the context in which it is uttered (this is somewhat inaccurate). In contrast, speaker meaning tends to require language users to know not only what the speaker says but also things about what the speaker intends to get across to his listeners. This textbook focuses on the literal meaning of sentences.

To summarize, one key element in symbolic logic is the *proposition*. A proposition is a sentence (or some abstract meaning expressed by a sentence) that is capable of being true or false. While generally it is complete sentences that express propositions, not all sentences are propositions since commands, questions, or exclamations are not capable of being true or false, and some one-word utterances, in context, express propositions. Finally, a single sentence can express multiple propositions depending upon whether or not we interpret its literal meaning or what the speaker means in uttering the sentence.

1.2 ARGUMENTS

A second key concept in logic is the notion of an *argument*. In this section, the concept of an argument is defined, and arguments are distinguished from nonarguments.

1.2.1 What Is an Argument?

In everyday speech, when someone argues, he or she is trying to prove a proposition. In logic, an *argument* is a series of propositions in which a certain proposition (a conclusion) is represented as following from a set of premises or assumptions.

> Argument   An argument is a series of propositions in which a certain proposition (a conclusion) is represented as following from a set of premises or assumptions.

To illustrate, consider the following argument:

1 If John is a crooked lawyer, then he will hide evidence.   Premise
2 John is a crooked lawyer.   Premise

3 Therefore, John will hide evidence.   Conclusion

The *premises* of an argument are propositions that are claimed to be true. In the above example, (1) and (2) are claimed to be true. The *conclusion* of an argument is the proposition or propositions claimed to follow from the premises. In the above example, (3) is claimed to follow from premises (1) and (2).

Later in the text, you will find some arguments that do not have premises. These arguments instead start from an assumption. *Assumptions* are propositions that are not claimed to be true but instead are supposed to be true for the purpose of argument. Below is an example of an argument that does not involve a premise but instead starts with an assumption (in order to indicate assumptions, they are indented).

1 | Assume that God exists.   Assumption
         | If God exists then there will be no evil in the world.
         | Thus, from above, there is no evil in the world.
         | But there is evil in the world!
5   Therefore, God does not exist.   Conclusion

The above argument is called a *reductio ad absurdum*, or a proof by contradiction. The argument begins with a proposition that is assumed to be true (an assumption) and uses the assumption to draw out a consequence that is absurd (i.e. , there is evil in the world; there is no evil in the world). Given that the assumed proposition leads to absurdity, the conclusion is that the assumption must be false. That is, God does not exist. Thus, not only can an argument begin from a set of premises and end in a conclusion but it can also begin with an assumption and terminate in a conclusion.

1.2.2 Identifying Arguments

One important practical skill is the ability to distinguish passages of text (or speech) that form an argument from ones that are not arguments. In everyday discourse, arguments tend to have two features:

(1) Arguments have *argument indicators* (words like *therefore, in conclusion*, etc.).
(2) The conclusion of an argument is claimed to follow from the premises/assumptions.

These two features are addressed in the next two subsections.

1.2.3 Argument Indicators

First, there tends to be a set of expressions that indicate arguments. These are known as *argument indicators*. Examples include *therefore, it follows that, since, hence, thus*, and so on.

Argument Indicators		
Therefore	*I infer that*	*Since*
So	*It follows that*	*Hence*
In conclusion	*For the reason that*	*Thus*
Consequently	*Inasmuch as*	*We deduce that*
It implies	*Ergo*	*We can conclude that*

To illustrate, consider the following example presented earlier:

If John is a crooked lawyer, then he will hide evidence. John is a crooked lawyer. Therefore, John will hide evidence.

In the above example, the argument indicator *therefore* indicates that *John will hide evidence* is the conclusion and follows from the other two propositions. That is,

1 If John is a crooked lawyer, then he will hide evidence.   Premise
2 John is a crooked lawyer.   Premise

3 Therefore, John will hide evidence.   Conclusion

In everyday speech, the parts of an arguments are expressed in a variety of ways. The conventional, standard, or ideal way of organizing an argument is as follows:

Standard Organization for Arguments		
First	*Second*	*Third*
(1) Premises/assumptions	Argument indicator	Conclusion

In the standard organization of arguments, the argument begins by laying out all of the premises or assumptions, then provides an argument indicator to signal that the conclusion is the next proposition in the series, and finally presents the conclusion of the argument.

Here is another example:

Ryan is either strong or weak. If Ryan is strong, then he can lift 200 lbs. Ryan can lift 200 lbs. Thus, Ryan is strong.

Notice how the argument begins with a set of propositions that are the premises, then has an argument indicator *thus*, and ends with a proposition that follows from the premises (the conclusion). That is,

1	Ryan is either strong or weak.	Premise
2	If Ryan is strong, then he can lift 200 lbs.	Premise
3	Ryan can lift 200 lbs.	Premise
4	Thus, Ryan is strong.	Conclusion

However, arguments are frequently organized in what might be thought of as nonstandard or nonideal ways. In some cases, the conclusion will occur at the beginning, followed by premises and the argument indicators. In other instances, there may be an initial premise, then an argument indicator and a conclusion, followed by a number of additional premises.

To illustrate, consider the following example of an argument where the argument indicator and the conclusion come before the premises:

We can conclude that John is the murderer. For John was at the scene of the crime with a bloody glove. And if John was at the scene of the crime with a bloody glove, then he is the murderer.

In the above example, the argument begins with the argument indicator, which is followed by the conclusion, and the following two sentences are premises.

1.2.4 Types of Arguments

A second feature of arguments is that their conclusion is claimed to follow from the premises/assumptions. The manner in which a conclusion follows from premises/ assumptions is diverse. For example, a proposition might be the best explanation for (and thus follow from) another proposition, or a proposition might lend strong support for the conclusion, or the truth of a set of propositions might guarantee the truth of another proposition. While the focus of this textbook is on the latter of these relations (more on this later), it is important to see in this section that the sense in which the conclusion of an argument follows from a set of premises/assumptions is different from the sense in which propositions are ordered in nonarguments like narratives or simple descriptions.

Consider the following example:

Imagine that you see a brown bag on a table. You reach inside and pull out a single black bean. You reach in again and pull out another black bean. You do this repeatedly until you have one hundred black beans in front of you. You might conclude that the next bean you pull out is black.

The above is a type of *inductive argument*. The premises provide some degree of support for the conclusion but certainly do not guarantee the truth of the conclusion.

We are tempted to say that because all of the beans pulled from the bag thus far have been black, the next bean will probably (but not necessarily) be black.

Consider another example:

> Imagine that there is a closed brown bag on a table and some black beans next to the bag. You are puzzled by where the beans came from and reason as follows:

(1) There is a closed bag on the table.
(2) There are some black beans next to the bag.
(3) Thus, those black beans came from that brown bag.

The above is a type of *abductive argument*. In the above case, while the truth of the premises does not guarantee the truth of the conclusion or even provide direct support for the conclusion, the conclusion would (perhaps best) explain the premises.

Finally, one last example:

> If John is a crooked lawyer, then he will hide evidence. John is a crooked lawyer. Therefore, John will hide evidence.

The above is a type of *deductively valid argument*. In the above case, the truth of the premises guarantees the truth of the conclusion. That is, it is necessarily the case that, if the premises are true, then the conclusion is true. In other words, it is logically impossible for the premises to be true and the conclusion to be false.

In all of the above examples of arguments, one proposition (the conclusion) *follows from* a set of premises/assumptions. This sense of following from differs importantly from other ways in which propositions are arranged. For example, in contrast to arguments where a proposition follows from another proposition, propositions can be ordered in a *temporal arrangement* or *narrative arrangement*, as in the case of a chronological list, a work of fiction, a witness's description of a brutal crime, or even a grocery list.

To illustrate, the following passage is not an argument:

> It was a sunny August afternoon when John walked to the store. On his way there, he saw a little, white rabbit hopping in a meadow. He smiled and wondered about the rabbit on the rest of his walk.

Not only are there no argument indicators in the above passage, but no proposition follows from any other. While we might be able to order these propositions, the ordering would be chronological or temporal. Here is another example:

> A mist is rising slowly from the fields and casting an opaque veil over everything within eyesight. Lighted up by the moon, the mist gives the impression at one moment of a calm, boundless sea, at the next of an immense white wall. The air is damp and chilly. Morning is still far off. A step from the bye-road, which runs along

the edge of the forest, a little fire is gleaming. A dead body, covered from head to foot with new white linen, is lying under a young oak-tree. (Anton Chekhov)

While in the above passage, a number of sentences express propositions, there are no argument indicators, and one sentence does not follow from the others in an inductive, abductive, or deductive sense. The above passage does not express an argument but rather provides a description of a particular scene.

To review, an *argument* is a series of propositions in which a certain proposition (a conclusion) is represented as following from a set of premises or assumptions. Arguments can be identified by (1) the presence of argument indicators, and (2) the fact that the conclusion follows from premises/assumptions in a way distinct from a temporal or narrative ordering.

Exercise Set #1

A. For each of the following sentences, determine whether it expresses a proposition:
   1. * I've failed over and over and over again in my life, and that is why I succeed. (Michael Jordan)
   2.   Be a yardstick of quality. (Steve Jobs)
   3. * Be thankful for what you have; you'll end up having more. (Oprah Winfrey)
   4.   All the religions of the world, while they may differ in other respects, unitedly proclaim that nothing lives in this world but Truth. (Mahatma Gandhi)
   5. * Decide that you want it more than you are afraid of it. (Bill Cosby)
   6.   All my life I have tried to pluck a thistle and plant a flower wherever the flower would grow in thought and mind. (Abraham Lincoln)
   7. * I have sacrificed all of my interests to those of the country. (Napoleon Bonaparte)
   8.   All religion, my friend, is simply evolved out of fraud, fear, greed, imagination, and poetry. (Edgar Allen Poe)
   9. * I have offended God and mankind because my work didn't reach the quality it should have. (Leonardo da Vinci)
   10.   You ask, What is our policy? I say it is to wage war by land, sea, and air. (Winston Churchill)
   11.   No man should escape our universities without knowing how little he knows. (Robert Oppenheimer)
   12.   How does a kid from Coos Bay, with one leg longer than the other, win races? (Steve Prefontaine)
   13.   Art is a lie that makes us realize truth. (Pablo Picasso)
   14.   How can we help President Obama? (Fidel Castro)
   15.   If you can't feed a hundred people, then feed just one. (Mother Teresa)

B. For each of the following passages, determine whether it is an argument. If the passage is not an argument, indicate what type of passage it is (e.g., a narrative, a list of statements, a description of events or facts, etc.). If the passage is an argument, indicate the presence of any argument indicators or other features that indicate that the passage is an argument (e.g., the argument is in schematic form).

1. * In recent events, it was announced that a Japanese couple was married by a robot. The robot, known as the I-Fairy, wed the happy couple on the rooftop of a restaurant in Tokyo.

2. Robots are not people. Marriage is a religious and legal arrangement between people. Therefore, robots should not be allowed to marry couples.

3. * The laws on the consumption of alcohol in this state are absurd and should be revised. Bottle shops can sell six-packs but not cases, beer distributors can sell cases but not six-packs, and no one can purchase alcohol unless they are at least twenty-one years of age, and it is before 3 a.m. In other countries, liquor laws are less strict, and they do not have the same problems with alcoholism and crimes related to drinking. Thus, the laws of this state should be revised.

4. "You are a fine artist, indeed," said Mr. Brown. "No," said Nancy, "I only paint a little and have never received any formal training." "Ha, training," Mr. Brown replied, "training is a mere trifle for someone as gifted as you."

5. * God exists because if God didn't exist, then there would be nothing at all. But since there is something, and that something had to come from somewhere, then God certainly exists.

6. There are many evil events and evil people in the world. For this reason, God does not exist. For if God did exist, and by definition God is good, then there would be no evil in the world. But since there is evil, God must not exist.

7. * There are many evil events and evil people in the world. Evil is caused by human freedom, and human freedom is a good that outweighs all of this evil. Ergo, although there is evil in the world, God does exist and is good.

8. There are many evil events and evil people in the world. However, even if some evil is caused by human freedom, and human freedom is a good, there is still evil that is not caused by human freedom (e.g., earthquakes). We can conclude that God does not exist or that God is not good.

9. * I believe in God. I attend a religious ceremony every Saturday and Sunday. I also am active in neighborhood events.

10. I don't believe in God. However, I generally spend some time involved in neighborhood events.

11. Conservatives believe that taxes should be kept low. However, taxes ought to be raised to cover important social programs. Thus, people who are conservative in their politics are wrongheaded.

12. Liberals believe that taxes should be high. Since higher taxes would hurt the economy, it follows that people who are liberal in their politics are wrongheaded.

13. I voted for a Republican in the last three elections. When I went to vote in the last election, there were many people voting for Democrats. This made me feel somewhat worried that the person I was voting for wouldn't win.

14. I voted for a Democrat in the last three elections. If I voted for a Democrat in the last three elections, then I will surely vote for one in the future. Hence, I will vote for a Democrat in the future.

15.   Yesterday, I witnessed a robbery. Today, I witnessed a murder. The police are suspicious because I was at the scene of a violent crime on two consecutive days.

Solutions to Starred Exercises from Exercise Set #1

A. For each of the following sentences, determine whether it expresses a proposition:
1. *   Yes.
3. *   The first part of the sentence is a command and so does not express a proposition, but the sentence after the semicolon does express a proposition.
5. *   No.
7. *   Yes.
9. *   Yes.

B. For each of the following passages, determine whether it is an argument:
1. *   No. It is a description of events.
3. *   Yes. The passage consists of sentences that express propositions, there is the argument indicator *thus*, and propositions are used to support a conclusion.
5. *   Yes. The passage consists of sentences that express propositions, and propositions are used to support a conclusion.
7. *   Yes. The passage consists of sentences that express propositions, propositions are used to support a conclusion, and there is an argument indicator *ergo*.
9. *   No. Although the passage consists mostly of propositions, they do not provide support for a conclusion. The passage is a list of beliefs and facts.

1.3 DEDUCTIVELY VALID ARGUMENTS

Let's take stock. An *argument* is a series of propositions in which a certain proposition (a conclusion) is represented as following from a set of premises or assumptions. There are roughly three kinds of arguments, and each has a different standard for how the premises (or assumptions) relate to the conclusion. Elementary symbolic logic is primarily concerned with the strongest standard, namely, with deductively valid arguments. The remaining part of this chapter aims to clarify the notion of a *deductively valid argument* and articulates a number of concepts and methods for identifying these types of arguments.

1.3.1 Deductive Validity Defined

In everyday discourse, *valid*, *deductively valid*, and *invalid* have a broader meaning than they do in logic. Generally, people say that an argument or proposition is valid if it is supported by evidence or convincing or if they agree with it. You hear expressions like 'That is a valid point.' However, in elementary symbolic logic, the words 'valid' and 'validity' have a narrower sense that should not be confused with how these terms are used in everyday conversation.

First, in elementary symbolic logic, validity only applies to arguments. That is, it is a property that applies to arguments and nothing else. It is inappropriate to say that propositions or sets of propositions that do not form an argument are valid. Consider the following two passages:

(1) It was a sunny August afternoon when John walked to the store. On his way there, he saw a little, white rabbit hopping in a meadow. He smiled and wondered about the rabbit on the rest of his walk.

(2) Everyone with a vivid imagination has the mind of a child. When John walked to the store, he had a number of highly unique, unusual, and colorful thoughts about a rabbit he saw. Therefore, John has the mind of a child.

Since (1) is not an argument, it is inappropriate to call it *deductively valid* or *invalid* since validity and invalidity only apply to arguments.

A second feature of deductive validity is that to classify an argument as deductively valid or invalid is to characterize a unique way in which the conclusion relates to the set of premises/assumptions. Premises or assumptions can relate to a conclusion in many ways, but to say an argument is valid is to pick a very specific kind of relation. Consider the following deductively valid argument:

1	John lives in Dallas or Philadelphia.	Premise
2	John does not live in Philadelphia.	Premise
3	Therefore, John lives in Dallas.	Conclusion

For the purpose of the above example, suppose that John does not live in Dallas or Philadelphia. In fact, suppose that John lives in San Francisco. Despite the fact that all of the propositions above are false, the conclusion still follows from the premises.

To see this more clearly, consider the following procedure:

First, assume that (1) and (2) are jointly true. That is, assume that (1) is true and that (2) is true.

Second, now that you have assumed (1) and (2) are both true, given the truth of (1) and (2), is it possible for (3) to be false? That is, if (1) and (2) are both true, can (3) be false?

The answer to the final question is no. It is necessarily the case that, if the premises are true, then the conclusion is true. When an argument is one that it is impossible for the premises to be true and the conclusion false, the argument is deductively valid.

To get a clearer hold of this definition, we turn to the definition of *deductive validity*. There are two ways of defining deductive validity. First, an argument is deductively valid if and only if, it is necessarily the case that, on the assumption the premises are true, the conclusion is true. Note that this does not mean that the premises are (in fact) true. It only means that it is necessary that if the premises are true, then it will be logically necessary that the conclusion is true. The second formulation is as follows. An argument is deductively valid if and only if it is logically impossible for the premises to be true and the conclusion to be false.

Validity   An argument is deductively valid if and only if, it is necessarily the case that if the premises are true, then the conclusion is true. That is, an argument is deductively valid if and only if it is logically impossible for its premises/assumptions to be true and its conclusion to be false.

In contrast, an argument is *deductively invalid* if and only if it is not valid. That is, an argument is deductively invalid if and only if it is possible for the premises/assumptions to be true and the conclusion to be false.

Invalidity   An argument is deductively invalid if and only if the argument is not valid.

To get a clearer understanding of deductive validity, let's focus on the second formulation. This formulation says that a deductively valid argument is one where the following two conditions are jointly impossible:

Condition 1: The premises are true.

Condition 2: The conclusion is false.

What does it mean to say that these two conditions are impossible? To say that something is logically impossible is to say that the state of affairs it proposes involves a logical contradiction.

Impossibility   Something is logically impossible if and only if the state of affairs it proposes involves a logical contradiction.

A proposition is a *logical contradiction* if and only if, no matter how the world is, no matter what the facts, the proposition is always false.

Contradiction   A proposition is a contradiction that is always false under every circumstance.

To illustrate, the notion of a logical impossibility, consider the following examples:

(1) John is 5'11, and John is not 5'11.
(2) Toronto is in Canada, and Toronto is not in Canada.
(3) Frank is the murderer, and Frank is not the murderer.

(1) to (3) are logical contradictions. No matter how we imagine the world, no matter what the circumstances are, (1) to (3) are always false. Since the state of affairs they propose involves a contradiction, each one of these is *logically impossible*. That is,

under no situation, circumstance, or way the world could be can John be two different heights, can Toronto be in Canada and not in Canada, or can Frank be the murderer and not be the murderer.

Returning now to the definition of a deductively valid argument, to say that an argument is *deductively valid* is to say that it would be impossible for the argument's premises to be true and its conclusion to be false. In other words, we would be uttering something contradictory (always false) if we were to say that a deductive argument's premises/assumptions were true and its conclusion was false.

1.3.2 Testing for Deductive Validity

The definition of deductive validity states that an argument is deductively valid if and only if it is logically impossible for its premises/assumptions to be true and its conclusion to be false. That is, an argument is deductively valid if and only if the following two conditions are jointly impossible:

Condition 1: The premises are true.

Condition 2: The conclusion is false.

Using the two conditions above, we can determine whether the argument is valid by asking the following question:

Is it logically impossible for the premises to be true (condition 1) and the conclusion to be false (condition 2)?

This is a difficult question to answer for all cases, so let's consider a more step-by-step method for determining validity. We will call this the *negative test for validity*. The negative test works as follows:

The Negative Test for Validity	
1	Is it possible for all of the premises to be true?
2	If the answer to (1) is yes, then assuming that the premises are true, is it possible for the conclusion to be false?
	If the answer to (1) is no, then the argument is valid.
Result	If the answer to (2) is yes, then the argument is not valid.
	If the answer to (2) is no, then the argument is valid.

To see how the negative test works, consider the following argument:

(1) All men are mortal.
(2) Barack Obama is a man.
(3) Therefore, Barack Obama is mortal.

In order to determine whether the above argument is valid, the first step is to look at the premises and to ask yourself whether it is possible for (1) and (2) to be true. That is, can both (1) and (2) be true at the same time? The answer is yes. Even though there might be some immortal man living on this planet, it seems that there is nothing that would make it impossible for both to be true at the same time.

The second step is to consider, given that the premises are assumed true, whether or not it is possible for the conclusion to be false. In other words, under the assumption that all men are mortal and Barack Obama is a man, is it possible for *Barack Obama is mortal* to be false? The answer is no since it is logically impossible for (1) and (2) to be true, and (3) to be false. According to the negative test, the argument is deductively valid.

Since an argument is valid if and only if it is impossible for the premises to be true and the conclusion to be false, a deductively valid argument can have any of the following:

(1) False premises and a true conclusion
(2) False premises and a false conclusion
(3) True premises and a true conclusion
(4) Some true and some false premises and a true conclusion
(5) Some true and some false premises and a false conclusion

It can never consist of

(6) true premises and a false conclusion.

Deductively Valid Arguments		
	False conclusion	*True conclusion*
False premises	Yes	Yes
True premises	No	Yes

In order to further clarify the notions of validity and invalidity, it will be helpful to look at some concrete examples using the negative test. Consider the following argument:

(1) All men are immortal.
(2) Barack Obama is a man.
(3) Therefore, Barack Obama is immortal.

Step 1 says to ask whether it is possible for all of the premises to be true. The answer is yes; even though (1) is in fact false, we can imagine that (1) and (2) are true. According to the negative test, if the answer to the first step is yes, then we need to move to step 2. Step 2 asks the following: Given that (1) and (2) are assumed true, can the conclusion be false? The answer is no; if all men are immortal and Barack Obama is a man, then it is impossible for *Barack Obama is immortal* to be false. Therefore, the above argument is valid.

Consider the next argument, which has a false premise.

(1) All men are rational.
(2) Some men are mortal.
(3) Therefore, some mortals are rational.

Is it possible for all of the premises to be true? Yes; even though (1) is in fact false, we can imagine that (1) and (2) are both true. Given that (1) and (2) are assumed true, can the conclusion be false? No; if all men are rational, and some men are mortal, then it is impossible for *Some mortals are rational* to be false. Therefore, the above argument is valid.

Consider the next argument, which has true premises and a true conclusion.

(1) Some horses are domesticated.
(2) All Clydesdales are horses.
(3) Therefore, all Clydesdales are domesticated.

Is it possible for all of the premises to be true? Yes; (1) and (2) are both true. Given that (1) and (2) are assumed true, can the conclusion be false? Yes; even if we assume that some horses are domesticated, and all Clydesdales are horses, it is logically possible for *All Clydesdales are domesticated* to be false, for we can imagine a wild Clydesdale. While all Clydesdales are in fact domesticated, nothing about the premises requires (3) to be true. Since it is possible for the premises to be true and the conclusion to be false, the argument is invalid.

Consider the next argument where there is a false conclusion.

(1) Some horses are domesticated.
(2) Some Clydesdales are horses.
(3) Therefore, all men are immortals.

Notice that in the above example, the conclusion is unrelated to the premises. Is it possible for all of the premises to be true? Yes; (1) and (2) are both true. Given that (1) and (2) are assumed true, can the conclusion be false? Yes; in fact, (3) is false. Since it is possible for the premises to be true and the conclusion to be false, the argument is invalid.

Finally, consider an example of an argument where all of the premises are false, but the conclusion is true.

(1) All humans are donkeys.
(2) James the donkey is human.
(3) James the donkey is a donkey.

Notice that (1) and (2) are both false, and (3) is true. Notice that it is necessarily the case that if (1) and (2) are assumed true, then (3) is true. That is, the above argument is valid.

Deductive arguments can either be valid or invalid, and if they are valid, they can be sound or unsound. An argument is *sound* if and only if it is both valid and all of its premises are true.

> Sound   An argument is sound if and only if it is valid and all of its premises are true.

One way of thinking about soundness is through the following formula:

$$\text{Validity} + \text{all true premises} = \text{sound argument}$$

An argument is *not sound* (or *unsound*) in either of two cases: (1) if an argument is invalid, then it is not sound; (2) if an argument is valid but has at least one false premise, then it is not sound.

> Unsound   An argument is unsound if and only if it is either invalid or at least one premise is false.

While the determination of whether an argument is valid or invalid falls within the scope of logic, the determination of the truth or falsity of the premises often falls outside logic. The reason for this is that the truth or falsity of a large number of propositions cannot be determined by their form. That is, if a premise is contingent (one whose truth or falsity depends upon the facts of the world), then its logical form does not tell us whether the premise is true or false, and while the argument may be valid, we will not be able to determine whether or not it is sound without empirical investigation.

1.4 SUMMARY

In this introductory chapter, the first goal has been to define the concept of a *proposition*. The second goal has been to separate the concept of *argument* from narrative and other discursive forms involving sentences and propositions. The third goal has been to acquire a basic understanding of the deductive validity. In a later chapter, deductive validity along with other logical properties will be defined more rigorously by using a more precise logical language. It is to this language that we now turn.

END-OF-CHAPTER EXERCISES

A. *Logical possibility of the premises.* Identify the premises of the following arguments and determine whether it is logically possible for the premises to be true.
    1. * You should only believe in something that is backed by science. Belief in God is not backed by science. Consequently, you should not believe in God.

2.   Drinking coffee or tea stains your teeth. You should never do anything to stain your teeth. Therefore, you should not drink coffee or tea.

3. * Some athletes are drug users. Some drug users are athletes. Thus, we can conclude that some people are both athletes and drug users.

4.   Democrats think we should raise taxes. Republicans think we should lower them. Hence, we should lower taxes.

5. * People can doubt many things, but you cannot doubt God exists. John is a person. John doubts that God exists. Therefore, John is a bad person.

6.   All politicians are crooks. John is a politician, but he is not a crook. Thus, everyone should vote for John.

7. * We can conclude that in order to become an elite distance runner, it is necessary that you learn to mid-foot strike when running. Elite distance runners tend to mid-foot strike when running. Amateur distance runners land on the heel of their feet when running.

8.   Frank cooked a delicious pizza. Liz ate the pizza and got food poisoning. Ergo, Frank's delicious pizza caused Liz to get food poisoning.

9.   John's fingerprints were on the murder weapon. John was not the murderer. Therefore, John is innocent of any crime.

10.   Alcohol is a dangerous substance. Marijuana is a dangerous substance. Marijuana is illegal, but alcohol is legal. For that reason, alcohol should be made illegal.

11.   Alcohol is a dangerous substance. Marijuana is a dangerous substance. Marijuana is illegal, but alcohol is legal. For that reason, marijuana should be made legal.

12.   John prayed to God for a new bike. John didn't get the bike. God exists. It follows that God was not listening to John's prayers.

13.   There is good reason to conclude that God does not like John. First, John prayed to God for a new bike; second, John didn't get the bike.

14.   Studies show that people who get one hug a day are happier, more productive people than people who get less than one hug a day. Studies show that people who get more than ten hugs a day are unhappier and less productive than people who get one and only one hug a day. For that reason, I infer that everyone should try to hug another person at least once a day.

15.   Studies show that people who get more than one hug a day are happier, more productive people than people who get less than one hug a day. Studies show that people who get more than ten hugs a day are unhappier and less productive than people who do not get at least one hug a day. Therefore, it is advisable to hug as many people as you can as much as you can.

B. *Validity or invalidity.* Using the negative test, determine whether the following arguments are valid or invalid:

1. * All men are mortal. Socrates is a man. Therefore, Socrates is a mortal.

2.   All fish are in the sea. Frank is a fish. Therefore, Frank is in the sea.

3. * Some monsters are friendly. Frank is a monster. Therefore, Frank is friendly.

4.   Some men are smokers. Some men ride bikes. Therefore, some men smoke and ride bikes.

5. * God is good. God is great. Therefore God is good and great.

6.   John is a nice person. Sarah is a nice person. Therefore, John and Sarah are nice people.

7. * John loves Sarah. Sarah loves John. Therefore, John and Sarah are married.

8.   John loves Sarah. Sarah loves John. Therefore, John and Sarah love each other.

9.   Democrats think that taxes should be raised, and Democrats are always right. Therefore, taxes should be raised.

10. Republicans think that taxes should be lowered, and Republicans are always right. Therefore, taxes should be lowered.

11. Murder is always wrong and should be illegal. Abortion is murder. Therefore, abortion is wrong and should be illegal.

12. Murder is always wrong and should be illegal. Abortion is not murder. Therefore, abortion is not wrong and should not be illegal.

13. Smoking causes cancer, which raises health-care costs. We should never do anything that raises health-care costs. We should never smoke.

14. Smoking causes cancer, which raises health-care costs. It is sometimes acceptable to do things that raise health-care costs. It is acceptable to smoke.

15. The government should not create any law that interferes with a person's basic human rights. Passing a law that makes smoking illegal interferes with a person's basic human rights. Therefore, the government should not create a law that makes smoking illegal.

C. *Conceptual questions.* Answer the following questions about valid arguments:

1. * Is it possible for a valid argument to be sound?

2.   Is it possible for a sound argument to be invalid?

3. * Is it possible for an argument with false premises to be sound?

4.   Is it possible for an argument with false premises to be valid?

5. * If an argument has two premises, and these premises cannot both be true, is the argument valid or invalid? Justify your answer.

Solutions to Starred Exercises from End-of-Chapter Exercises

A. *Logical possibility of the premises*

1. * Logically possible.

3. * Logically possible.

5. * Not logically possible.

7. * Logically possible.

B. *Validity or invalidity*

1. * Valid.

3. * Invalid. While some monsters are friendly, and Frank is a monster, it is possible that Frank is an unfriendly monster.

5. * Valid.

7. * Invalid. Two people can love each other without being married.

C. *Conceptual questions*

    1. * Yes. It is possible for a valid argument to be sound, provided the argument has true premises.

    3. * No. It is not possible for an argument with false premises to be sound because all of the premises in a sound argument must true.

DEFINITIONS

Proposition	A proposition is a sentence (or something expressed by a sentence) that is capable of being true or false.
Argument	An argument is a series of propositions in which a certain proposition (a conclusion) is represented as following from a set of premises or assumptions.
Validity	An argument is deductively valid if and only if, it is necessarily the case that on the assumption the premises are true, the conclusion is true. That is, an argument is deductively valid if and only if it is logically impossible for its premises/assumptions to be true and its conclusion to be false.
Invalidity	An argument is deductively invalid if and only if the argument is not valid.
Impossibility	Something is logically impossible if and only if the state of affairs it proposes involves a logical contradiction.
Contradiction	A proposition is a contradiction that is always false under every circumstance.
Sound	An argument is sound if and only if it is valid and all of its premises are true.
Unsound	An argument is unsound if and only if it is either invalid or at least one premise is false.

Chapter Two

Language, Syntax, and Semantics

In the Introduction to this book, *logic* was defined as a science that aims to identify principles for good and bad reasoning. *Symbolic logic* was defined as a branch of logic that represents how we ought to reason through the use of a formal language. In this chapter, you will be introduced to just such a logical language. This language is called the *language of propositional logic* (or PL for short). The major goals of this chapter are the following:

(1) Learn all of the symbols of PL.
(2) Learn the syntax (or grammar) of PL.
(3) Be able to translate from English into PL, and vice versa.
(4) Obtain a clear grasp of truth functions and truth-functional operators.

2.1 TRUTH FUNCTIONS

Many propositions in English are built up from one or more propositions. For example, consider the following three propositions:

(1) John went to the store.
(2) Liz went to the game.
(3) John went to the store, and Liz went to the game.

Notice that (3) is built up from (1) and (2) by placing *and* between them. The term *and* connects the sentence *John went to the store* to the sentence *Liz went to the game* to form the complex proposition in (3). A number of other terms are used to generate complex propositions from other propositions. For example,

(4) John went to the store, or Liz went to the game.
(5) If John went to the store, then Liz went to the game.
(6) John went to the store if and only if (iff) Liz went to the game.

Terms like *and, or, if . . ., then. . .*, and *if and only if* are called *propositional connectives*. These are terms that connect propositions to create more complex propositions.

> Propositional   A propositional connective is a term (e.g., *and, or, if . . . then . . .,*
>   connective   *. . . if and only if . . .*) that connects or joins propositions to create
>   more complex propositions.

In addition to propositional connectives, there is an important class of terms that operate on a single proposition to form a more complex proposition. For example,

(1) John went to the store.
(7) It is not the case that John went to the store.
(8) It is known that John went to the store.
(9) It is suspected that John went to the store.

Since the terms above do not connect propositions to form complex propositions, it would be misleading to call them propositional connectives. Instead, we call terms similar to the above that work (or operate) on single propositions to form more complex propositions, as well as the propositional connectives, *propositional operators.*

> Propositional   A propositional operator is a term (e.g., *and, or, it is not the case that*)
>   operator   that operates on propositions to create more complex propositions.

In this text, we focus on the truth-functional use of propositional operators. A propositional operator is used truth-functionally insofar as the truth value of the complex proposition is entirely determined by the truth values of the propositions that compose it. Propositional operators that are used truth-functionally are called *truth-functional operators.*

> Truth-functional   A truth-functional operator is a propositional operator (e.g.,
>   operator   *and, or, it is not the case that*) that is used in a truth-functional
>   way.

Before the notion of a truth-functional operator is clarified, it is helpful to distinguish between two different types of propositions. An *atomic proposition* is a proposition without any truth-functional operators.

>   Atomic   An atomic proposition is a proposition without any truth-func-
> proposition   tional operators.

Examples include the following:

 (1) John went to the store.
 (2) Liz went to the game.
(10) Vic plays soccer.

These propositions do not contain terms like *and, or,* and *it is not the case that* that work on propositions to create more complex propositions.

In contrast to atomic propositions there are *complex propositions*. These are propositions that contain at least one truth-functional operator.

> Complex   A proposition is complex when it has at least one truth-functional
> proposition   operator.

Examples include the following:

(4) John went to the store, or Liz went to the game.
(5) If John went to the store, then Liz went to the game.
(6) John went to the store if and only if Liz went to the game.
(7) It is not the case that John went to the store.

To get a clearer idea of the truth-functional use of propositional operators, it is helpful to consider a number of different types of functions. In general, a function associates a value or values (known as the *input*) with another value (known as the *output*). Consider the following function, which we will call the *lightening-color function*. This function takes a color as input and generates a lighter version of that color as output. We will define this function as follows:

Lightening-color function = df. Given a color as input, generate a lighter color as output.

Thus, the lightening-color function takes a particular color as an input (e.g., brown) and yields a lighter version of that color as output (e.g., light brown).

Input (colors)	*Lightening-Color Function*	**Output** (colors)
grey *brown*	⟶ ⟶	*lightgrey* *lightbrown*

In the case of the lightening-color function, the output color is entirely determined by the input color. Here is a slightly more complicated function that takes two colored items (red or blue) as input and produces an emotion (happy or sad) as output:

Color-emotion function = df. If the color-value input of both of the colored items is *blue*, then the output is *happy*. If the color-value input of either of the colored items is *red*, then the output is *sad*.

Input (colors)	**Input** (colors)	*Color-Emotion Function*	**Output** (colors)
blue	*blue*	⟶	*happy*
blue	*red*	⟶	*sad*
red	*blue*	⟶	*sad*
red	*red*	⟶	*sad*

In the above example, suppose that someone's emotions are determined according to the color-emotion function, and suppose that we present that person with two different pieces of clothing, a blue shirt and a blue pair of pants. The color-emotion function says that if the color-value input of both items is *blue*, then this individual will be happy. Alternatively, if we hand him or her a pair of blue pants and a red shirt, the color-emotion function says that if the color-value input of either of the items is *red*, then the individual will be sad.[1]

Let us now return to the truth-functional use of propositional operators. In order to distinguish between a propositional operator that is used truth-functionally and one that is not, it is important to define a truth-value function. A *truth-value function* is a kind of function where the truth-value output is entirely determined by the truth-value input.

> Truth   A truth function is a kind of function where the truth-value output is
> function   entirely determined by the truth-value input.

In order to illustrate, consider a type of function that takes the truth value of two propositions as input and determines a truth value for a complex proposition as output. This truth-value function is defined as follows:

Special truth function = df. If the truth-value input of both of the propositions is true, then the complex proposition is true. If the truth-value input of either of the propositions is false, then the complex proposition is false.

The *special truth function* takes the truth value of two propositions and then, using the truth-functional rule above, determines the value for a complex proposition that is composed of these propositions. The truth value of the output proposition is entirely determined by the truth values of the input propositions.

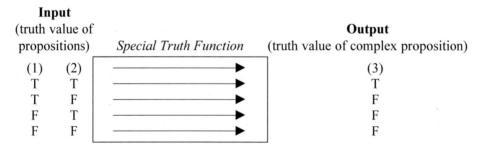

Input (truth value of propositions)		*Special Truth Function*	**Output** (truth value of complex proposition)
(1)	(2)		(3)
T	T		T
T	F		F
F	T		F
F	F		F

To illustrate, take the propositions discussed earlier:

(1) John went to the store.
(2) Liz went to the game.

Now take a complex proposition that is composed of (1) and (2) and connected by the propositional operator *and*:

(3) John went to the store, and Liz went to the game.

Let's further suppose that the propositional operator *and* is being used in a truth-functional way and that *and* corresponds to the *special truth function* expressed above. If this is the case, then the truth value of (3) should be entirely determined by the truth values of (1) and (2). To illustrate, suppose that (1) is true, and (2) is true. According to the *special truth function*, if the truth-value input of both of the propositions is true, then the complex proposition is true. Thus, (3) is also true. And it is entirely determined by the input. Alternatively, suppose that (1) is true, and (2) is false. According to the *special truth function*, if the truth-value input of either of the proposition is false, then the complex proposition is false. Thus, (3) is false. In plain English, the truth value of *John went to the store and Liz went to the game* is entirely determined by the truth value of *John went to the store*, the truth value of *Liz went to the game*, and the *special truth function* expressed by *and*.

To summarize, recall that many propositions in English are built up from one or more propositions or involve propositional operators. A *propositional operator* is a term (e.g., *and, or, it is not the case that*) that works on propositions to create more complex propositions. Some propositional operators are used in ways that correspond to truth functions. A *truth function* is a function where the truth-value output of a proposition is entirely determined by the truth-value input. Propositional operators that correspond to truth functions are called *truth-functional operators*. Complex propositions involving truth-functional operators are propositions whose truth values are determined entirely by the truth values of the propositions that compose them.

2.2 THE SYMBOLS OF PL AND TRUTH-FUNCTIONAL OPERATORS

In this section, we introduce the formal language of propositional logic (PL). We start by introducing the symbols that make up PL:

1 Uppercase Roman (unbolded) letters with or without subscripted integers ('A_1,' 'A_2,' 'A_3,' 'B,' 'C,'. . . , 'Z') for atomic propositions
2 Truth-functional operators ($\vee$, $\rightarrow$, $\leftrightarrow$, $\neg$, $\wedge$)
3 Parentheses, braces, and brackets to indicate the scope of truth-functional operators

An *atomic proposition* is a proposition that does not contain any truth-functional operators. In the language of PL, single, uppercase Roman letters are used to abbreviate atomic propositions. For example, take the following proposition:

(1) John is grumpy.

This atomic proposition can be abbreviated in the language of PL by any single, capital Roman letter of our choosing. Thus, *John is grumpy* can be abbreviated as follows:

J

The particular letter used to represent the proposition is unimportant (e.g., 'Q' or 'W' or 'A') provided that once a letter is chosen, it is used consistently and unambiguously. Thus, *John is grumpy* could have been abbreviated in PL as follows:

G

English Statement	Translation into PL
Mary is a zombie.	M
John is running from the law.	J
A crazy ape escaped from the zoo.	Z
I saw a monster yesterday.	Y
There are three elephants in my bedroom.	E

To ensure that any simple English proposition can be represented in PL, integers can be subscripted to uppercase Roman letters (e.g., A_1, A_2, A_3, A_4, B_1, B_2, B_3, B_4, Z_1, Z_2, . . ., Z_{30}).This will ensure that the symbols available in PL to represent English propositions are infinite.

As noted in the previous section, more complex propositions are built up by adding propositional operators to them. Consider the following complex proposition:

(2) John is grumpy, and Liz is happy.

To represent (2) in PL, we could abbreviate (2) as 'J.' However, the problem with abbreviating (2) as 'J' is that this would cover over the fact that (2) is a complex proposition composed of two propositions and a truth-functional operator. In abbreviating English propositions in PL, we want to represent as much of the underlying truth-functional structure as possible. In order to do this, we introduce a number of new symbols ($\land$, $\lor$, $\rightarrow$, $\leftrightarrow$, $\neg$) into PL that abbreviate various truth-functional operators (*and, or, if . . ., then . . ., . . . if and only if . . ., not*) that are found in English. Our method for explaining these truth-functional operators in PL will not be to consider all of the various propositional operators that occur in English one by one and then show how they can be abbreviated by one of our new truth-functional symbols. Instead, we will define five truth-functional operators in PL ($\lor$, $\rightarrow$, $\leftrightarrow$, $\neg$, $\land$), and then offer some general suggestions as to how they relate to the truth-functional use of propositional operators in English.

2.2.1 Conjunction

In the language of PL, where '**P**' is a proposition and '**Q**' is a proposition, a proposition of the form

P$\land$Q

is called a *conjunction*. The '$\land$' symbol is a truth-functional operator called the *caret*. Each of the two propositions that compose the conjunctions are called the proposition's *conjuncts*. The proposition to the left of the caret is called the *left conjunct*, while the proposition to the right of the caret is called the *right conjunct*. That is,

Left Conjunct   Right Conjunct

P   ∧   Q

The caret symbolizes the following truth function:

Conjunction = df. If the truth-value input of both of the propositions is true, then the complex proposition is true. If the truth-value input of either of the proposition is false, then the complex proposition is false.

In other words, a conjunction is only true in the case when both of the conjuncts are true. We can also represent this truth function in terms of truth-functional input and output as follows:

Proposition	Input		Output
	P	**Q**	**P∧Q**
Truth value	T	T	T
Truth value	T	F	F
Truth value	F	T	F
Truth value	F	F	F

The best translations into English of '**P∧Q**' are sentences that make use of a . . . *and* . . . structure, such as '**P** and **Q**.' For example, consider the following sentence:

(1) John is grumpy, and Liz is happy.

(1) can be abbreviated by letting 'G' stand for *John is grumpy*, 'L' stand for *Liz is happy*, and the caret for the truth-functional operator *and*. Thus, (1) can be abbreviated as follows:

G∧L

Determining the truth value of a complex proposition in English involving the truth-functional use of *and* is the same as determining the truth value of a complex proposition of the form '**P∧Q**,' which involves the ∧ operator. That is, when someone utters, *John is grumpy, and Liz is happy*' this proposition is true if and only if both of the conjuncts are true.

While *and* in English often behaves similarly to the caret, conjunction is also represented in English in a number of different ways. Consider a few below:

(2) Both John and Liz are happy.
(3) Although Liz is happy, John is grumpy.
(4) Liz is happy, but John is grumpy.

These can be expressed in PL as follows:

(2*) H∧L
(3*) L∧G
(4*) L∧G

2.2.2 Negation

In the language of PL, where '**P**' is a proposition, a proposition of the form

$$¬\mathbf{P}$$

is called a *negation*. The symbol for negation represents the following truth function:

Negation = df. If the truth-value input of the proposition is true, then the complex proposition involving '¬' is false. If the truth-value input of the proposition (atomic or complex) is false, then the complex proposition involving '¬' is true.

That is,

	Input	Output
Proposition	**P**	**¬P**
Truth value	T	F
Truth value	F	T

In other words, the negation function changes the truth value of the proposition it operates upon. If 'M' is true, then '¬M' is false. And if 'M' is false, then '¬M' is true. To put this in plain English, if a proposition is true, adding '¬' to it changes it to false. If the proposition is false, adding '¬' to it changes it to true.

The best translations into English of '¬P' are sentences involving the use of *not* or *it is not the case that* (e.g.,'**not-P**'). For example, consider the following propositions:

(1) Liz is not happy.
(2) It is not the case that John is grumpy.
(3) It is false that Mary is a zombie.

In PL, (1) to (3) are expressed as follows:

(1*) ¬L
(2*) ¬G
(3*) ¬Z

Again, it is important to remember that one goal in translating an English proposition into PL is to try to represent as much of the underlying structure as possible. Consider the following English proposition:

(4$_E$) John is not tall.

You might be tempted to translate (4$_E$) with a single letter. For example,

(4) J

The problem with this translation is that it does not translate the *not* in (4$_E$) with the '¬' operator. A more comprehensive translation of (4$_E$) is the following:

(4*) ¬J

The main reason for not translating (4$_E$) as (4) is that it would make the analysis of English propositions in terms of their truth functions pointless. To see this more clearly, consider the following proposition:

(5$_E$) John is not tall, and Frank is not an orphan.

One way of translating (5$_E$) is by a single letter, since (5$_E$) is a proposition. Thus,

(5) J

Again, if we were to translate (5$_E$) as (5), we would no longer be considering the underlying truth-functional nature of English, and so there would be no reason to have truth-functional operators in PL at all. Thus, one goal in translation from English to PL is to represent as much of the underlying truth-functional nature of English with the use of truth-functional operators.

Before moving to a consideration of the remaining truth-functional operators (∨, →, and ↔), it is instructive to pause and consider some features of the syntax of PL.

2.2.3 Scope Indicators and the Main Operator

Parentheses (( )), brackets ([ ]), and braces ({ }) are used for the purpose of indicating the scope of the operators. For ease of reference, these will be referred to as *scope indicators*.

Scope indicators are used to indicate the order of operations of truth-functional operators. To see this more clearly, consider that in mathematics, (2 × 3) + 3 has a different value than 2 × (3 + 3). In the first equation, we are instructed to multiply first and then add. In the second equation, the parentheses instruct us to add first and then multiply.

Let's consider an example of how scope indicators function by considering the use of parentheses and the two truth-functional operators (∧, ¬) already introduced. Consider the following propositions:

(1) ¬M
(2) ¬(M∧J)
(3) ¬M∧J

The scope of '¬' is the proposition (atomic or complex) that occurs to the right of the negation. In the absence of parentheses, '¬' simply operates on the propositional letter to its immediate right. For example, in the case of (1), '¬' applies to 'M.' In order to indicate that '¬' applies to a complex proposition, parentheses are used. For instance, in the case of (2), '¬' operates not merely on 'M' and not merely on 'J' but on the complex proposition 'M∧J' contained within the parentheses. This proposition is importantly different from one like (3), where '¬' applies not to the conjunction 'M∧J' but only to 'M.'

Thus, '¬' has more (or wider) scope in (2) than in (3) since in (2) '¬' applies to the complex proposition ('M∧J') and in (3) it only applies to the atomic proposition ('M'). Conversely, '∧' has more (or wider) scope in (3) but less scope in (2) since in (3) '∧' operates on '¬M' and 'J,' whereas in (2) it is being operated on by '¬.' Notice further that in (2), '¬' has '∧' in its scope, whereas in (3), '∧' has '¬' in its scope.

(2) ¬(M∧J)   The '¬' operates on the conjunction 'M∧J.'
(3) ¬M∧J   The '∧' operates on '¬M' and 'J.'

The operator with the greatest scope is known as the *main operator*. The main operator is the operator that has all other operators within its scope.

> Main   The main operator of a proposition is the truth-functional operator
> operator   with the widest or largest scope.

Thus, in the case of (2), since '¬' has '∧' in its scope, '¬' is the main operator. In the case of (3), since '∧' has '¬' in its scope, '∧' is the main operator. Consider a few more examples:

(4) ¬¬P
(5) ¬(P∧¬Q)
(6) ¬¬P∧¬¬Q

In the case of (4), the leftmost '¬' has the most scope since it has '¬P' within its scope. In the case of (5), the leftmost '¬' has the most scope since it has '∧' and the rightmost '¬' in its scope. In the case of (6), '∧' has the most scope since it has '¬¬P' and '¬¬Q' in its scope.

The main operator also determines how we classify a given proposition. Since the operator with the greatest scope in (1) is '¬' (negation), and it operates on '∧' (conjunction), (1) is classified as a *negated conjunction*. In the case of (2), since the operator with the greatest scope is '∧' (conjunction), and it operates on two propositions (one of the conjuncts being negated), we can say that (2) is a *conjunction with a negated conjunct*.

(2) ¬(M∧J)   negated conjunction
(3) ¬M∧J   conjunction with a negated conjunct

Finally, as noted, scope indicators indicate the order of operations of truth-functional operators. This is important because the order of operations of the truth-functional operators has an effect on the truth value of the complex proposition. In other words, just as in mathematics, where the use of scope indicators in (2 × 3) + 3 and 2 × (3 + 3) has an effect on the sum (or product) of the equation, scope indicators in '¬(M∧J)' and '¬M∧J' have an effect on the truth value of the complex proposition.
    Consider the following:

(2) ¬(M∧J)
(3) ¬M∧J

    In (2), the order of operations is first to use the truth-functional rule corresponding to '∧' to determine a truth value for 'M∧J' and then to use the truth function corresponding to '¬' to determine the truth value for '¬(M∧J).' In other words, in determining the truth value of a complex proposition, we move from the truth-functional operators with the least scope to the truth-functional operators with the most scope.
    Consider the following procedure for (2), supposing that 'M' is true and 'J' is false. Begin by writing 'T' (for true) or 'F'(for false) underneath the corresponding propositional letter:

¬(	M	∧	J)
	T		F

Next, start with the truth-functional operator with the least scope and assign a truth value to the complex proposition that results from the input of the propositions being operated upon. In short, use the truth values for 'M' and 'J' and the truth-functional rule for conjunction to determine the truth value of 'M∧J.' Once determined, write the corresponding truth value underneath '∧':

¬(	M	∧	J)
	T	F	F

Finally, move to the truth-functional operator with the next-least scope—this is '¬'—and use the truth values of the input proposition(s) to determine the truth value of the complex proposition:

¬(	M	∧	J)
T	T	F	F

Thus, given the above truth values assigned to 'M' and 'J,' '¬(M∧J)' is true.

Now consider the same process for (3). First, start by assigning truth values to the propositional letters:

¬	M	∧	J
	T		F

In this case, since '¬' has the least scope, we determine the truth value of '¬M' first:

¬	M	∧	J
F	T		F

Then we determine the truth value of the entire proposition:

¬	M	∧	J
F	T	F	F

Thus, given the above truth values, '¬M∧J' is false. This shows that the order of operations can affect the truth value of complex propositions. For when 'M' is true and 'J' is false, then (2) is true and (3) is false.

The difference in the order of operations of (2) and (3) also influences how (2) and (3) are translated into English. For the purpose of translation, let 'M' = *Mary is a zombie* and 'J' = *John is running*. Translating (2) and (3) into English gives us the following:

(2_E) It is not both the case that Mary is a zombie and John is running.
(3_E) Mary is not a zombie, and John is running.

Notice that both translations aim at capturing the scope of the operators. In (2), '¬' has wide scope and operates on complex proposition 'M∧J.' This is reflected in (2_E), where *it is not both the case that* operates on the complex sentence *Mary is a zombie and John is running*. This differs from (3) and (3_E), where '∧' has wide scope. In the case of (3), '∧' connects two conjuncts, '¬M' and 'J.' This is reflected in (3_E), where the use of *and* operates on the sentence *Mary is not a zombie* and the sentence *John is running*.

To keep the order of operations clear and to maximize readability, three different scope indicators are employed: parentheses (( )), then brackets ([ ]), and then braces ({ }).

Using Parentheses, Brackets, and Braces		
Single proposition	P	No parentheses needed
Single proposition with negation	¬P	No parentheses needed

Two propositions	P∧Q	No parentheses needed
Two propositions with negation	¬P∧Q	No parentheses needed
Two propositions with two negations	¬P∧¬Q	No parentheses needed
Compound proposition that is negated	¬(P∧Q)	Parentheses needed
Three propositions	(P∧Q)∧R	Parentheses needed
More than three propositions	(P∧Q)∧(R∧S)	Parentheses needed
	[(P∧Q)∨R]∨S	Then Brackets
	{[(P∧Q)∧R]∧S}∧M	Then Braces

Finally, only the scope of '¬' and '∧' has been discussed. What about the remaining truth-functional operators (∨, →, ↔)?

Remember that the scope of '¬' (negation) is the proposition (atomic or complex) to its immediate right, and the scope of '∧' (conjunction) includes the propositions (atomic or complex) to its immediate left and right. Truth-functional operators that apply to one and only one proposition are called *unary operators*. The truth-functional operator for negation, '¬,' is a unary operator. The remaining truth-functional operators (∧, ∨, →, ↔) are *binary operators* in that they apply to two propositions (atomic or complex). Since '∧,' '∨,' '→,' and '↔' always connect two propositions (atomic or complex), these operators are called *connectives* because they are operators that connect two propositions.

| Unary operator | ¬ |
| Binary operator (connective) | ∧, ∨, →, ↔ |

Before moving forward, consider the following complex example:

(6) {[(¬P→Q)↔R]∨S}∧M

In (6), the rightmost '∧' has the greatest scope (and so is the main operator) because it contains all other operators in its scope. That is, it operates on two propositions: '[(¬P→Q) ↔ R]∨S' and 'M.' Next, '∨' has the next most scope since it has '↔' and '→' in its scope. The '∨' operates on two propositions: '(¬P→Q)↔R' and 'S.' Next, '↔' has the next most scope. The '↔' operates on two propositions: '¬P→Q' and 'R.' Next, '→' has the next most scope. The '→' operates on two propositions: '¬P' and 'Q.' Finally, '¬' has the least scope. The '¬' operates on only one proposition: 'P.'

Proposition	Main Operator
¬P∧Q	∧
¬(P∧Q)	¬
(¬P∧Q)→W	→
(¬P∧Q)→(W∨R)	→

[(P∧Q)↔R]∨S	∨
S∨[(P∧Q)∧R]	∨
M↔W	↔
¬(¬M∧W)	¬
¬(¬R→S)∧(F→P)	∧
¬[(R→¬M)→M]	¬
¬¬Q→¬S	→
(W∧¬P)∨¬(P∧¬S)	∨

2.3 SYNTAX OF PL

Before moving on to the remaining truth-functional operators, we turn to a formulation of the syntax (or grammar) of PL. In English, there are right and wrong ways to combine sentences to generate new sentences. When an expression is not put together correctly, we say that it is 'ungrammatical.' Consider the following example:

(1) John is tall.
(2) Vic is short.
(3) John is tall, Vic is short, and

While (1) and (2) are sentences, (3) is not since it is an ungrammatical expression. The same is true in PL. The next few subsections articulate the syntax of PL. In these subsections, we articulate what it means for an expression to be a well-formed formula (grammatically correct) of PL. But before this, we need to make a crucial distinction so that we can talk about PL without confusion.

2.3.1 Metalanguage, Object Language, Use, and Mention

A distinction can be drawn between the language we are talking about and the language we are using to talk about another language. The former language is called the *object language*. The object language is the language that we are talking about. It is the language under investigation or discussion. The language that we use to talk about another language is called the *metalanguage*. So far, we have been dealing with two languages: English and PL. We have been using English (the metalanguage) in order to discuss, characterize, and investigate PL (the object language).

In addition to the object language/metalanguage distinction, we can distinguish between two different uses of a language. Typically, when we use words, we do so to characterize something about the world or our mental state. For example, in the following sentence, the word 'John' refers to an individual in the world who, it is claimed, looks tired:

(1) John is looking rather sluggish today.

In cases like (1), where we use language to express something about the world or our mental state, we will say that an expression is being *used*. That is, the proper name 'John' is being used in (1) to refer to John. In contrast to (1), when we use language not to talk about something in the world but to talk about language itself, we will say that an expression is being *mentioned*. Here is an example:

(2) 'John' has four letters in it.

In (2), the proper name 'John' is being mentioned and not used because the expression 'John' in (2) refers to the term John and not a real, living person John.

As you may have noticed from the examples above, the use/mention distinction is commonly marked by the use of single quotation marks. We make use of two different explicit methods to indicate that a term is being mentioned rather than used. First, single quotation marks indicate that a term is being mentioned, and the absence of quotation marks indicates that it is being used. For example,

'John' has four letters
'P∧Q' is a conjunction
'(P→Q)∧S' is a complex proposition

Second, sometimes we will mark the use/mention distinction simply by putting expressions on display. For example,

$$\neg P$$
$$P \land Q$$
$$(P \to Q) \land S$$

are all instances where expressions in PL are being mentioned rather than used.

2.3.2 Metavariables, Literal Negation, and the Main Operator

It is frequently the case that we want to talk about propositions in PL in a very general way. That is, we want to use the metalanguage to talk about not one and only one expression in an object language but about any expression of a particular type (e.g., all conjunctions or all negated propositions). Here is an example: Suppose that you want to say that for any proposition in PL (atomic or complex), if you put a '¬' in front of it, you will get a negation. Now let's suppose that you try to do this as follows:

If 'P' is a proposition, then '¬P' is a negation.

If this is your formulation of such a rule, you will fall miserably short of your goal. Why? Well, because your goal is to say that for *any* proposition in PL (not just the atomic proposition 'P'), if you put a '¬' in front of it, you form a negation. That is, your goal is not to express something about a particular proposition but to characterize a general feature of any proposition in PL.

In order to do achieve this end, you will need *metalinguistic variables* (or *meta-variables*). A metavariable is a variable in the metalanguage of PL (not an actual part of the language of PL) that is used to talk about expressions in PL. In other words, metavariables are variables in the metalanguage that allow us to make general statements about the object language like the one we are currently aiming at. In order to clearly demarcate metavariables from propositions that belong to PL, we will represent metavariables using bold, uppercase letters (with or without numerical subscripts) (e.g., '**P**,' '**Q**,' '**R**,' '**Z**$_1$').

Before we look at a number of examples involving the use of metalinguistic variables, at least two things should be pointed out. First, metavariables are part of the *metalanguage* of PL. This means that they will be used to talk about PL and are not part of PL itself. Second, metavariables are variables for expressions in the object language. To illustrate, consider the following mathematical expression: for any positive integer *n*, if *n* is odd, then *n* + 2 is odd. In this example, *n* does not stand for some particular positive integer but is instead a variable for positive integers. Likewise, our metalinguistic variables are variables, but they are variables for expressions in PL.

Let's consider a few examples that illustrate the fact that metavariables provide a very general means of talking about the object language. First, let's consider our earlier effort, where we wanted to say that for any proposition in PL (atomic or complex), if you put a '¬' in front of it, you get a negation.

If '**P**' is a proposition in PL, then '¬**P**' is a negation.

In the above example, note that '**P**' is metavariable. '**P**' is not part of the vocabulary of PL. Instead, it is used to refer to any proposition in PL (e.g.,'A,''¬A,''A→B,''¬A→B,'etc.). Since the above statement makes use of the metavariable '**P**,' it captures the general statement that if you place a negation in front of a proposition, you get a negation.

Consider a second example:

If '**P**' is a proposition in PL, and '**Q**' is a proposition in PL, then '**P**∧**Q**' is a proposition in PL.

In the above example, again note that '**P**' and '**Q**' are metavariables. They do not stand for some particular proposition in PL (e.g.,'P' or 'Q') and are not part of the language. Instead, they are being used to refer to any proposition in PL (e.g., 'A,' '¬A,' 'A→B,' '¬A→B,' and so on). Again, since the above statement makes use of the metavariables '**P**' and '**Q**,' it expresses conjunctions in PL.

A final example:

If '**P**' is a proposition in PL consisting of a proposition, followed by a caret, and followed by another proposition, then '**P**' is a conjunction.

The above example is another way to characterize conjunctions in PL using meta-variables. This time, instead of using two metavariables, we make use of only one.

2.3.3 The Language and Syntax of PL

Here we characterize the language of PL in a more rigorous way. First, the language consists of uppercase Roman (unbolded) letters with or without subscripted integers:

$$A_1, A_2, A_3, B, C,. . ., Z$$

Second, there are five truth-functional operators:

$$\lor, \rightarrow, \leftrightarrow, \neg, \land$$

Third, there are parentheses, braces, and brackets to indicate the scope of truth-functional operators

$$(), [], \{\}$$

Other, complex propositions are formed by combining the above three elements in a way that is determined by the grammar of PL. A syntactically correct proposition in PL is known as a *well-formed formula* (wff, pronounced 'woof'). The rules that determine the grammatical and ungrammatical ways in which the elements of PL can be combined are known as *formation rules*.

The formation rules for PL are as follows:

1   Every propositional letter (e.g., 'P,' 'Q,' 'R') is a wff.
2   If '**P**'is a wff, then '¬**P**' is a wff.
3   If '**P**' and '**R**' are wffs, then '(**P**∧**R**)' is a wff.
4   If '**P**' and '**R**' are wffs, then '(**P**∨**R**)' is a wff.
5   If '**P**' and '**R**' are wffs, then '(**P**→**R**)' is a wff.
6   If '**P**' and '**R**' are wffs, then '(**P**↔**R**)' is a wff.
7   Nothing else is a wff except what can be formed by repeated application of rules (1)–(6).

Rule (1) specifies that every uppercase Roman (unbolded) letter (with or without subscripted integers) is an atomic proposition. Rules (2) to (6) specify how complex propositions are formed from simpler propositions. Finally, rule (7) specifies that further complex propositions in PL can only be formed by repeated uses of rules (1) to (6).

The formation rules listed above provide a method for determining whether or not an expression is an expression in PL. That is, the formation rules are rules for constructing well-formed (or grammatically correct) formulas, and so, if a formula cannot be created using the rules, then that formula is not well formed (not grammatically correct). To illustrate how these rules can be used to determine whether or not a proposition is a wff, consider whether the following expression is a wff:

$$P \rightarrow \neg Q$$

Begin by showing that all of the atomic letters that compose P→¬Q are wffs. That is, by rule (1) every propositional letter (e.g., 'P,''Q,''R') is a wff; thus, it follows that 'P' and 'Q' are wffs. Next, move to more complex propositions. That is, by rule (2) if '**P**' is a wff, then '¬**P**' is a wff; thus, it follows that since 'Q' is a wff, then '¬Q' is a wff. Finally, by rule (5), if '**P**' and '**R**' are wffs, then '(**P**→**R**)' is a wff; thus, it follows that since 'P' is a wff and '¬Q' is a wff, then 'P→¬Q' is a wff. More compactly,

1	P and Q are wffs.	Rule 1
2	¬Q is a wff.	Line 1 + rule 2
3	P→¬Q is a wff.	Line 1, 2 + rule 5

Thus, using formation rules (1) to (7), it was shown that 'P→¬Q' is a wff. Next, consider a slightly more complex example. That is, show that 'P→(R∨¬M)' is a wff.

First, rule (1) states that every propositional letter (e.g., 'P,''Q,''R') is a wff. Thus, in the case of 'P→(R∨¬M),' it follows, by rule (1),that 'P,''R,' and 'M' are wffs. Next, by rule (2), if '**P**' is a wff, then '¬**P**' is a wff. Thus, since 'M' is a wff, then '¬M' is also a wff. Rule (4) states if '**P**' and '**R**' are wffs, then '(**P**∨**R**)' is a wff. Thus, in the case of 'P→(R∨¬M),' since 'R' and '¬M' are wffs, then '(R∨¬M)' is a wff. Finally, rule (5) states that if '**P**' and '**R**' are wffs, then '(**P**→**R**)' is a wff. Thus, in the case of 'P→(R∨¬M),' since 'P' and '(R∨¬M)'are wffs, then so is 'P→(R∨¬M).'

More compactly, 'P→(R∨¬M)' is a wff because

1	P, R, M are wffs.	Rule 1
2	¬M is a wff.	Line 1 + rule 2
3	R ∨¬Mis a wff.	Line 1, 2 + rule 4
4	P → (R ∨¬M) is a wff.	Line 1–3 + rule 5

Thus, using formation rules (1) to (7), 'P→(R∨¬M)' is shown to be a wff.

2.3.4 Literal Negation and the Main Operator

One useful reason for the introduction of metavariables is that it allows for specifying the *literal negation* of any proposition in PL. The literal negation of a proposition '**P**' is its corresponding negated form'¬**P**.'

Literal negation   If '**P**'is a proposition in PL, the literal negation of a proposition '**P**' is the proposition of the following form '¬**P**.'

To illustrate, consider the following propositions

(1) P
(2) ¬Q
(3) W→R
(4) W∧R
(5) P∧(R∨S)

The literal negations (or negated forms) of the above propositions are formed by placing a negation before the entire proposition. Thus,

(1*) ¬P
(2*) ¬¬Q
(3*) ¬(W→R)
(4*) ¬(W∧R)
(5*) ¬(P∧(R∨S)) or ¬[P∧(R∨S)]

Another useful reason for the introduction of metavariables is that they allow for a more succinct specification of the main operator of a proposition.

1   If '**P**' is an atomic proposition, then '**P**' has no truth-functional operators and so has no main operator.
2   If '**Q**' is a proposition, and if '**P**' is of the form '¬**Q**,' then the main operator of '**P**' is the negation that occurs before '**Q**.'
3   If '**Q**' and '**R**' are both propositions, and if '**P**' is of the form '**Q**∧**R**,' '**Q**∨**R**,' '**Q**→**R**,' or '**Q**↔**R**,' then the main operator of '**P**' is the truth-functional operator that occurs between '**Q**' and '**R**.' These are '∧,' '∨,' '→,' and '↔,' respectively.

To illustrate, take the following propositions in PL:

(1) P
(2) S∧M
(3) (P∧Q)→¬Z

In the case of (1), 'P' is an atomic proposition and so has no main operator. In the case of (2), 'S∧M' is a proposition of the form '**Q**∧**R**' where '**Q**' is the proposition 'S' and '**R**' is the proposition 'M.' Thus, the main operator is the truth-functional operator between '**Q**' and '**R**,' which is the caret. In the case of (3), '(P∧Q)→¬Z' has the form '**Q**→**R**,' where '**Q**' is the proposition '(P∧Q)' and '**R**' is the proposition '¬Z.' Thus, the main operator is the truth-functional operator between '**Q**' and '**R**,' which is the arrow.

Exercise Set #1

A. Translate the following English sentences into PL. Let 'J' = *John is tall*, 'F' = *Frank is tall*, 'L' = *Liz is happy*, and 'Z' = *Zombies are coming to get me*.
   1. * John is tall and Frank is tall.
   2.   Frank is tall.
   3. * Liz is happy.
   4.   Zombies are coming to get me and Liz is happy.
   5. * Zombies are not coming to get me and Liz is not happy.
   6.   John is tall and Liz is happy.
   7. * Liz is not happy and Frank is tall.
   8.   Liz is happy and Liz is not happy.

    9. * John is tall and zombies are coming to get me.
  10.   Liz is happy and zombies are not coming to get me.

B. Abbreviate the following sentences with uppercase letters for propositions and truth-functional operators ∧ and ¬. Capture as much of the logical form as possible.
    1. * John is not a murderer
    2.   John is not a murderer, but Frank is a murderer.
    3. * Mary is an excellent painter, and John is a fantastic juggler.
    4.   Two plus two equals four, four plus four equals eight, and eight plus eight equals sixteen.
    5.   John is not a nice guy, Frank is not a good man, and Mary is not a talented musician.
    6.   It is not both the case that John is the murderer and Mary is the murderer.
    7. * John is not the murderer and Mary is the murderer.
    8.   John is not the murderer; nor is Mary.
    9.   Two plus two does not equal four, four plus four does not equal eight, and eight plus eight does not equal sixteen.
  10. * John is a great juggler and not a great juggler.

C. State whether each of the following is a well-formed formula (wff, pronounced 'woof'). If it is wff, determine the main operator for each.
    1. * J∧(Q∨R)
    2.   J∧(Q∧¬R)
    3. * J∧¬(Q∧R)
    4.   J∨¬(¬Q∧¬R)
    5. * ¬J¬∧(R∨R)
    6.   ¬J¬∧(R∨R)
    7.   ¬J∧R∨R
    8.   ¬J∧R∨R∧R
    9.   ¬¬J∧R
  10.   (J↔R)∨(R↔R)
  11.   (J→R)∨(R↔R)
  12.   ¬[(J→¬R)∨(R↔R)]
  13.   ¬(J↔R)∧¬(R↔R)
  14.   (J↔R)∧¬(R↔R)
  15.   (J↔R)→¬(R↔R)
  16.   ¬(J↔R)∧¬¬(R↔R)
  17.   J↔[R∨(R↔R)]
  18.   ¬(J↔¬R)∨¬(¬R↔¬R)
  19. * (J↔R)∨(R↔R)
  20.   (J↔R)∨¬(¬R∧¬R)

D. Using formation rules (1) to (7), show that the following propositions are well-formed.
    1. * P
    2.   P∧Q
    3. * P∧¬Q

4.   (P∧¬Q)→R
5. * ¬(P→¬Q)
6.   ¬¬(P↔Q)
7. * ¬[P∧(Q∨R)]
8.   (P→Q)∨(P∧R)
9.   (P↔S)→[(¬R→S)∨¬T]

Solutions to Starred Exercises in Exercise Set #1

A.
    1. * J∧F
    3. * L
    5. * ¬Z∧¬L
    7. * ¬L∧F
    9. * J∧Z

B.
    1. * ¬J, where 'J' = *John is a murderer.*
    3. * M∧J, where 'M' = *Mary is an excellent painter*, and 'J' = *John is a fantastic juggler.*
    7. * ¬J∧M, where 'J' = *John is the murderer*, and 'M' = *Mary is the murderer.*
  10. * J∧¬J, where 'J' = *John is a great juggler.*

C.
    1. * Wff; the '∧' in 'J∧(Q∨R).'
    3. * Wff; the leftmost '∧' in 'J∧¬(Q∧R).'
    5. * Not a wff.
  19. * Wff; the '∨' in '(J↔R)∨(R↔R).'

D.
    1. * 'P' is a wff. Proof: by rule (1), 'P' is a wff.
    3. * 'P∧¬Q' is a wff. Proof: by rule (1), 'P' and 'Q' are wffs. By rule (2), if 'Q' is a wff, then '¬Q' is a wff. Finally, by rule (3), if 'P' and '¬Q' are wffs, then 'P∧¬Q' is a wff.
    5. * '¬(P→¬Q)' is a wff. Proof: by rule (1), 'P' and 'Q' are wffs. By rule (2), if 'Q' is a wff, then '¬Q' is a wff. By rule (5), if 'P' and '¬Q' are wffs, then 'P→¬Q' is a wff. Finally, by rule (2), if 'P→¬Q' is a wff, then '¬(P→¬Q)' is a wff.
    7. * '¬[P∧(Q∨R)]' is a wff. Proof: by rule (1), 'P,''Q,' and 'R' are wffs. By rule (4), if 'Q' and 'R' are wffs, then 'Q∨R' is a wff. By rule (3), if 'P' and 'Q∨R' are wffs, then 'P∧(Q∨R)' is a wff. Finally, by rule (2), if 'P' is a wff, then '¬[P∧(Q∨R)]' is a wff.

2.4 DISJUNCTION, CONDITIONAL, BICONDITIONAL

In this section, we return to a discussion of the remaining truth-functional operators found in PL. These are '∧,''→,' and '↔.'

2.4.1 Disjunction

In the language of PL, where '**P**' is a proposition and '**Q**' is a proposition, a proposition of the form

$$\mathbf{P \lor Q}$$

is called a *disjunction*. The '∨' symbol is a truth-functional operator called the *wedge* (or *vee*). Each of the two propositions that compose the disjunction are called the proposition's *disjuncts*. The proposition to the left of the wedge is called the *left disjunct*, while the proposition to the right of the caret is called the *right disjunct*. That is,

Left Disjunct   Right Disjunct
P   ∨   Q

The '∨' symbolizes the following truth function:

Disjunction = df. If the truth-value input of either of the propositions is true, then the complex proposition is true. If the truth-value input of both of the propositions is false, then the complex proposition is false.

In other words, a disjunction is true if either (or both) of the disjuncts are true and false only when both of the disjuncts are false. This function can be represented as follows:

	Input		Output
Proposition	**P**	**Q**	**P∨Q**
Truth value	T	T	T
Truth value	T	F	T
Truth value	F	T	T
Truth value	F	F	F

The best translations into English of '**P∨Q**' are sentences involving the inclusive use of *or* as in '**P** or **Q**.' Consider the following proposition:

(2_E) Mary is a zombie, or John is a mutant.

Provided *or* is used inclusively, we generally understand (2_E) to be true if either (or both) of the simpler sentences that compose (2_E) are true. By translating *Mary is a zombie* as 'M,' and *John is a mutant* as 'J,' and by treating the English use of *or* in terms of '∨,' we can translate (2) into the following proposition in PL:

(2) M∨J

Although complex translation is not the focus of this text, it should be noted that while '∨' is almost exclusively represented in English by the word *or*, '∨' (in conjunction with other truth-functional operators) is used to translate a number of other English expressions. For example, let 'Z' = *Mary is a zombie* and 'M' = *Mary is a mutant*. Now consider the following sentence.

(3$_E$) Mary is neither a zombie nor a mutant.

One way to translate (3) is by using '∨,' '¬,' and a scope indicator:

(3) ¬(Z∨M)

(3) says that it is not the case that 'Z' or 'M.'

Finally, it should be noted that not every instance of *or* is equivalent to the truth-functional operation represented by the wedge. In English, the word *or* can also be used exclusively to mean that one and only one of the propositions is true. Here are two examples.

(4$_E$) Michael Jordan or Kobe Bryant is the greatest basketball player ever.
(5$_E$) Either Julia Child or Jeff Smith is the greatest TV chef ever.

In (4$_E$) and (5$_E$), the connective *or* is interpreted exclusively. For (4$_E$) or (5$_E$) to be true, one or the other of the simpler sentences (but not both) has to be true. Either one or the other is the greatest, but both are not the greatest. Thus, the sentences are elliptical in that we could add *not both* to both (4$_E$) and (5$_E$). That is,

(4$_E$*) Michael Jordan or Kobe Bryant is the greatest basketball player ever, not both.
(5$_E$*) Either Julia Child or Jeff Smith is the greatest TV chef ever, not both.

The missing *not both* is implied by the fact that 'greatest' usually conveys that one and only one person or thing is the greatest. Thus, (4$_E$*) and (5$_E$*) can only be true if one and only one of the constitutive propositions is true. That is, (4$_E$*) will be false if both Michael Jordan and Kobe Bryant are the greatest, and (5$_E$*) will be false if both Julia Child and Jeff Smith are the greatest.

This use of *or* is distinct from the inclusive sense of *or*, where the complex proposition can be true even if both of the constitutive propositions are true. Consider the following disjunctions:

(6$_E$) Mary will visit John, or Mary will have lunch.
(7$_E$) Mary is a zombie, or John is a mutant.

Suppose in the case of (6$_E$) that Mary visits John and has lunch, and in the case of (7$_E$), Mary is a zombie, and John is a mutant. In these cases, nothing prompts us to add *not both* to (6$_E$) and (7$_E$). Thus, (6$_E$) and (7$_E$) are true provided either of the disjuncts are true (even if both are true).

Since 'v' is a part of PL, whenever we translate an English proposition involving *or* into PL, we need to be sure that *or* is used inclusively. If *or* is used exclusively, then we can translate the proposition with 'v,' but we will also need to append a translation to indicate that not both of the disjuncts are true. For example, reconsider (4_E*).

(4_E*) Michael Jordan or Kobe Bryant is the greatest basketball player ever, not both.

A translation of (4_E*) is (M∨K)∧¬(M∧K). This is done by translating the *or* as if it were inclusive but then adding *and not both*. That is,

(4_E*)	Michael Jordan or Kobe Bryant is the greatest	and	not both.
(4_E*)	M∨K	∧	¬(M∧K)

Another way to do the same thing would be to introduce a new truth-functional operator into the language of PL. We might introduce '⊕' as the truth-functional operator that stands for the exclusive sense of *or*. The operator could be defined in terms of the following truth function:

	Input		**Output**
Proposition	**P**	**Q**	**P ⊕ Q**
Truth value	T	T	F
Truth value	T	F	T
Truth value	F	T	T
Truth value	F	F	F

However, since there is already a way to express the exclusive use of *or* in PL by adding *and not both* to a disjunction, the '⊕' will not be included in PL.

2.4.2 Material Conditional

In the language of PL, where '**P**' is a proposition and '**Q**' is a proposition, a proposition of the form

$$\mathbf{P} \rightarrow \mathbf{Q}$$

is called a *conditional* (the *material conditional*). The '→' symbol is a truth-functional operator called the *arrow*. The proposition to the left of the arrow is called the *antecedent*, while the proposition to the right of the arrow is called the *consequent*. That is,

Antecedent   Consequent
P   →   Q

The '→' symbolizes the following truth function:[2]

Conditional = df. If the truth-value input of the proposition to the left of the '→' is true and the one to the right is false, then the complex proposition is false. For all other truth-value inputs, the complex proposition is true.

This function can be represented as follows:

	Input		Output
Proposition	**P**	**Q**	**P→Q**
Truth value	T	T	T
Truth value	T	F	F
Truth value	F	T	T
Truth value	F	F	T

The best translations into English of 'P→Q' are sentences that make use of an *if. . ., then. . .* structure, such as 'if **P**, then **Q**.'

For example,

(1_E) If John is in Toronto, then he is in Canada.
(2_E) If Mary is a zombie, then John is a mutant.
(3_E) If hell freezes over, then men are pigs.

These propositions are symbolized as follows:

(1_{PL}) T→C
(2_{PL}) M→J
(3_{PL}) H→P

In addition to *if. . ., then . . .* statements, there are a number of other ways to express conditionals.

English Proposition	Symbolic Representation
If Mary is a zombie, then John is a zombie	M→J
If Mary is not a zombie, John is not a zombie	¬M→¬J
In the case that Mary is a zombie, John is a zombie.	M→J
Mary being a zombie means that John is a zombie.	M→J
On the condition that Mary is a zombie, John is a zombie.	M→J
Only if John is a zombie, Mary is a zombie	M→J

There is question about which uses of the *if. . ., then . . .* construction in English correspond to the '→' truth function. At this point, we'll ignore the philosophical and logical debate concerning this issue and translate *if. . ., then . . .* and equivalent constructions by using '→.'[3] Later on in the text, we'll give a defense of why truth-functional uses of *if . . ., then . . .* constructions should behave like the truth function presented above, but for now, try to commit the above truth function to memory.

However, it should be noted here that there are two different ways that an *if . . ., then . . .* construction can be used in English: (1) a truth-functional way, and (2) a non-truth-functional way. Truth-functional uses of *if . . ., then . . .* are uses where the truth value of the complex proposition are determined by the truth value of the component propositions. So, *If John is in Toronto, then he is in Canada* is true depending upon the values of *John is in Toronto* and *John is in Canada*. However, we can use *if . . ., then . . .* statements in a non-truth-functional way. Perhaps the most evident example concerns *causal statements*.

Consider the following two causal statements:

(4) If John prays before his big logic exam, then he will receive an A.
(5) If John jumps up, then (assuming normal conditions) he will come down.

Assume that in the case of (5), the antecedent and consequent are true. If the causal use of *if . . ., then. . .* is truth-functional, then we have the following:

	Input			Output
Proposition	U	→	D	
Truth value	T		T	T

This is exactly what the truth-functional use of '→' tells us should happen.

However, assume that in the case of (4), the antecedent and consequent are true. John prays before his exam and receives an A. If the causal use of *if . . ., then . . .* is truth-functional, then we have the following:

	Input			Output
Proposition	J	→	S	
Truth value	T		T	T

However, what if the cause of John getting an A on the exam was not his praying but that he cheated? If this is the case, (4) is false even if the antecedent and consequent are both true. This is because (4) asserts that John's prayer caused him to get an A, and what caused John to get an A was his cheating. Thus, our input and output conditions are as follows:

	Input			Output
Proposition	J	→	S	
Truth value	T		T	F

Therefore, although many statements expressing causation take the *if . . ., then . . .* form, knowing the truth value of the components does not sufficiently determine the truth value of the complex proposition. Causal statements do not correspond to the truth-functional use of '→' since *truth-functional operators* uniquely determine the truth value of a complex proposition in virtue of the truth values of the components.

2.4.3 Material Biconditional

In the language of PL, where '**P**' is a proposition and '**Q**' is a proposition, a proposition of the form

$$\mathbf{P \leftrightarrow Q}$$

is called a *biconditional* (the *material biconditional*). The '↔' symbol is a truth-functional operator called the *double arrow*.[4] The proposition to the left of the double arrow is called the *left-hand side of the biconditional*, while the proposition to the right of the double arrow is called the *right-hand side of the biconditional*. That is,

$$\text{Left-Hand Side} \qquad \text{Right-Hand Side}$$
$$\text{P} \qquad \leftrightarrow \qquad \text{Q}$$

The truth function corresponding to '↔' is the following:

Biconditional = df. If the truth value input of the propositions are identical, then the complex proposition is true. If the truth value inputs differ, the complex proposition is false.

This truth-functional operator can be represented as follows:

	Input			Output
Proposition	**P**	↔	**Q**	
Truth value	T		T	T
Truth value	T		F	F
Truth value	F		T	F
Truth value	F		F	T

The best translations into English of '**P**↔**Q**' are sentences that make use of an . . .*if and only if . . .* structure, such as '**P** if and only if **Q**.' For example, *Mary is a zombie if and only if she was infected by the T-virus* or *John will win the election if and only if he campaigns in southern states.* By abbreviating *Mary is a zombie* as '**M**' and *Mary was infected by the T-virus* as '**T**,' and by using the symbolic representation for the double arrow, the above complex English proposition is abbreviated as follows:

English Proposition	Abbreviation
Mary is a zombie.	M
Mary was infected by the T-virus.	T
Mary is a zombie if and only if Mary was infected by the T-virus.	M↔T

While the material biconditional is traditionally represented in English by the words *if and only if*, it can be represented by other means.

English Proposition	Abbreviation
Mary is a zombie *just in the case that* John is a vampire	M↔J
Mary is a zombie *if and only if* John is a vampire	M↔J

2.5 ADVANCED TRANSLATION

In the previous sections, the emphasis has been on using the following truth-functional operators (¬, ∧, ∨, →, ↔) to translate the following English expressions (*not, and, or, if. . ., then . . .,* and *if and only if*). As it stands, your ability to translate from English into propositional logic and from propositional logic into English is limited to these expressions. This section considers a number of additional English expressions and suggests various ways to translate these into the language of propositional logic.

In this section, the following proposition types are considered:

neither **P** nor **Q**
not both **P** and **Q**
P only if **Q**
P even if **Q**
not-P unless **Q** or **P** unless **Q**

First, a 'neither **P** nor **Q**' proposition is true if and only if both '**P**'and '**Q**'are false. Thus, it can be translated as '¬**P**∧¬**Q**.' Consider the following propositions:

(1$_E$) Neither John nor Liz plays guitar.
(2$_E$) Neither Barack nor George is a good president.

(1$_E$) says *John does not play guitar* and *Liz does not play guitar*, and so (1$_E$) can be translated as follows:

(1) ¬J∧¬L

Likewise, (2$_E$) says that Barack is not a good president and George is not a good president. Thus, (2$_E$) is best translated as a conjunction where each of the conjuncts is negated:

(2) ¬B∧¬G

Second, a 'not both **P** and **Q**' proposition is true so long as '**P**'and '**Q**'are not jointly true. Thus, 'not both **P** and **Q**' is true in three different cases. First, it is true when '**P**' is true and '**Q**' is false. Second, it is true when '**Q**' is true and '**P**' is false. Third, it is true when '**P**' and '**Q**'are both false. Given that 'not both **P** and **Q**' is true in three different cases and false only when '**P**' is true and '**Q**' is true, the best way to translate 'not both **P** and **Q**' is a negated conjunction, that is,¬(**P**∧**Q**).

To illustrate, consider the following propositions:

(3$_E$) Frank did not kiss both Corinne and George.
(4$_E$) George did not eat both the hamburger and the hot dog.

Start translating (3$_E$) by isolating the two propositions that compose it. These are

(3$_{E1}$) Frank kissed Corinne.
(3$_{E2}$) Frank kissed George.

When (3$_E$) states that Frank did not kiss both Corinne and George, this means that while he may have kissed one of them, it is not the case that Frank kissed Corinne *and* George. Thus, (3$_E$) is best translated as a negated conjunction:

(3) ¬(C∧G)

Likewise, (4$_E$) receives a similar treatment. It may be the case that George ate the hamburger but not the hot dog, or George may have eaten the hot dog but not the hamburger, but (4$_E$) says that he did not eat them both.

(4) ¬(B∧D)

Third, it is tempting to translate '**P** only if **Q**' as '**Q**→**P**' since you may think that *if* signifies the antecedent like it does in 'if **P** then **Q**.' But the translation of '**P** only if **Q**' as '**Q**→**P**' should be avoided. In discussing '**P** only if **Q**,' it is helpful to consider two different explanations for why '**P** only if **Q**' should be translated as '**P**→**Q**' rather than '**Q**→**P**.' The first way involves getting clearer on the distinction between a *necessary condition* and a *sufficient condition*. '**P**' is a sufficient condition for '**Q**' when the truth of '**P**' guarantees the truth of '**Q**.' By contrast, '**P**' is a necessary condition for '**Q**' when the falsity of '**P**' guarantees the falsity of '**Q**.'

In the material conditional '**P**→**Q**,' '**P**'is a sufficient condition for '**Q**,' while '**Q**'is a necessary condition for '**P**.' To see this more clearly, consider the following argument:

If Toronto is the largest city in Canada ('P'), then Toronto is the largest city in Ontario ('Q').
Toronto is the largest city in Canada ('P').
Therefore, Toronto is largest city in Ontario ('Q').

Notice that the truth of 'P' in the above argument guarantees the truth of 'Q.' Thus, 'P' is sufficient for 'Q.' In contrast, consider the following argument:

If Toronto is the largest city in Canada ('P'), then Toronto is the largest city in Ontario ('Q').
Toronto is not the largest city in Ontario ('¬Q').
Therefore, Toronto is not the largest city in Ontario ('¬P').

Notice that the falsity of 'Q' in the above argument (as represented by the second premise) guarantees the falsity of 'P' (as represented by the conclusion). Thus, 'Q' is necessary for 'P.' Now consider that 'P only if Q' says that in order for 'P' to be true, 'Q' needs to be true. That is, 'P only if Q' says that 'Q' is a necessary condition for 'P.' Thus, 'P only if Q' should be translated as 'P→Q.'

A second way to explain why 'P only if Q' should be translated as 'P→Q' begins by considering the conditions under which 'P only if Q' is false. 'P only if Q' is false in just one case, namely, where 'P' is true and 'Q' is false. Thus, in looking for a translation of 'P only if Q,' we want to use truth-functional operators that makes 'P only if Q' true in every case, except when 'P' is true and 'Q' is false. In looking at the truth table definitions (see below), we see that 'P→Q' is false just in the case that 'P' is true and 'Q' is false, and it is true in all others. Thus, 'P only if Q' is best translated as 'P→Q.'

Truth Table Definitions for Propositional Operators							
P	**¬P**	**P**	**Q**	**P∧Q**	**P∨Q**	**P→Q**	**P↔Q**
T	F	T	T	T	T	T	T
F	T	T	F	F	T	F	F
		F	T	F	T	T	F
		F	F	F	F	T	T

To illustrate, consider the following propositions:

(5_E) Ryan will let Daniel live only if Daniel wins the lottery.
(6_E) Stock prices will go up only if people buy more stocks.
(7_E) Only if people buy more stocks will stock prices go up.

Since 'P only if Q' should be translated as 'P→Q,' (5_E) is best translated as follows:

(5) R→D

Other *only if* propositions should be translated similarly. That is, the proposition that immediately comes after the *only if* is the consequent of the condition while the other proposition is the antecedent. Thus, in the case of (6_E),

(6) U→B

Likewise, in the case of (7_E),

(7) U→B

Fourth, 'P even if Q' is true if and only if 'P' is true. That is, 'P even if Q' says 'P regardless of Q,' and so the truth (or falsity) of 'P even if Q' entirely depends upon whether 'P' is true and is independent of whether 'Q' is true or false. Given that this is the case, there are two ways to translate 'P even if Q.' First, if the goal of a translation into a formal language is merely to capture the conditions under which a proposition is true, then we can disregard 'Q' altogether and translate 'P even if Q' as simply 'P.' However, if the goal of a translation is to preserve what is expressed, then we can translate 'P even if Q' as 'P∧(Q∨¬Q)'

To illustrate, consider the following sentences:

(8_E) Corinne is a good worker even if her employer is incompetent.
(9_E) Stock prices will go up even if people buy more stocks.

(8_E) says that Corinne is a good worker regardless of whether her employer is incompetent. Since the truth or falsity of (8_E) does not depend upon whether or not her employer is incompetent, the truth or falsity of (8_E) turns entirely on whether or not Corinne is a good worker. Thus, (8_E) is best translated simply as follows:

(8) C

Likewise, (9_E) is best translated as follows:

(9) U

Fifth, and finally, one of the most difficult expressions to translate is 'P unless Q' because there are two seemingly conflicting ways to translate the expression. First, consider the following proposition:

(11_E) You will not win the lottery unless you acquire a ticket.

Before beginning, notice that (11_E) is not 'P unless Q' but 'not-P unless Q.' In thinking about the meaning of (11_E), let's consider the conditions under which (11_E) is true and false (beginning with the uppermost row).

You win the lottery.	You acquire a lottery ticket.	You will not win the lottery unless you buy a ticket.
T	T	T
T	**F**	**F**
F	T	T
F	F	T

(11_E) is true if you acquired a ticket and won the lottery. Congratulations! Second, (11_E) is false if did not acquire a ticket and you did win the lottery. Third, (11_E) is true if you acquired a ticket and did not win the lottery. (11_E) doesn't say you will win the lottery if you buy a ticket; it only says that acquiring a ticket is a precondition for winning. If buying a ticket were a sufficient condition for winning the lottery, then everyone would play! Finally, (11_E) is true if you did not acquire a ticket and did not win the lottery. Did you expect to win the lottery without acquiring a ticket? Get real!

If we let 'P' stand for *You will win the lottery* and 'Q' stand for *You will acquire a lottery ticket*, a translation of (11_E) will be a proposition that is false just in the case that 'P' is true and 'Q' is false (and true in all others). Since '¬P∨Q' is false only when 'P' is true and 'Q' is false, '¬P∨Q' is a translation of (11_E). Thus, we can translate propositions like '**not-P** unless **Q**' as '¬P∨Q.' Some further examples include the following:

> You will not graduate unless you complete your coursework.
> You are not happy unless you smile.
> You will not live unless you drink this antidote.

Other cases of '**P** unless **Q**' seem to say something stronger. Consider the following proposition:

(12_E) John is at the party unless Liz called him.

In thinking about the meaning of (12_E), let's consider the conditions under which (12_E) is true and false.

John is at the party.	Liz called John.	John is at the party unless Liz called him.
T	T	?
T	F	T
F	T	T
F	F	F

Skipping row 1 and beginning with row 2, (12_E) is true if John is at the party and Liz did not call him. (12_E) says that John is pretty much assured to be at the party, and the only thing that is going to stop him from being there is Liz's call. Moving to row 3, (12_E) is true if John is not at the party and Liz did call him. Again, (12_E) says that the only thing that is going to keep John from being at the party is Liz's call, and so, if Liz called him, and he is not at the party. (12_E) makes good on what it says. Moving to row 4, (12_E) is false if John is not at the party and Liz did not call. Part of what (12_E) says is that if Liz does not call, then John will be there. So, in the case that Liz did not call and John is not at the party, (12_E) is false.

Thus far, our analysis of (12_E) does not differ too much from our analysis of (11_E). However, what is problematic about (12_E), and cases of '**P** unless **Q**' in general, is how to treat row 1. In one reading of (12_E), (12_E) is true if John is at the party and Liz called John (she may have called John to tell him to have a great time at the party). I find a variety of readings of this sort to be unnatural and to depend upon ambiguous (and sometimes unarticulated) aspects of the sentence. For example, consider the following:

I will take the job unless I get another offer.

Suppose I take the job, although I did get another offer, but the offer was not as good. Thus, '**P** unless **Q**' is true at row 1 (i.e., when '**P**'is true and '**Q**' is true). But the rationale for this reading is built upon ambiguity, for above sentence really says,

I will take the job unless I get another [better] offer.

In that case, '**P** unless **Q**' is false at row 1 (i.e., when '**P**' is true and '**Q**' is true). Therefore, I think a more natural reading of (12_E) is that (12_E) is false if Liz called John and John is at the party.

John is at the party.	Liz called John.	John is at the party unless Liz called him.
T	T	F
T	F	T
F	T	T
F	F	F

If we let '**P**' stand for *John is at the party* and '**Q**' stand for *Liz called John*, a translation of (12_E) will be a proposition that is true in just two cases: (1) where '**P**' is true and '**Q**' is false, and (2) where '**P**' is false and '**Q**' is true. Since '$\neg(P\leftrightarrow Q)$' is true just in these cases, '$\neg(P\leftrightarrow Q)$' is a translation of (12_E). Thus, we can translate propositions like '**P** unless **Q**' using the exclusive disjunction '**P**⊕**Q**,' which is equivalent to '$(P\lor Q)\land\neg(P\land Q)$' or '$\neg(P\leftrightarrow Q)$.' Some further examples include the following:

John won the race unless Liz beat him.
Liz will be a great dancer unless she is an automobile accident.
Vic will be a successful mathematician unless he decides to be an artist.

Knowing when to translate '**P** unless **Q**' as '**P**∨**Q**' as opposed to '$\neg(P\leftrightarrow Q)$' is a somewhat complicated affair. We suggest that as a rule of thumb, '**P** unless **Q**' is translated as '$\neg(P\leftrightarrow Q)$' while '**not-P** unless **Q**' is best translated as '**P**∨**Q**,' but it is best to proceed in a step-by-step fashion like the method used above.

English Sentence	Translation into PL
neither **P** nor **Q**	¬P∧¬Q
not both **P** and **Q**	¬(P∧Q)
P only if **Q**	P→Q
P even if **Q**	**P** or P∧(Q∨¬Q)
not P unless **Q**	P∨Q
P unless **Q**	P∨Q∧¬(P∧Q)or ¬(P↔Q)

END-OF-CHAPTER EXERCISES

A. Identify the main operator of each of the following:
   1. * M∧¬Q
   2. ¬¬M∧¬¬Q
   3. * ¬(M∧¬Q)
   4. ¬M∨¬Q
   5. * ¬(¬M∨¬Q)
   6. ¬M→¬Q
   7. (¬M→¬Q)∧¬(M∧¬Q)
   8. ¬[(P→Q)→(¬R→¬M)]
   9. * ¬{¬[¬(R ∨ S) → P]∨¬[W → (S ∨ Q)]}

B. Determine whether the following sets of propositions contain a proposition and its literal negation. If the set does, identify which propositions these are.
   1. * A, B, C, ¬A
   2. A, B, ¬C, A
   3. * A∧B, ¬A∧B
   4. A↔B, A↔¬B, D∧B, ¬(A↔B)
   5. * A→B, ¬A∨B, ¬A→B, ¬A→¬B
   6. R, T, R∨¬Z, ¬R∨P
   7. * A∧B, ¬A∧¬B, ¬(A∧B)
   8. A, ¬¬A
   9. * P→(Q∧R), ¬P→(Q∧R)
   10. P→(Q∧R), ¬[P→(Q∧R)]

C. Write out the literal negation of the following propositions.
   1. * P
   2. Q∧R
   3. * P→Q
   4. Q↔S
   5. * P∧¬M
   6. P→(L∧T)
   7. * P∧(R↔S)

    8.   (Z∨M)∧¬(Z↔P)

    9. * (L→M)∧¬(¬Z↔¬P)

   10.   ¬Q

D. *Basic translation.* Translate the following English expressions into a symbolic propositional logic expression. Make sure to capture as much of these expressions as possible with the propositional operators.

    1. * John is robbing the store, and Mary is in the getaway car.

    2.   John is not a happy man, and Mary is a happy women.

    3. * John will go to the store, or he will buy a new car, or he will run from the law.

    4.   If John is a zombie, then Mary should run, or Mary should fight.

    5.   If John is not a zombie, then Mary should run or fight.

    6.   If John is hungry or a zombie, then Mary should flee.

    7. * If Mary left the store two hours ago, and John left one hour ago, and Frank leaves now, then John will not arrive at the store before Mary.

    8.   Frank is hungry if and only if John stole his sandwich.

E. Translate the following symbolic propositional logic expressions into English. Use the following: 'J' = *John is a zombie*, 'M' = *Mary is a mobster*, 'F' = *Frank is a fireman*. If you are having difficulty, first identify and translate the main operator, then translate the component sentences, and finally make sure to pay attention to the scope of negation.

    1. * J→M

    2.   ¬J→F

    3. * F→¬J

    4.   (F∨J)→¬M

    5.   ¬(F∨J)→¬J

    6.   J↔¬M

    7.   [(J∧M)∧F]→(¬M∧¬J)

    8.   (M↔J)→(J∨¬F)

    9. * M∨(F∨¬J)

   10.   F∨(M∨¬J)

F. For each proposition, identify the number of different ways that they can be represented using metalinguistic variables.

    1. * A↔B

    2.   (A→B)↔C

    3. * ¬(T→R)

    4.   S∧R

    5. * (P↔Q)→W

    6.   A∨¬B

    7. * ¬¬P

    8.   ¬¬P∧Q

    9. * ¬(W→¬R)

   10.   ¬(¬P↔¬E)→¬Q

G. *Advanced translation.* Translate the following English expressions into symbolic propositional logic expressions. Make sure to capture as much of these expressions as possible with the propositional operators.

    1. * Even if you're on the right track, you'll get run over if you just sit there. (Will Rogers)

    2.   Human progress is neither automatic nor inevitable. (Martin Luther King Jr.)

    3. * Marriage is neither heaven nor hell; it is simply purgatory. (Abraham Lincoln)

    4.   The moral virtues, then, are produced in us neither by nature nor against nature. (Aristotle)

    5. * Happiness is neither virtue nor pleasure. (William Butler Yeats)

    6.   Truth stands, even if there be no public support. (Mahatma Gandhi)

    7. * Every author in some way portrays himself in his works, even if it be against his will. (Goethe)

    8.   To be wronged is nothing unless you continue to remember it. (Confucius)

    9. * America is never wholly herself unless she is engaged in high moral principle. (George H. W. Bush)

  10.   Man is not free unless government is limited. (Ronald Reagan)

  11.   All murderers are punished unless they kill in large numbers and to the sound of trumpets. (Voltaire)

  12.   Art is permitted to survive only if it renounces the right to be different and integrates itself into the omnipotent realm of the profane.

  13.   We will be remembered only if we give to our younger generation a prosperous and safe India (Abdul Kalam)

  14.   People will generally accept facts as truth only if the facts agree with what they already believe. (Andy Rooney)

  15.   You can be a famous poisoner or a successful poisoner, but not both. (Clive Anderson)

H. *Conceptual questions*

    1.   For the sake of simplicity, every proposition is assumed to be either true or false (the principle of bivalence). Can you think of any propositions that violate this principle? What sort of truth value, if any, would you assign to them if they are declarative sentences that assert something is the case?

Solutions to Starred Exercises in End-of-Chapter Exercises

A.

    1. * ∧.
    3. * The leftmost ¬.
    5. * The leftmost ¬.
    9. * The leftmost ¬.

B.

    1. * Yes; A, ¬A.
    3. * No.

5. * No.
7. * Yes; A∧B, ¬(A∧B).
9. * No.

C.

1. * ¬P.
3. * ¬(P→Q).
5. * ¬(P∧¬M).
7. * ¬[P∧(R↔S)].
9. * ¬[(L→M)∧¬(¬Z↔¬P)].

D.

1. * J∧M, where 'J' = *John is robbing the store*, and 'M' = *Mary is in the getaway car.*
3. * (S∨B)∨L or S∨(B∨L), where 'S' = *John will go to the store*, 'B' = *John will buy a new car*, and 'L' = *John will run from the law.*
7. * [(M∧J)∧F]→¬A, where 'M' = *Mary left the store two hours ago*, 'J' = *John left one hour ago*, 'F' = *Frank leaves now*, and 'A' = *John will arrive at the store before Mary.*

E.

1. * J→M; if John is a zombie, then Mary is a mobster.
3. * F→¬J; if Frank is a fireman, then John is not a zombie.
9. * M∨(F∨¬J); Mary is a mobster, or Frank is a fireman, or John is *not* a zombie.

F.

1. * **P,P↔Q.**
3. * **P, ¬P,¬(P→Q).**
5. * **P, P→Q, (P↔Q)→W.**
7. * **P, ¬P, ¬¬P.**
9. * **P, ¬P, ¬(P→Q), ¬(P→¬Q).**

G.

1. * S∧(T∨¬T).
   Key: 'T' = *You are on the right track*; 'S' = *You will get run over if you just sit there.*
3. * (¬N∧¬L)∧P
   Key: 'N' = *Marriage is heaven*; 'L' = *Marriage is hell*; 'P' = *Marriage is purgatory.*
5. * ¬V∧¬P
   Key: 'V' = *Happiness is a virtue*; 'P' = *Happiness is a pleasure.*
7. * A∧(W∨¬W)
   Key: 'A' = *Every author in some way portrays himself in his works*; 'W' = *Portraying oneself in one's work is against one's will.*
9. * ¬A∨M
   Key: 'A' = *America is wholly herself*; 'M' = *America is engaged in high moral principle.*

DEFINITIONS

Propositional connective	A propositional connective is a term (e.g., *and, or, if . . ., then . . ., . . .if and only if. . .*) that connects or joins propositions to create more complex propositions.
Propositional operator	A propositional operator is a term (e.g., *and, or, it is not the case that*) that operates on propositions to create more complex propositions.
Truth-functional operator	A truth-functional operator is a propositional operator (e.g., *and, or, it is not the case that*) that is used in a truth-functional way.
Atomic proposition	An atomic proposition is a proposition without any truth-functional operators.
Complex proposition	A complex proposition is a proposition that has at least one truth-functional operator.
Truth function	A truth function is a kind of function where the truth-value output is entirely determined by the truth-value input.
Main operator	The main operator of a proposition is the truth-functional operator with the widest or largest scope.

GUIDE TO TRANSLATION FROM ENGLISH TO PL

English Sentence	Translation into PL
P and **Q**	**P**∧**Q**
not **P**	¬**P**
P or **Q**	**P**∨**Q**
if **P**, then **Q**	**P**→**Q**
P if and only if **Q**	**P**↔**Q**
neither **P** nor **Q**	¬**P**∧¬**Q**
not both **P** and **Q**	¬(**P**∧**Q**)
P only if **Q**	**P**→**Q**
P even if **Q**	**P** or **P**∧(**Q**∨¬**Q**)
not P unless **Q**	**P**∨**Q**
P unless **Q**	(**P**∨**Q**)∧¬(**P**∧**Q**) or ¬(**P**↔**Q**)

NOTES

1. Functions are abundant in mathematics. They typically associate a quantity (the input) with another quantity (the output). For example, $f(x) = 2x$ is a function that associates with any positive integer (the input) an integer twice as large (the output).

2. The truth function represented by '→' is sometimes represented as ⊃, also known as the *horseshoe*.

3. If you are interested, see J. Bennett, *A Philosophical Guide to Conditionals* (Oxford: Clarendon Press, 2003); D. Sanford, *If P, Then Q: Conditionals and the Foundations of Reasoning* (New York: Routledge, 1989); J. Etchemendy, *The Concept of Logical Consequence* (Cambridge, MA: Harvard University Press, 1990).

4. Symbolically, the material biconditional is sometimes represented by '≡,' also known as the *tribar*.

Chapter Three

Truth Tables

Thus far, we have articulated the symbols and syntax of PL. The primary goal of this chapter is to explain more fully the semantics of PL by (1) revisiting the notion of a valuation (truth-value assignment), (2) articulating a mechanical method that shows how the truth value of complex proposition '**P**' in PL is determined by the truth value of the propositions that make up '**P**,' and (3) using this method to determine whether certain logical properties belong to propositions, sets of propositions, and arguments. This mechanical method will give us a determinate yes or no answer as to whether a proposition is contingent, contradictory, or tautological; as to whether sets of propositions are consistent or inconsistent; as to whether a pair of propositions are equivalent or nonequivalent; and as to whether arguments are valid or invalid. Such a method is known as a *decision procedure*.

3.1 VALUATIONS (TRUTH-VALUE ASSIGNMENTS)

The key semantic concept in PL is a *valuation* (or *truth-value assignment*). A valuation in PL is an assignment of a truth value ('T' or 'F') to a proposition.

> Valuation   A valuation in PL is an assignment of a truth value ('T' or 'F') to a proposition.

There are two things to note about the above definition. First, we stipulate that a valuation can only assign a value of 'T' or 'F' to a proposition. This is an idealization since it is an open question whether or not there are additional truth values (e.g., 'I' for indeterminate). Second, previously when we wanted to say that a particular proposition has a certain truth value, we expressed this as follows:

'A' is true.

From now on, we will make use of a notational abbreviation to represent this same fact. That is, we will use an italicized lowercase letter '*v*' in order to represent that 'A' is true:

$$v(A) = T$$

The above says that 'A' is assigned the truth value of 'T.'

A key feature of the *syntax* of PL is that every proposition in PL can be generated using formation rules (see 2.3.3.). In addition, the truth value of any complex proposition in PL is determined by the truth value of the propositional letters that make it up and the use of the following truth table definitions:

P	¬P		P	R	P∧R	P∨R	P→R	P↔R
T	F		T	T	T	T	T	T
F	T		T	F	F	T	F	F
			F	T	F	T	T	F
			F	F	F	F	T	T

Truth Table Definitions for Propositional Operators

Using the truth table definitions, we can determine how the truth value of a complex proposition is determined by the truth values of the atomic propositions that make it up. For example, if 'A' is true, and 'B' is false, then using the above truth table definition, 'A∧B' is false. We will consider how this works in two steps.

The first step is to see that the truth value of a complex well-formed formula (wff, pronounced 'woof') can be determined, provided the truth values are assigned to the atomic propositions composing the complex wffs. For example, consider the following proposition:

(1$_E$) Mary is a zombie, and John is not a mutant.

Translate *Mary is a zombie* as 'Z' and *John is a mutant* as 'J.' Next, insert the appropriate symbolic operators to reflect the truth-functional syntax of English. In the above example, (1$_E$) can be translated as follows:

(1$_{PL}$) Z∧¬J

Since 'Z' and 'J' are both propositions, they have a fixed truth value. Assume that $v(Z) = T$, $v(J) = F$. Using the truth values of the atomic propositions and the truth-functional definitions, the truth value of the complex proposition 'Z∧¬J' can be determined. This is done in two steps.

(1) Write the appropriate truth value underneath each proposition.

(2) Starting with the truth-functional operator with the least scope and proceeding to the truth-functional operator with the most scope, use the appropriate truth-functional definition to determine the truth value of the complex proposition.

Starting with step 1, start by writing the truth values below each atomic proposition.

$$\begin{array}{ccc} \mathbf{Z} & \wedge & \neg \ \mathbf{J} \\ \mathbf{T} & & \quad \mathbf{F} \end{array}$$

Moving to step 2, starting with the truth-functional operator with the least amount of scope and proceeding to the operators with more scope, assign truth values to complex propositions until a truth value is assigned to the main operator. This procedure will thus require knowledge of the corresponding truth-functional rules associated with each truth-functional operator (see table above).

In the above example, '¬' has the least amount of scope and operates on 'B.' The truth-functional rule for '¬' says that if the truth-value input is 'F,' then the truth-value output is 'T.' In the above example, $v(J) = F$, and so the truth value for '¬J' is $v(\neg J) = T$. Represent this determination by writing a 'T' under '¬' to the immediate left of 'B.'

$$\begin{array}{ccc} \mathbf{Z} & \wedge & \neg \ \mathbf{J} \\ \mathrm{T} & & \mathbf{T} \ \mathrm{F} \end{array}$$

Next, proceed to the operator with the next-least scope. In the above example, this is '∧.' The '∧' operates on 'Z' and '¬J,' where $v(Z) = T$ and $v(\neg J) = T$. The truth-functional definition for '∧' determines that the complex proposition is true.

$$\begin{array}{ccc} \mathbf{Z} & \wedge & \neg \ \mathbf{J} \\ \mathrm{T} & \mathbf{T} & \mathrm{T} \ \mathrm{F} \end{array}$$

Using the truth values of the atomic propositions and the truth-functional rules, the truth value of 'Z∧¬J' has been determined, that is, $v(Z \wedge \neg J) = T$.

Consider a slightly more complex example:

(2$_E$) If Mary is not a zombie or John is a mutant, then we are doomed.

Translate *Mary is a zombie* as 'Z,' *John is a mutant* as 'J,' and *We are doomed* as 'D.' Next, insert the appropriate symbolic operators to reflect the truth-functional syntax of English. (2$_E$) can be translated as follows:

(2$_{PL}$) (¬Z∨J)→D

Assume that $v(Z) = T$, $v(J) = F$, and $v(D) = F$. Following step 1, write the appropriate truth value below each proposition.

$(\neg$	**Z**	$\lor$	**J)**	$\rightarrow$	**D**
	T		**F**		**F**

Following step 2, start with truth-functional operators with the least amount of scope and assign truth values to complex propositions until a truth value is assigned to the main operator. In the above example, '$\neg$' has the least amount of scope. Thus, given the truth function associated with '$\neg$' and $v(Z) = T$, the truth value for the complex proposition '$\neg Z$' is $v(\neg Z) = F$. This is represented by writing an 'F' under '$\neg$' to the left of 'Z.'

$(\neg$	**Z**	$\lor$	**J)**	$\rightarrow$	**D**
F	T		**F**		**F**

Continue to the operator with the next-least scope. In the above example, this is '$\lor$.' Thus,

$(\neg$	**Z**	$\lor$	**J)**	$\rightarrow$	**D**
F	T	**F**	F		**F**

In the above example, note that 'F' is written underneath '$\lor$' because '$\lor$' operates on (or connects) two false propositions, that is, $v(\neg Z) = F$ and $v(J) = F$. Now, look for the operator with the next-least scope. In our case this is '$\rightarrow$,' which is also the main operator. The '$\rightarrow$' operates on $v(D) = F$ and the complex proposition $v(\neg Z \lor J) = F$.

$(\neg$	**Z**	$\lor$	**J)**	$\rightarrow$	**D**
F	T	F	F	**T**	F

Using the notion of scope and the truth-functional definition associated with the various operators, the truth value of '$(\neg Z \lor J) \rightarrow D$' has been determined to be true. This is represented as 'T' under the main operator.

Exercise Set #1

A. Determine the truth value of the following complex propositions.
1. * $A \land \neg B$, where $v(A) = T$, $v(B) = T$
2. $\neg A \rightarrow \neg B$, where $v(A) = T$, $v(B) = F$
3. * $A \leftrightarrow \neg B$, where $v(A) = T$, $v(B) = T$
4. $(A \lor B) \land C$, where $v(A) = F$, $v(B) = T$, $v(C) = T$
5. * $(\neg A \rightarrow B) \rightarrow C$, where $v(A) = T$, $v(B) = T$, $v(C) = T$.
6. $A \rightarrow (B \land \neg C)$, where $v(A) = T$, $v(B) = T$, $v(C) = T$.
7. * $\neg B \rightarrow (A \land \neg A)$, where $v(A) = T$, $v(B) = T$, $v(C) = T$.

8. (A↔¬B)∨C, where *v*(A) = T, *v*(B) = T, *v*(C) = T.
9. * [(A→B)→(B→C)]∨A, where *v*(A) = T, *v*(B) = T, *v*(C) = T.
10. [(A∧¬B)∨¬(C→D)]↔Q, where *v*(A) = T, *v*(B) = T, *v*(C) = F, *v*(D) = T, *v*(Q) = F

Solutions to Starred Exercises in Exercise Set #1

A.
1. * *v*(A∧¬B) = F
3. * *v*(A↔¬B) = F
5. * *v*((¬A→B)→C) = T
7. * *v*(¬B→(A∧¬A)) = T
9. * *v*([(A→B)→(B→C)]∨A) = T

3.2 TRUTH TABLES FOR PROPOSITIONS

In the previous section, truth table definitions were used to determine the truth value of complex propositions in the case where the propositional letters that compose these propositions were assigned valuations. In these cases, given the truth values of the propositional letters, we were able to determine the truth value of the complex propositions. Namely, if *John is tall* is true, and *Liz is happy* is true, the proposition *John is tall, and Liz is happy John is tall, and Liz is happy* is true.

However, the function of a truth table is much more general for it can be used to give a description of the different ways in which truth values can be assigned to propositional letters. To see this more clearly, let's take a simple case involving the proposition 'P∧Q.' Notice that the proposition 'P∧Q' consists of two propositional letters, 'P' and 'Q.' Now, it might be the case that 'P' is true and 'Q' is true. Or, it might be the case that 'P' is true and 'Q' is false. Or it might be the case that 'P' is false and 'Q' is true. Or it might be the case that 'P' is false and 'Q' is false. A truth table will take the different ways in which the propositional letters of 'P∧Q' might be valuated and determine the truth value of 'P∧Q' on that basis.

To represent these different scenarios using a truth table, start by constructing a table with three columns and five rows, where 'P∧Q' is placed in the upper-right-most cell, and the atomic propositions 'P' and 'Q' are placed in the two columns to the left.

P	Q	P∧Q

Next, we want to represent the different ways that 'P' and 'Q' can be evaluated. To do this, start by writing 'T,' 'T,' 'F,' 'F' under the leftmost 'P,' and alternating 'T,''F,''T,''F' under 'Q.'[1]

P	Q	P∧Q
T	T	
T	F	
F	T	
F	F	

Now that the truth table is set up, the procedure for computing the truth value of the complex proposition 'P∧Q' is the same as computing the truth value for complex propositions discussed in the previous section. We can follow the same two-step procedure we followed earlier:

(1) Write the appropriate truth value underneath each propositional letter.
(2) Starting with the truth-functional operator with the least scope and proceeding to the truth-functional operator with the most scope, use the appropriate truth-functional rule to determine the truth value of the complex proposition.

Thus, starting from the first row, we write a 'T' under every 'P' in 'P∧Q' and a 'T' under every 'Q' in 'P∧Q'

P	Q		P	∧	Q
T	T		T		T
T	F				
F	T				
F	F				

Next, moving to the second row, write 'T' under every 'P' and 'F' under every 'Q.'

P	Q		P	∧	Q
T	T		T		T
T	F		T		F
F	T				
F	F				

Continue the process until all of the data concerning the truth values of the propositional letters is under the complex proposition.

P	Q	P	∧	Q
T	T	T		T
T	F	T		F
F	T	F		T
F	F	F		F

Next, the truth value of the complex proposition 'P∧Q' is determined using the truth values of the propositional letters plus the truth table definitions. Since 'P∧Q' is a conjunction, we fill out the table accordingly:

P	Q	P	∧	Q
T	T	T	T	T
T	F	T	F	F
F	T	F	F	T
F	F	F	F	F

Let's consider a more complicated example involving '(P∨¬P)→Q.' Start by constructing a table with three columns and five rows and put '(P∨¬P)→Q' in the upper-right box.

		(P∨¬P)→Q

Next, notice that the propositional letters that compose '(P∨¬P)→Q' are 'P' and 'Q.' So, write 'P' in the uppermost left box and 'Q' to the right of 'P.'

P	Q	(P∨¬P)→Q

Next, we need to consider all possible combinations of truth and falsity for the compound expression. Start by writing 'T,' 'T,' 'F,' 'F' under the leftmost 'P.' Now alternate 'T,''F,''T,''F' under 'Q.'

P	Q	(P∨¬P)→Q
T	T	
T	F	
F	T	
F	F	

Now that the truth table is set up, the procedure for computing the truth value of the complex propositional form is essentially the same as computing the truth value for complex propositions. We can follow the same two-step procedure we followed earlier:

(1) Write the appropriate truth value underneath each propositional letter.
(2) Starting with the truth-functional operator with the least scope and proceeding to the truth-functional operator with the most scope, use the appropriate truth-functional rule to determine the truth value of the complex proposition.

Thus, first look at the truth value for 'P' in row 1. It is 'T.' Now move right across the row, inserting 'T' wherever there is a 'P.' Do the same for rows 2, 3, and 4.

	P	Q	(P	∨	¬	P)	→	Q
1.	T	T	T			T		
2.	T	F	T			T		
3.	F	T	F			F		
4.	F	F	F			F		

Now do this for 'Q' using the 'Ts' and 'Fs' that occur below it.

	P	Q	(P	∨	¬	P)	→	Q
1.	T	T	T			T		T
2.	T	F	T			T		F
3.	F	T	F			F		T
4.	F	F	F			F		F

Now that all of the truth values have been transferred from the left side of the table, the next step is to determine the truth value of the more complex propositions within the table. In order to do this, assign truth values to propositions that have the least scope, moving to the expression that has the most scope (i.e., the main operator). Since the main operator of '(P∨¬P)→Q' is '→,' the proposition has a conditional form. The operator with the least scope is '¬.' Thus, start with negation by writing the appropriate truth value below the negation (¬).

	P	Q	(P	∨	¬	P)	→	Q
1.	T	T	T		F	T		T
2.	T	F	T		F	T		F
3.	F	T	F		T	F		T
4.	F	F	F		T	F		F

Second, the wedge (∨) has the next-least scope in '(P∨¬P)→Q.' Using the truth values under the '¬' in '¬P' and the truth values under the non-negated 'P,' determine the truth value for the disjunction '(P∨¬P)' and write the truth value under '∨.'

	P	Q	(P	∨	¬	P)	→	Q
1.	T	T	T	T	F	T		T
2.	T	F	T	T	F	T		F
3.	F	T	F	T	T	F		T
4.	F	F	F	T	T	F		F

The next step is to finish the table by writing the correct truth value under the main operator of the proposition. The truth value under the main operator will determine the truth value for the propositional form. In order to do this, look at the truth value under the main operator in '(P∨¬P)' and the truth value of 'Q.' One is written under '∨,' the other is written under 'Q.'

	P	Q	(P	∨	¬	P)	→	Q
1.	T	T	T	T	F	T		T
2.	T	F	T	T	F	T		F
3.	F	T	F	T	T	F		T
4.	F	F	F	T	T	F		F
	1	2	3	4	5	6	7	8

Look at row 1 of the truth table. We see from column 4 that the disjunction is 'T,' and the truth value of 'Q' is 'T.' What does the truth table definition say about '*v*(P→Q)' when *v*(P) = T and *v*(Q) = T? The conditional is true, so write 'T' underneath the '→' in row 1 of column 7.

	P	Q	(P	∨	¬	P)	→	Q
1.	T	T	T	T	F	T	T	T
2.	T	F	T	T	F	T		F
3.	F	T	F	T	T	F		T
4.	F	F	F	T	T	F		F
	1	2	3	4	5	6	7	8

What does it say about 'v(P→Q)' when v(P) = T and v(Q) = F? It says v(P→Q) = F. Write 'F' in line 2, column 6 under the '→.'

	P	**Q**	**(P**	**∨**	**¬**	**P)**	**→**	**Q**
1.	T	T	T	T	F	T	T	T
2.	T	F	T	**T**	F	T	**F**	**F**
3.	F	T	F	T	T	F		T
4.	F	F	F	T	T	F		F
	1	2	3	4	5	6	7	8

Repeat this process until you have the following:

	P	**Q**	**(P**	**∨**	**¬**	**P)**	**→**	**Q**
1.	T	T	T	T	F	T	**T**	T
2.	T	F	T	T	F	T	**F**	F
3.	F	T	F	T	T	F	**T**	T
4.	F	F	F	T	T	F	**F**	F
	1	2	3	4	5	6	7	8

This is a complete truth table. The truth value in the column under the main operator of '(P∨¬P)→Q' indicates the truth value of the proposition given a specific valuation of its atomic parts.

Exercise Set #2

A. Construct and complete truth tables for the following propositions.
    1.   ¬P∨¬Q
    2.   ¬(P∧¬Q)
    3. * (P↔Q)∧(P→¬Q)
    4.   P→(¬P∧¬Q)
    5. * (P↔Q)∧(P↔¬Q)
    6.   ¬¬P∧(P↔¬Q)
    7. * (P→Q)→(P∨¬Q)
    8.   ¬P→¬Q
    9. * ¬¬(¬P∧¬¬Q)
    10.   ¬P∧(¬Q∨R)
    11. * [(P→Q)∧(R→P)]∧¬(R→Q)
    12.   ¬P→(¬Q↔ R)
    13. * [(P∧Q)∧(R∧P)]∧¬(R∧Q)

Solutions to Starred Exercises in Exercise Set #2

A.

3. * (P↔Q)∧(P→¬Q)

P	Q	(P	↔	Q)	∧	(P	→	¬	Q)
T	T	T	T	T	F	T	F	F	T
T	F	T	F	F	F	T	T	T	F
F	T	F	F	T	F	F	T	F	T
F	F	F	T	F	T	F	T	T	F

5. * (P↔Q)∧(P↔¬Q)

P	Q	(P	↔	Q)	∧	(P	↔	¬	Q)
T	T	T	T	T	F	T	F	F	T
T	F	T	F	F	F	T	T	T	F
F	T	F	F	T	F	F	T	F	T
F	F	F	T	F	F	F	F	T	F

7. * (P→Q)→(P∨¬Q)

P	Q	(P	→	Q)	→	(P	∨	¬	Q)
T	T	T	T	T	T	T	T	F	T
T	F	T	F	F	T	T	T	T	F
F	T	F	T	T	F	F	F	F	T
F	F	F	T	F	T	F	T	T	F

9. * ¬¬(¬P∧¬¬Q)

P	Q	¬	¬	(¬	P	∧	¬	¬	Q)
T	T	F	T	F	T	F	T	F	T
T	F	F	T	F	T	F	F	T	F
F	T	T	F	T	F	T	T	F	T
F	F	F	T	T	F	F	F	T	F

11. * [(P→Q)∧(R→P)]∧¬(R→Q)

P	Q	R	[(P	→	Q)	∧	(R	→	P)]	∧	¬	(R	→	Q)
T	T	T	T	T	T	T	T	T	T	F	F	T	T	T
T	T	F	T	T	T	T	F	T	T	F	F	F	T	T
T	F	T	T	F	F	F	T	T	T	F	T	T	F	F
T	F	F	T	F	F	F	F	T	T	F	F	F	T	F
F	T	T	F	T	T	F	T	F	F	F	F	T	T	T
F	T	F	F	T	T	T	F	T	F	F	F	F	T	T
F	F	T	F	T	F	F	T	F	F	F	T	T	F	F
F	F	F	F	T	F	T	F	T	F	F	F	F	T	F

                     Chapter Three

13. *   [(P∧Q)∧(R∧P)]∧¬(R∧Q)

P	Q	R	[(P	∧	Q)	∧	(R	∧	P)]	∧	¬	(R	∧	Q)
T	T	T	T	T	T	T	T	T	T	F	F	T	T	T
T	T	F	T	T	T	F	F	F	T	F	T	F	F	T
T	F	T	T	F	F	F	T	T	T	F	T	T	F	F
T	F	F	T	F	F	F	F	F	T	F	T	F	F	F
F	T	T	F	F	T	F	T	F	F	F	F	T	T	T
F	T	F	F	F	T	F	F	F	F	F	T	F	F	T
F	F	T	F	F	F	F	T	F	F	F	T	T	F	F
F	F	F	F	F	F	F	F	F	F	F	T	F	F	F

3.3 TRUTH TABLE ANALYSIS OF PROPOSITIONS

The precise syntax of PL, truth table definitions and scope indicators allow for the use of a *decision procedure*. A decision procedure is a mechanical method that determines in a finite number of steps whether a proposition, set of propositions, or argument has a certain logical property.

> Decision   A decision procedure is a mechanical method that determines in a
> procedure   finite number of steps whether a proposition, set of propositions, or
>   argument has a certain logical property.

What logical properties are we interested in? With respect to propositions, we would like to know whether a proposition is always true (*tautological*), always false (*contradictory*), or neither always true nor always false (*contingent*), independent of how truth values are assigned to propositional letters. Truth tables provide us with a decision procedure for determining whether a proposition has one of these three properties; that is, it is a mechanical method that will give us a yes or no answer as to the question, Is 'P' contingent? Is 'P∨¬P' a tautology? and so on.

3.3.1 Tautology

In rough terms, a tautology is a proposition that is always true. Sentences like 'John is tall or not tall,' 'Mary is the murderer or she isn't,' or 'It is what it is' are all examples of tautologies. More precisely, a proposition '**P**' is a *tautology* if and only if (iff) '**P**' is true under every valuation. That is, a proposition '**P**' is a tautology if and only if it is true no matter how we assign truth values to the atomic letters that compose it.

Using a completed truth table, we can determine whether or not a proposition is a tautology simply by looking to see whether there are only 'Ts' under the proposition's main operator (or in the case of no operators, under the propositional letter).

Tautology   A proposition '**P**' is a tautology if and only if '**P**' is true under every valuation. A truth table for a tautology will have all 'Ts' under its main operator (or in the case of no operators, under the propositional letter).

As an illustration, consider the following truth table for 'P→(Q→P)':

P	Q	P	→	(Q	→	P)
T	T	T	**T**	T	T	T
T	F	T	**T**	F	T	T
F	T	F	**T**	T	F	F
F	F	F	**T**	F	T	F

Notice that there is a single line of 'Ts' under the main operator in the above table. This means that under every combination of valuations of 'P' and 'Q,' the complex proposition 'P→(Q→P)' is true, and therefore 'P→(Q→P)' is a tautology.

As a second example, consider the following truth table for 'P∧(Q→P)':

P	Q	P	∧	(Q	→	P)
T	T	T	**T**	T	T	T
T	F	T	**T**	F	T	T
F	T	F	**F**	T	F	F
F	F	F	**F**	F	T	F

Notice that under the main operator of 'P∧(Q→P),' there are two 'Fs,' indicating that under some valuation, 'P∧(Q→P)' is false. Given our definition that a proposition is a tautology if and only if it is true under every valuation, the proposition 'P∧(Q→P)' is not a tautology.

3.3.2 Contradiction

In rough terms, a *contradiction* is a proposition that is always false. Sentences like 'John is tall and not tall,' 'Mary is the murderer, and she isn't,' or 'A is not A' are examples of contradictions. More precisely, a proposition '**P**' is a contradiction if and only if '**P**' is false under every valuation. That is, a proposition '**P**' is a contradiction if and only if it is false no matter how we assign truth values to the propositional letters that compose it.

Using a completed truth table, we can determine whether or not a proposition is a contradiction simply by looking to see whether there are only 'Fs' under the proposition's main operator (or in the case of no operators, under the propositional letter).

Contradiction   A proposition '**P**' is a contradiction if and only if '**P**' is false under every valuation. A truth table for a contradiction will have all 'Fs' under its main operator (or in the case of no operators, under the propositional letter).

As an illustration, consider the following truth table for '¬P∧(Q∧P)':

P	**Q**	**¬**	**P**	**∧**	**(Q**	**∧**	**P)**
T	T	F	T	**F**	T	T	T
T	F	F	T	**F**	F	F	T
F	T	T	F	**F**	T	F	F
F	F	T	F	**F**	F	F	F

Notice that there is a single line of 'Fs' under the main operator in the above table. This means that under every combination of valuations of 'P' and 'Q,' the complex proposition '¬P∧(Q∧P)' is false, and therefore '¬P∧(Q∧P)' is a contradiction.

As a second example, consider the truth table for 'P∧(Q→P).'

P	**Q**	**P**	**∧**	**(Q**	**→**	**P)**
T	T	T	**T**	T	T	T
T	F	T	**T**	F	T	T
F	T	F	**F**	T	F	F
F	F	F	**F**	F	T	F

Notice that under the main operator of 'P∧(Q→P),' there are two 'Ts,' indicating that under some valuation, 'P∧(Q→P)' is true. Given our definition that a proposition is a contradiction if and only if it is false under every valuation, the above truth table shows that proposition 'P∧(Q→P)' is not a contradiction.

3.3.3 Contingency

In the previous two sections, a tautology was defined as a proposition that is true under every valuation, while a contradiction was defined as a proposition that is false under every valuation. This leaves one final case. In rough terms, a *contingency* is a proposition whose truth value depends on how it is valuated. Sentences like 'John is tall,' 'Mary is the murderer,' and 'Politicians are trustworthy' are all examples of contingencies. More precisely, a contingency is a proposition '**P**' that is neither always true (a tautology) nor always false (a contradiction) under every valuation.

Using a completed truth table, we can determine whether or not a proposition is a contingency simply by looking to see whether there is at least one 'F' and at least one 'T' under the proposition's main operator (or in the case of no operators, under the propositional letter).

Contingency   A proposition 'P' is a contingency if and only if 'P' is neither always false under every valuation nor always true under every valuation. A truth table for a contingency will have at least one 'T' and at least one 'F' under its main operator (or in the case of no operators, under the propositional letter).

As a very simple illustration, consider the truth table for 'P.'

P	P
T	T
F	F

Notice that the truth table for 'P' has one 'F' under it and one 'T' under it. Thus, it is a contingency since it is neither always true nor always false.

Next, consider the truth table for 'P→Q.'

P	Q	P	→	Q
T	T	T	T	T
T	F	T	F	F
F	T	F	T	T
F	F	F	T	F

Notice that 'P→Q' is neither true under every valuation nor false under every valuation. That is, 'P→Q' is neither always true nor always false. This is evident from the fact that 'P→Q' is true when 'P' is true and 'Q' is true (as row 1 indicates), and 'P→Q' is false when 'P' is true and 'Q' is false (as row 2 indicates). Thus, 'P→Q' is a contingency.

Finally, consider the truth table for 'P∧(Q→P).'

P	Q	P	∧	(Q	→	P)
T	T	T	**T**	T	T	T
T	F	T	**T**	F	T	T
F	T	F	**F**	T	F	F
F	F	F	**F**	F	T	F

We considered the above proposition and its corresponding truth table earlier when discussing contradictions and tautologies. In asking whether the proposition was a tautology, we used a truth table of this proposition to show that 'P∧(Q→P)' is not a tautology. In asking whether the proposition was a contradiction, we used a truth table of this proposition to show that 'P∧(Q→P)' is not a contradiction.

Notice that under the main operator of 'P∧(Q→P),' there are two 'Ts,' indicating that under some valuation 'P∧(Q→P)' is true. In addition, notice that under the main operator of 'P∧(Q→P),' there are two 'Fs,' indicating that under some valuation 'P∧(Q→P)' is false. Thus, we know that 'P∧(Q→P)' is neither always true nor always false, and therefore it is a contingency.

Exercise Set #3

A. Construct truth tables for the following propositions, then explain whether the proposition is a tautology, contradiction, or contingency.
   1. * R→¬R
   2.   (P→Q)∧(¬Q→¬P)
   3. * (¬P∧¬Q)∧P
   4.   P∨¬(¬M∨M)
   5. * (R∧R)∧R
   6.   (P→W)∧(P↔W)
   7.   ¬(P∧W)∧¬(¬P∨W)
   8.   ¬(M∨W)∧Q
   9.   Q→Q
   10. (R∨R)∨R

Solutions to Starred Exercises in Exercise Set #3

A.
   1. * R→¬R; contingent.

R	R	→	¬	R
T	T	F	F	T
F	F	T	T	F

   3. * (¬P∧¬Q)∧P; contradiction.

P	Q	(¬	P	∧	¬	Q)	∧	P
T	T	F	T	F	F	T	F	T
T	F	F	T	F	T	F	F	T
F	T	T	F	F	F	T	F	F
F	F	T	F	T	T	F	F	F

   5. * (R∧R)∧R; contingency

R	(R	∧	R)	∧	R
T	T	T	T	T	T
F	F	F	F	F	F

3.4 TRUTH TABLE ANALYSIS OF SETS OF PROPOSITIONS

A complete truth table allows for analysis of various properties of propositions, sets of propositions, and arguments. In this section, we consider how to analyze sets of propositions for logical equivalence and consistency with truth tables.

3.4.1 Equivalence

In rough terms, two propositions are logically equivalent if and only if they always have the same truth value. For example, the proposition expressed by *John is a married man* will always have the same truth value as the proposition expressed by *John is not an unmarried man*. More precisely, a pair of propositions 'P' and 'Q' is *logically equivalent* if and only if 'P' and 'Q' have identical truth values under every valuation. That is, 'P' and 'Q' are equivalent if and only if, no matter how we assign truth values to the propositional letters that compose them, there will never be a case where 'P' is true and 'Q' is false or a case where 'P' is false and 'Q' is true.

Using a completed truth table, we can determine whether or not a pair of propositions 'P' and 'Q' is logically equivalent by looking to see whether there is a row on the truth table where one of the 'P' has a different truth value than 'Q.'

> Equivalence A pair of propositions 'P' and 'Q' is logically equivalent if and only if 'P' and 'Q' have identical truth values under every valuation. In a truth table for an equivalence, there is no row on the truth table where one of the pair 'P' has a different truth value than the other 'Q.'

A truth table provides a decision procedure for showing logical equivalence by allowing for a comparison of truth values at each row of a truth table. For example, take the following two propositions: 'P→Q' and 'Q∨¬P.' Start by putting both of these expressions into a truth table.

P	Q	(P	→	Q)	(Q	∨	¬	P)
T	T	T	**T**	T	T	**T**	F	T
T	F	T	**F**	F	F	**F**	F	T
F	T	F	**T**	T	T	**T**	T	F
F	F	F	**T**	F	F	**T**	T	F

The truth table above shows that 'P→Q' is logically equivalent to 'Q∨¬P' because whenever $v(P{\to}Q) = T$, then $v(Q{\lor}\neg P) = T$, and whenever $v(P{\to}Q) = F$, then $v(Q{\lor}\neg P) = F$.

Another way to determine whether two propositions are logically equivalent is to join two propositions together using the double arrow (↔) and then determine whether the resulting biconditional forms a tautology. To see why this is the case, first consider that two propositions are logically equivalent if and only if they always have the same truth value. Second, note that a biconditional is true whenever both sides of the biconditional have the same truth value. Finally, if a biconditional is always true (i.e., a tautology), then each side of the biconditional always has the same truth value as the other. Thus, if a biconditional is a tautology, then the left and right sides of the biconditional are logically equivalent to each other.

This may be somewhat difficult to grasp, so it is helpful to consider this idea using a truth table. Consider whether 'P→Q' and 'Q∨¬P' are equivalent.

P	Q	P	→	Q		Q	∨	¬	P
T	T	T	**T**	T		T	**T**	F	T
T	F	T	**F**	F		F	**F**	F	T
F	T	F	**T**	T		T	**T**	T	F
F	F	F	**T**	F		F	**T**	T	F

Notice that the truth values of 'P→Q' and 'Q∨¬P' do not differ, and so 'P→Q' and 'Q∨¬P' are logically equivalent. However, now consider the truth table where these two propositions are joined together using the double arrow (↔).

P	Q	(P	→	Q)	↔	(Q	∨	¬	P)
T	T	T	T	T	**T**	T	T	F	T
T	F	T	F	F	**T**	F	F	F	T
F	T	F	T	T	**T**	T	T	T	F
F	F	F	T	F	**T**	F	T	T	F

Notice that a biconditional is true if and only if each side of the biconditional has the same truth value. And notice that in the case of logically equivalent propositions, like those of 'P→Q' and 'Q∨¬P,' the truth values are the same for every row of the truth table. Since they have the same truth value for every row of the truth table, '(P→Q)↔(Q∨¬P)' determines a tautology.

Thus, we can determine whether two propositions are logically equivalent by placing the double arrow between these propositions and testing to see whether the proposition is a tautology. If the proposition is a tautology, then the propositions are logically equivalent. If the proposition is not a tautology, then the propositions are not logically equivalent.

3.4.2 Consistency

In rough terms, a set of propositions is *logically consistent* if and only if it is logically possible for all of them to be true. For example, the following propositions can all be true:

John is a bachelor.
Frank is a bachelor.
Vic is a bachelor.

More precisely, a set of propositions '{**P,Q, R,** . . . **Z**}'is logically consistent if and only if there is at least one valuation where all of propositions '**P**,''**Q**,''**R**,'. . .'**Z**' in the set are true. That is, '{**P, Q, R,** . . . **Z**}' is logically consistent if and only if there is at least one way of assigning truth values to the propositional letters such that '**P**,''**Q**,''**R**,'. . .'**Z**' are jointly true.

Using a completed truth table, we can determine whether or not a set of propositions '{**P, Q, R,** . . . **Z**}' is logically consistent by looking to see whether there is at least one row on the truth table where '**P**,''**Q**,''**R**,'. . .'**Z**' are all true.

Consistency   A set of propositions '{**P, Q, R,** . . ., **Z**}'is logically consistent if and only if there is at least one valuation where '**P**,' '**Q**,''**R**,'. . ., '**Z**' are true. A truth table shows that a set of propositions is consistent when there is at least one row in the truth table where '**P**,' '**Q**,''**R**,'. . .,'**Z**' are all true.

A truth table provides a decision procedure for demonstrating logical consistency by allowing for a comparison of truth values at each row of a truth table. Using a truth table, there is a logical consistency when there is at least one row where there is a 'T' under all the main operators (or under a propositional letter where there are no truth-functional operators). Consider the following three propositions: 'P→Q,' 'Q∨P,' 'P↔Q.'

	P	**Q**	(**P**	→	**Q**)	(**Q**	∨	**P**)	(**P**	↔	**Q**)
1.	T	T	T	**T**	T	T	**T**	T	T	**T**	T
2.	T	F	T	**F**	F	F	**T**	T	T	**F**	F
3.	F	T	F	**T**	T	T	**T**	F	F	**F**	T
4.	F	F	F	**T**	F	F	**F**	F	F	**T**	F

To determine whether the above three propositions are logically consistent requires that there be at least one row where 'T' is located under the main connective for each proposition. A 'T' is not located under the main operator for lines 2, 3, and 4, but a 'T' is located under the main operator for line 1. Therefore, 'P→Q,' 'Q∨P,' and 'P↔Q' are logically consistent.

Remember that we are concerned with logical properties of propositions. A set of propositions may be materially or factually inconsistent yet remain logically consistent. For example, consider a case where empirical science informed us that $v(P{\rightarrow}Q)$ = F and $v(Q{\vee}P)$ = T. If this happened, then 'P→Q' and 'Q∨P' would form a factually inconsistent set since they are not both true. However, at the level of logical or semantic analysis, they *could* both be true since it could be the case that $v(P{\rightarrow}Q)$ = T and $v(Q{\vee}P)$ = T.

If a set of propositions is not consistent, then the set of propositions is *inconsistent*. An inconsistent set of propositions is a set of propositions where there is no valuation where all of the propositions in the set can be true.

> Inconsistency   A set of propositions '{**P, Q, R, . . ., Z**}' is logically inconsistent if and only if there is no valuation where '**P**,' '**Q**,' '**R**,'. . .,'**Z**' are jointly true. A truth table shows that a set of propositions is inconsistent when there is no row on the truth table where '**P**,' '**Q**,' '**R**,'. . .,'**Z**' are all true.

To illustrate, compare the following two propositions: '(P∨Q)' and '¬(Q∨P).'

	P	Q	(P	∨	Q)	¬	(Q	∨	P)
1.	T	T	T	T	T	F	T	T	T
2.	T	F	T	T	F	F	F	T	T
3.	F	T	F	T	T	F	T	T	F
4.	F	F	F	F	F	T	F	F	F

The above truth table shows that there is no row where both propositions are true under the same truth valuations. Therefore, these two propositions are inconsistent.

When an inconsistent set of propositions is conjoined to form a conjunction, a *contradiction* is formed. That is, given two separate yet inconsistent propositions, the conjunction of these two propositions forms a contradiction.

	P	Q	(P	∨	Q)	∧	¬	(Q	∨	P)
1.	T	T	T	T	T	**F**	F	T	T	T
2.	T	F	T	T	F	**F**	F	F	T	T
3.	F	T	F	T	T	**F**	F	T	T	F
4.	F	F	F	F	F	**F**	T	F	F	F

3.5 THE MATERIAL CONDITIONAL EXPLAINED (OPTIONAL)

In chapter 2, it was noted that there are difficulties assimilating the truth function associated with '→' to the use of *if . . ., then . . .* in English. It was briefly explained how various uses of the English *if . . ., then . . .* do not correspond to the truth-functional '→' (e.g., causal statements). No justification was given for why '→' should receive the following evaluation:

P	Q	P→Q
T	T	T
T	F	F
F	T	T
F	F	F

With a better understanding of truth tables and logical properties defined in terms of truth-functionality, a more compelling case can be made for why 'if P then Q' corresponds to 'P→Q.' There are two main strategies for justifying the claim that truth-functional uses of *if . . ., then . . .* in English correspond with the truth function associated with the arrow.

The first strategy involves considering all of the possible truth functions and eliminating those that don't correspond to an intuitive understanding of what conditionals mean. The other involves considering certain intuitions about valid inference.

First, consider all of the possible truth functions.

P	Q								
T	T	T	T	T	F	T	T	T	F
T	F	T	F	T	T	F	T	T	F
F	T	F	T	T	T	T	F	T	T
F	F	F	F	T	T	T	T	F	T
		1	2	3	4	5	6	7	8

P	Q								
T	T	T	F	F	F	F	F	T	F
T	F	F	T	T	F	T	F	F	F
F	T	F	T	F	T	F	F	F	F
F	F	T	F	T	F	F	T	F	F
		9	10	11	12	13	14	15	16

If there is a truth-functional use of *if . . ., then . . .*, then it must correspond to at least one of these since they exhaust all of the possible candidates. We can eliminate a number of these possibilities using basic intuitions about *if . . ., then . . .* statements. First, one intuition is that if both the antecedent and the consequent are true, then the conditional is true. This means that (4), (8), (10), (11), (12), (13), (14), and (16) can be eliminated as possible candidates. This leaves the following truth functions:

T	T	T	T	T	T
T	F	T	F	T	T
F	T	T	T	F	T
F	F	T	T	T	F
1	2	3	5	6	7

T	T
F	F
F	F
T	F
9	15

Second, intuitively, we think that a conditional is false whenever the antecedent is true and the consequent is false. This means that we should eliminate those truth functions that valuate the complex proposition as true when the antecedent is true and the consequent is false. Thus, we can eliminate (1), (3), (6), and (7) since they claim that 'P→Q' is true whenever $v(P) = T$ and $v(Q) = F$. This leaves (2), (5), (9), and (15) as possible candidates for the truth-functional use of *if. . ., then . . . .*

Third, consider the truth functions expressed by (2), (5), (9), and (15).

P	**Q**	**Q**	**¬P∨Q**	**P↔Q**	**P∧Q**
T	T	T	T	T	T
T	F	F	F	F	F
F	T	T	T	F	F
F	F	F	T	T	F
		2	5	9	15

Consider (2). Generally, we think that 'P→Q' and truth-functional uses of 'if P then Q' say something more than simply 'Q.' That is, *If John is in Toronto, then John is in Canada* says something more than *John is in Canada*. Thus, (2) should be eliminated. Consider (9). The truth function in (9) can be expressed by the biconditional 'P↔Q.' However, there is a difference between biconditionals and conditionals. In the case of the former, 'P↔Q' is logically equivalent to 'Q↔P.' However, this is not the case with conditionals. For example, consider the following conditional:

(1) If John is in Toronto, then he is in Canada.

This is not logically equivalent to

(2) If John is in Canada, then he is in Toronto.

because (1) can be true while (2) is false. Namely, (2) is false provided John is in Canada but not in Toronto. This allows us to eliminate (9).

Finally, consider (15). The truth function described in (15) can be represented by the conjunction 'P∧Q.' If *if. . ., then . . .* statements are represented by the '∧' function, then every truth-functional use of 'if P then Q' can be replaced by a statement of the form 'P and Q.' However, this does not seem to be the case. Consider the following propositions:

(3) If John is in Toronto, then he is in Canada.
(4) John is in Toronto, and he is in Canada.

Clearly, (3) and (4) do not say the same thing. (4) is true if and only if John is both in Toronto and in Canada. However, we think that (3) is true if John is in Canada but not Toronto, e.g., if he were in Vancouver. Thus, the only truth function that represents truth-functional uses of *if. . ., then . . .* is represented in (5), and this column corresponds to the truth function expressed by '→.'

A second justification for why truth-functional uses of *if. . ., then . . .* correspond to the '→' function depends upon our understanding of logical properties. For suppose that instead of treating truth-functional uses of 'if P then Q' in terms of '→,'they were defined as the following truth function '→*':

P→*Q
T
F
F
F

If this were the case, then the following two propositions would be logically inconsistent:

(5) If John is in Toronto, then he is in Canada.
(6) John is not in Toronto.

The inconsistency can be expressed in a truth table as follows:

P	Q	P→*Q	¬P
T	T	T	F
T	F	F	F
F	T	F	T
F	F	F	T

However, we do not regard (5) and (6) as inconsistent for suppose that John wants to convey to his friend Liz that Toronto is in Canada. But also assume that John and Liz are having this conversation, *not* in Toronto but in Chicago. John says to Liz, "If I'm in Toronto, then I'm in Canada." What John says is true even though he is not in Toronto, he's never been to Toronto, and never plans on going to Toronto. Even if the antecedent of (5) is false, the conditional should be true for, to be this different, John simply denies that he can be both in Toronto and in Canada, *for Toronto is in Canada*!

3.6 TRUTH TABLE ANALYSIS OF ARGUMENTS

A complete truth table allows for analysis of various properties of propositions, sets of propositions, and arguments. In this section, we consider how to analyze arguments to determine whether or not they are valid or invalid.

3.6.1 Validity

In this section, we use the truth table method to determine whether an argument is *valid* or *invalid*. An argument is valid if and only if it is impossible for the premises to be true and the conclusion to be false. In chapter 1, the negative test was used to determine whether or not arguments were valid. This test asked you to imagine whether it is possible for the premises to be true and the conclusion to be false. The negative test, however, has some limitations since it depends on an individual's psychological capacities, and this capacity is taxed when dealing with extremely long arguments. Truth tables provide an easier, more reliable, and purely mechanical method for determining validity. Using a truth table, an argument is valid in PL if and only if there is no row of the truth table where the premises are true and the conclusion is false. If there is a row where the premises are true and the conclusion is false, then the argument is invalid.

Before providing some examples of truth tables that illustrate arguments that are valid (or invalid), it will be helpful to introduce an additional symbol to represent arguments. This symbol is the single turnstile ($\vdash$). Although later in this text, the turnstile will stand for something more specific, temporarily we use it to indicate the presence of an argument: the propositions to the left of the turnstile are the *premises*, while the proposition to the right of the turnstile is the *conclusion*. For example, the turnstile in

$$\mathbf{P \wedge R \vdash R}$$

indicates the presence of an argument where '**P∧R**' is the premise and '**R**' is the conclusion. Likewise, the turnstile in

$$\mathbf{P \wedge R, \ Z \rightarrow Z, \ \neg(P \wedge Q) \vdash R}$$

indicates the presence of an argument where '**P∧R**,' '**Z→Z**,' and '**¬(P∧Q)**' are premises, and '**R**' is the conclusion. Lastly, the turnstile in

$$\vdash \mathbf{R}$$

indicates the presence of an argument that has no premises but has '**R**' as a conclusion.

Validity	An argument '**P**, **Q**, . . ., **Y** ⊢ **Z**' is valid in PL if and only if it is impossible for the premises to be true and the conclusion false. A truth table shows that an argument is valid if and only if there is no row of the truth table where the premises are true and the conclusion is false.
Invalidity	An argument '**P**, **Q**, . . ., **Y** ⊢ **Z**' is invalid in PL if and only if it is possible for the premises to be true and the conclusion false. A truth table shows that an argument is invalid if and only if there is a row of the truth table where the premises are true and the conclusion is false.

To illustrate, consider the following truth table for the following argument: 'P→Q, ¬Q ⊢ ¬P.'

P	Q	(P	→	Q)	¬	Q	⊢	¬	P
T	T	T	**T**	T	**F**	T		**F**	T
T	F	T	**F**	F	**T**	F		**F**	T
F	T	F	**T**	T	**F**	T		**T**	F
F	F	F	**T**	F	**T**	F		**T**	F

Notice that there is *no row* where the premises 'P→Q' and '¬Q' are true and '¬P' is false. Even though there is a row (the bottom row) where 'P→Q' and '¬Q' are jointly true, this is not a row where '¬P' is also false. Thus, 'P→Q, ¬Q ⊢ ¬P' is valid.

Next, consider the argument 'P→Q, Q ⊢ P' and its truth table.

P	Q	(P	→	Q)	Q	⊢	P
T	T	T	**T**	T	T		T
T	F	T	**F**	F	F		T
F	T	F	**T**	T	T		F
F	F	F	**T**	F	F		F

Notice that in the truth table above, there is a row where the premises are true and the conclusion is false. This is row 3. Thus, P→Q, Q ⊢ P is invalid.

3.6.2 Validity and Inconsistency

Another way to determine whether an argument is valid is by analyzing whether the set of propositions consisting of the premises and the negation of the conclusion is inconsistent. If the set is *inconsistent*, then the argument is valid. If it is consistent, then the argument is invalid. To see this more clearly, consider the following valid argument: 'P→Q, P ⊢ Q.' Using a truth table, this argument is shown to be valid since there is no line in the truth table where all of the premises are true and the conclusion is false.

		Premises					Conclusion
P	Q	P	→	Q	P ⊢		Q
T	T	T	T	T	T		T
T	F	T	F	F	T		F
F	T	F	T	T	F		T
F	F	F	T	F	F		F

Notice that in considering whether or not 'P→Q, P⊢Q' is valid, what is being analyzed is whether the following truth-value assignment is possible:

$$v(P{\to}Q) = T, v(P) = T, \text{ and } v(Q) = F.$$

If it is not possible, then 'P→Q, P⊢Q' is valid. If it is possible, then 'P→Q, P⊢Q' is invalid. However, notice that if $v(Q) = F$, then $v(\neg Q) = T$. This suggests another way of determining validity, that is, determining whether the following truth-value assignment is possible:

$$v(P{\to}Q) = T, v(P) = T, \text{ and } v(\neg Q) = T.$$

That is, another way of checking for validity is by asking the following question:

Is it possible for both the premises and the negation of the conclusion to be true?

If the answer to this question is no, then the argument is valid. If the answer is yes, then the argument is invalid.

An equivalent way of asking the same question is the following:

Are the premises and the negation of the conclusion logically consistent (i.e., all true under the same truth-value assignment)?

If the premises and the negation of the conclusion are not logically consistent, then the argument is valid. If the premises and the negation of the conclusion are logically consistent, then the argument is invalid.

To see this more clearly, consider again the valid argument 'P→Q, P⊢Q' but which determines whether the argument is valid by determining whether '{P→Q, P, ¬Q}' is inconsistent.

		Premises				Negation of the Conclusion
P	Q	P	→	Q	P	¬Q
T	T	T	T	T	T	F
T	F	T	F	F	T	T
F	T	F	T	T	F	F
F	F	F	T	F	F	T

Notice that in the above table, there is no line on the truth table where 'P→Q,' 'P,' and '¬Q' are all true. That is, '{P→Q, P, ¬Q}' is inconsistent. In saying that '{P→Q, P, ¬Q}' is inconsistent, we are saying that it is impossible for the premises 'P→Q' and 'P' to be true and the negation of the conclusion '¬Q' to be true. Thus, 'P→Q, P ⊢ Q' is valid.

To consider this more generally, compare the definitions for *validity* and *inconsistency*. An argument is valid if and only if it is impossible for the premises '**P**,''**Q**,'. . ., '**Y**' to be true and the conclusion '**Z**' to be false. This is just another way of saying that an argument is valid if and only if it is impossible for the propositions '{P, Q, . . ., Y, ¬Z}' all to be true. Notice, however, that if it is impossible for the propositions '{P, Q, . . .,Y, ¬Z}'to all be true, then the propositions '{P, Q, . . ., Y, ¬Z}' are inconsistent. Thus, validity can be defined in terms of inconsistency.

3.7 SHORT TRUTH TABLE TEST FOR INVALIDITY

A truth table is capable of showing that an argument is invalid by graphically showing that there is a way of assigning truth values to propositional letters that would make the premises of an argument true and the conclusion false. The process of filling out the truth table, however, is quite time-consuming. One way of shortening the truth table test for invalidity is by a process called *forcing*. Rather than beginning the test by assigning truth values to propositional letters, then using the truth-functional operator rules to determine the truth values of complex propositions, and then analyzing the argument to see if it is valid or invalid, the forcing method begins by assuming that the argument is invalid and working backward to assign truth values to propositional letters.

For example, consider the following argument: 'P→Q, R∧¬Q ⊢ Q.' Begin by assuming the argument is invalid, which involves assigning 'T' to propositions that are premises, and 'F' to the conclusion.

P	Q	R	(P	→	Q)	(R	∧	¬	Q)	⊢	Q
				T			T				F

Next, we work backward from our knowledge of the truth-functional definition and the truth values assigned to the complex propositions. For example, we know that if 'R∧¬Q' is true, then both of the conjuncts 'R' and '¬Q' are true.

P	Q	R	(P	→	Q)	(R	∧	¬	Q)	⊢	Q
				T		**T**	**T**	**T**			F

In addition, if $v(¬Q) = T$, then $v(Q) = F$.

P	Q	R	(P	→	Q)	(R	∧	¬	Q)	⊢	Q
	F			T		T	T	T	**F**		F

Using this forcing strategy, we have determined that 'Q' is false and 'R' is true, and so we can assign all other 'Rs' the value of true and 'Qs' the value of false.

P	Q	R	(P	→	Q)	(R	∧	¬	Q)	⊢	Q
	F	T		T	F	T	T	T	F		F

Finally, we know that for 'P→Q' to be true and 'Q' to be false, 'P' must be false. Thus,

P	Q	R	(P	→	Q)	(R	∧	¬	Q)	⊢	Q
F	F	T	F	T	F	T	T	T	F		F

And so, using the forcing method, we have shown that 'P→Q, R∧¬Q⊢Q' is invalid. This method began by assuming the argument was invalid (i.e., the premises were assumed to be true and the conclusion to be false). From there, we worked backward using the truth values and the truth-table definitions to obtain a consistent assignment of valuations to the propositional letters.

END-OF-CHAPTER EXERCISES

A. Construct a truth table for the following propositions. Determine whether they are contradictions, tautologies, or contingencies.
    1. * A→(B→A)
    2.   A∧¬A
    3. * ¬(A∧¬A)
    4.   ¬(R∨¬R)
    5.   P→[P∧(¬P→R)]
    6.   P↔(Q∨¬P)
    7.   Q↔(Q∨P)
    8.   R∧[(S∨¬T)↔(R∧T)]
    9.   Q↔¬¬Q
    10.   P∨¬(Q→¬P)

B. Construct a truth table for the following sets of propositions. Determine whether the set is consistent or inconsistent.
    1. * A↔C, ¬C∨¬A
    2.   ¬(A∧¬A), A∨¬A
    3. * P∨S, S∨P
    4.   P→(R∨¬R), (R∨¬R)→P
    5.   P↔R, ¬R↔¬P, (P→R)∧(R→P)
    6.   P, ¬¬P
    7.   P, (P∨Q)∨P
    8.   P→Q, ¬P∨¬Q, P

9.   P↔Q, ¬P↔¬Q
10.   P→Q, Q→P, ¬P→Q

C. Construct a truth table for the following pairs of propositions. Determine whether the pairs are equivalent.
   1. * (A↔B), (A→B)∧(B→A)
   2.   ¬(A∧¬A), A∨¬A
   3. * (P→R)∧(R→P), P↔R
   4.   P→(S∧M), ¬P∨(S∧M)
   5.   ¬P→¬R, R∧(¬P∧R)
   6.   P, ¬¬P
   7.   R, ¬(R→S)→S
   8.   R∨¬P, P→R
   9.   P→Q, ¬Q→¬P
  10.   S∧T, T∧S, ¬T∨¬S

D. Construct a truth table for the following arguments. Determine whether the argument is valid or invalid.
   1. * A ⊢ A
   2.   (A∧B)∧¬A ⊢ B
   3.   (¬M∨¬P)→M, ¬M∨¬P ⊢ M
   4. * P→M, ¬P ⊢ ¬M
   5.   A∨S, ¬S ⊢ A
   6.   P↔M ⊢ (P→M)∧(M → P)
   7. * A∨B ⊢ ¬(B∨¬B)
   8.   P, P∨¬Q ⊢ P∧¬Q
   9.   ¬(P∧Q) ⊢ ¬P∨¬Q
  10.   ¬(P∧Q) ⊢ ¬P∧¬Q

E. Determine the truth value for each of the following propositions where *v*(P) = T, *v*(W) = F, and *v*(Q) = T. Note that the truth value of some propositions can be determined even if their truth-value input is not known.
   1. * P∧¬W
   2.   (P→W)→¬W
   3. * (P∨W)↔¬Q
   4.   P∨¬R
   5. * ¬(P∨R)↔(¬P∧¬R)
   6.   ¬(P→Q)→W
   7. * ¬S∨S
   8.   ¬(S∧¬S)
   9. * W→W
  10.   (P∧¬P)→W

F. The main purpose of truth tables is their use as a *decision procedure*. A decision procedure is a mechanical test that can be used to determine whether a proposition, set of propositions, or argument has or does not have a certain logical property. Thus, provided we can translate an English sentence into the language of propositional logic, we don't have to think about whether a proposition is logically contingent, or whether two propositions are equivalent, or whether an argument is valid. We can simply calculate this with a truth table! This might not seem that important when considering simple arguments like *John is tall and a fisherman; therefore, John is tall*. It is extremely important, however, when considering longer arguments about issues we take to be important (e.g., a scientist's argument for why relativity theory is false, a politician's argument for why we should raise taxes, or your significant other's argument for why you should meet his or her parents). Just like arithmetical equations require a lot of energy to solve by counting or calculating by hand, more complex propositions, sets of propositions, and arguments require a lot of mental effort to analyze. Just as arithmetical calculators make our lives easier, so do truth-functional calculators. In thinking about the practical use of truth tables, answer the following questions associated with the three scenarios below.

Example 1: Suppose a political candidate, John McSneaky, claims that once elected, he will do five things and asserts them by uttering the following propositions: 'P,' 'Q,' 'R,' 'S,' and 'T.' Now, you do not know if John McSneaky is lying to you or not. It may be the case that he has no intention of doing any of the things that he asserts he will do. Generally, a number of factors go into determining whether or not we believe what John McSneaky says. But if '{P, Q, R, S, T}' are logically inconsistent, then what do we know about John McSneaky's claims? If John McSneaky made one hundred claims during his campaign, how might a truth-functional calculator help? Is there any reason why you might still vote for John?

Example 2: Suppose that an esteemed scientist, Mary McSneaky, asserts the following propositions: 'P,' 'Q,' 'R,' 'S,' and 'T.' Now, you do not know if Mary McSneaky is correct or incorrect since she is a specialist in her field and you are not. However, if '{P, Q, R, S, T}' are logically inconsistent, then we definitely know one thing: everything Mary McSneaky's asserts cannot be true since it is logically impossible for 'P,' 'Q,' 'R,' 'S,' and 'T' to be true at the same time. Suppose that a rival scientist points this out to her. What does this mean for Mary's future research? Will all of her work have to be abandoned? How might a truth-functional calculator help her in the future?

Example 3: Suppose you are a jury member, John McSneaky has lost the election (see example 1), and he is suspected of having turned to a life of crime. He is arrested for murder, and the prosecutor asserts the following argument: 'P,' 'Q,' therefore 'R.' You are instructed by the judge to vote guilty if and only if the prosecutor's conclusion (that John is the murderer) is true and his argument is valid. As you sit and listen to the prosecutor, he is extremely persuasive, good-looking, and comforting to witnesses who swear that John is the killer. He seems like an all-around good guy, and all of his premises are true. All of your fellow jury members are convinced that John McSneaky is guilty! He must die! What might keep you from issuing a guilty verdict, and how might the use of a truth-functional calculator help you to make your decision?

Solutions to Starred Exercises in End-of-Chapter Exercises

A.

1. * A→(B→A); tautology.

A	B	A	→	(B	→	A)
T	T	T	T	T	T	T
T	F	T	T	F	T	T
F	T	F	T	T	F	F
F	F	F	T	F	T	F

3. * ¬(A∧¬A); tautology.

A	¬	(A	∧	¬	A)
T	T	T	F	F	T
F	T	F	F	T	F

B.

1. * A↔C, ¬C∨¬A; consistent.

A	C	A	↔	C		¬	C	∨	¬	A
T	T	T	**T**	T		F	T	**F**	F	T
T	F	T	**F**	F		T	F	**T**	F	T
F	T	F	**F**	T		F	T	**T**	T	F
F	F	F	**T**	F		T	F	**T**	T	F

3.* P∨S, S∨P; consistent.

P	S	P	∨	S		S	∨	P
T	T	T	**T**	T		T	**T**	T
T	F	T	**T**	F		F	**T**	T
F	T	F	**T**	T		T	**T**	F
F	F	F	**F**	F		F	**F**	F

C.

1. * (A↔B), [(A→B)∧(B→A)]; equivalent.

A	B	A	↔	B		(A	→	B)	∧	(B	→	A)
T	T	T	**T**	T		T	T	T	**T**	T	T	T
T	F	T	**F**	F		T	F	F	**F**	F	T	T
F	T	F	**F**	T		F	T	T	**F**	T	F	F
F	F	F	**T**	F		F	T	F	**T**	F	T	F

3. * (P→R)∧(R→P), P↔R; equivalent.

P	R	(P	→	R)	∧	(R	→	P)		P	↔	R
T	T	T	T	T	**T**	T	T	T		T	**T**	T
T	F	T	F	F	**F**	F	T	T		T	**F**	F
F	T	F	T	T	**F**	T	F	F		F	**F**	T
F	F	F	T	F	**T**	F	T	F		F	**T**	F

D.

1. * A ⊢ A; valid.

A	⊢	A
T		T
F		F

4. * P→M, ¬M ⊢ ¬P; valid.

P	M	P	→	M		¬	M	⊢	¬	P
T	T	T	T	T		F	T		F	T
T	F	T	F	F		T	F		F	T
F	T	F	T	T		F	T		T	F
F	F	F	T	F		T	F		T	F

7. * A∨B ⊢ ¬(B∨¬B); invalid.

A	B	(A	∨	B)	⊢	¬	(B	∨	¬	B)
T	T	T	T	T		F	T	T	F	T
T	F	T	T	F		F	F	T	T	F
F	T	F	T	T		F	T	T	F	T
F	F	F	F	F		F	F	T	T	F

E.

1. * $v(P∧¬W) = T$
3. * $v[(P∨W)↔¬Q] = F$
5. * $v[¬(P∨R)↔(¬P∧¬R)] = T$
7. * $v(¬S∨S) = T$
9. * $v(W→W) = T$

DEFINITIONS

Valuation	A valuation in PL is an assignment of a truth value ('T' or 'F') to a proposition.
Decision procedure	A decision procedure is a mechanical method that determines in a finite number of steps whether a proposition, set of propositions, or argument has a certain logical property.

Tautology	A proposition '**P**' is a tautology if and only if '**P**' is true under every valuation. A truth table for a tautology will have all 'Ts' under its main operator (or in the case of no operators, under the propositional letter).
Contradiction	A proposition '**P**' is a contradiction if and only if '**P**' is false under every valuation. A truth table for a contradiction will have all 'Fs' under its main operator (or in the case of no operators, under the propositional letter).
Contingency	A proposition '**P**' is a contingency if and only if '**P**' is neither always false under every valuation nor always true under every valuation. A truth table for a contingency will have at least one 'T' and at least one 'F' under its main operator (or in the case of no operators, under the propositional letter).
Equivalence	A pair of propositions '**P**' and '**Q**' is equivalent if and only if '**P**' and '**Q**' have identical truth values under every valuation. In a truth table for an equivalence, there is no row on the truth table where one of the pair '**P**' has a different truth value than the other '**Q**.'
Consistency	A set of propositions '{**P, Q, R,**. . ., **Z**}'is logically consistent if and only if there is at least one valuation where '**P**,''**Q**,''**R**,'. . .,'**Z**' are true. A truth table shows that a set of propositions is consistent when there is at least one row on the truth table where '**P**,''**Q**,''**R**,' . . .,'**Z**' are all true.
Inconsistency	A set of propositions '{**P, Q, R,**. . ., **Z**}' is logically inconsistent if and only if there is no valuation where '**P**,''**Q**,''**R**,'. . .,'**Z**' are jointly true. A truth table shows that a set of propositions is inconsistent when there is no row on the truth table where '**P**,''**Q**,''**R**,'. . .,'**Z**' are all true.
Validity	An argument '**P, Q**,. . ., **Y** ⊢ **Z**' is valid in PL if and only if it is impossible for the premises to be true and the conclusion false. A truth table shows that an argument is invalid if and only if there is no row of the truth table where the premises are true and the conclusion is false.
Invalidity	An argument '**P, Q**,. . ., **Y** ⊢ **Z**' is invalid in PL if and only if it is possible for the premises to be true and the conclusion false. A truth table shows that an argument is valid if and only if there is a row of the truth table where the premises are true and the conclusion is false.

NOTE

1. The number of rows you need to construct is determined by the number of variables. If you have one variable (e.g., 'p'), then you will only need two rows, one for 'T' and one for 'F.' If you have two variables, you will need four rows. If you have three variables, you'll need eight. You will not need to construct tables with more than three variables, but the general expression is 2^n where n equals the number of variables. For example, where $n=30$ variables, there will be 1,073,741,824 rows.

Chapter Four

Truth Trees

The major goal of this chapter is to introduce you to the truth-tree decision procedure. In the previous chapter, truth tables were employed to mechanically determine various logical properties of propositions, sets of propositions, and arguments. As a decision procedure, truth tables have the advantage of giving a complete and graphical representation of all of the possible truth-value assignments for propositions, sets of propositions, and arguments. However, this decision procedure has the disadvantage of becoming unmanageable in cases involving more than three distinct propositional letters. The goal of this chapter is to introduce a more economical decision procedure called the *truth-tree method*. The truth-tree method is capable of yielding the exact same information as the truth table method. In addition, the complexity of the truth-tree method is not a function of the number of distinct propositional letters. Thus, whereas determining whether '(R∧¬M)∨(Q∨W)' is contingent requires producing a complex sixteen-row truth table, the truth-tree method will prove to be much more economical wise. Thus, trees provide a simpler method for testing propositions, sets of propositions, and arguments for logical properties.

In addition, truth trees will be useful in a later chapter where a more expressive logical language is introduced. In that language, the truth-table method will be unsuitable because predicate logic is not a truth-functional language. However, it will be possible to make use of the truth-tree method as a *partial decision procedure*.

4.1 TRUTH-TREE SETUP AND BASICS IN DECOMPOSITION

The truth-tree method consists of the following three-step procedure:

(1) Set up the truth tree for decomposition.
(2) Decompose any propositions into a nondecomposable form using decomposition rules.
(3) Analyze the completed truth tree for a specified logical property.

In order to acquire a clearer idea of this three-step procedure, it will be necessary (1) to introduce some vocabulary to classify types of propositions and to talk about trees, (2) to examine how truth trees are constructed and how propositions are decomposed, and (3) to discuss how to analyze a tree for various logical properties.

4.1.1 Truth-Tree Setup

A truth tree consists of three columns: (1) a column consisting of numbers for propositions found in column 2, (2) a column for propositions that are undergoing decomposition, and (3) a column that identifies which proposition has been decomposed and justifies how that proposition was decomposed.

1	2	3
Numbering	Propositions	Justification

Step 1 involves setting up the truth tree by vertically stacking and numbering each proposition. To illustrate, consider the following two propositions:

$$(R \wedge \neg M), R \wedge (W \wedge \neg M)$$

Begin your setup by vertically stacking each proposition (one under the other), numbering each proposition, and then justifying why the proposition is in on that line.

1	$R \wedge \neg M$	P
2	$R \wedge (W \wedge \neg M)$	P

The order of stacking does not matter so long as all of the propositions are stacked and numbered, and a 'P' (for 'proposition') is written along the right-hand side to indicate that this particular proposition is part of the set of propositions provided.

4.1.2 Truth-Tree Decomposition

Once the set of propositions has been stacked, step 2 is to *decompose* propositions that are in decomposable form. There are nine proposition types that can undergo decomposition.

Nine Decomposable Proposition Types	
Conjunction	$P \wedge R$
Disjunction	$P \vee R$
Conditional	$P \rightarrow R$
Biconditional	$P \leftrightarrow R$
Negated conjunction	$\neg(P \wedge R)$

Negated disjunction	$\neg(P \lor R)$
Negated conditional	$\neg(P \rightarrow R)$
Negated biconditional	$\neg(P \leftrightarrow R)$
Double negation	$\neg\neg P$

Decomposing a complex proposition '**P**' consists of graphically (1) *stacking*, (2) *branching*, or (3) *stacking and branching* propositions according to certain rules called *decomposition rules*. Before giving an articulation of the decomposition rules for PL, it is important to see, in a very general and abstract way, how these rules are formulated.

Take the complex propositional form '**P#Q**,' where '**P#Q**' is a decomposable proposition, and '#' is an arbitrarily chosen truth-functional operator. As mentioned above, '**P#Q**' can be decomposed by stacking '**P**' and '**Q**,'

P#Q   P
  P   Stacking rule
  Q   Stacking rule

by branching '**P**' and '**Q**,'

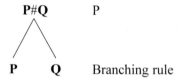

P#Q   P

P   **Q**   Branching rule

or by branching and stacking '**P**' and '**Q**,'

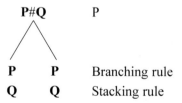

P#Q   P

P   **P**   Branching rule
Q   **Q**   Stacking rule

Whether a proposition '**P**' stacks, branches, or branches and stacks depends on the conditions under which '**P**' is true. *Stacking* occurs when the proposition being decomposed is true under only one truth-value assignment.

> Stacking rule   A stacking rule is a truth-tree rule where the condition under which a proposition '**P**' is true is represented by stacking. A stacking rule is applied to propositions that are true under one truth-value assignment.

To see this more clearly, consider the following truth functions:

P	Q	P#Q	P#Q	P#Q	P#Q
T	T	**T**	F	F	F
T	F	F	**T**	F	F
F	T	F	F	**T**	F
F	F	F	F	F	**T**
		1	2	3	4

In each of the above examples, 'P#Q' is true only under one truth-value assignment. Each of these can be represented graphically by a stacking rule. For example, suppose the truth function represented by column 1 for 'P#Q' is to be decomposed. This truth function states that 'P#Q' is true if and only if (iff) $v(P) = T$ and $v(Q) = T$. This can be represented graphically by writing 'P' and 'Q' directly under 'P#Q' in the tree.

P#Q	P
P	
Q	

To see this more clearly, consider the above diagram in an expanded form (read the diagram below from left to right)

The proposition	**P#Q**	is true
if and only if	**P**	is true
and	**Q**	is true

Consider the truth function represented by column 2 for 'P#Q.' Notice that under this valuation of 'P#Q,' $v(P\#Q) = T$ if and only if $v(P) = T$ and $v(Q) = F$. In this case, we cannot represent the conditions under which 'P#Q' is true by writing

The proposition	**P#Q**	is true
if and only if	**P**	is true
and	**Q**	is true—**NO!**

because $v(P\#Q) = T$ only if $v(Q) = F$. But since $v(Q) = F$ if and only if $v(\neg Q) = T$, we can represent the truth function represented by column 2 for 'P#Q' as follows:

The proposition	**P#Q**	is true
if and only if	**P**	is true
and	**¬Q**	is true.

Thus, in order to represent column 2 for 'P#Q,' we stack 'P' and ¬Q under 'P#Q' as follows:

P#Q	P
P	
¬Q	

The above tree represents that $v(\textbf{P\#Q}) = \text{T}$ in one and only one case. That is, $v(\textbf{P\#Q}) = \text{T}$ if and only if $v(\textbf{P}) = \text{T}$ and $v(\neg\textbf{Q}) = \text{T}$. This procedure can be used to represent the remaining truth functions in columns 3 and 4.

Branching occurs when the proposition being decomposed is false only under one truth-value assignment.

Branching rule   A branching rule is a truth-tree rule where the condition under which a proposition 'P' is true is represented by branching. A branching rule is applied to propositions that are false only under one truth-value assignment.

Consider the following truth functions:

P	Q	P#Q	P#Q	P#Q	P#Q
T	T	T	T	T	F
T	F	T	T	F	T
F	T	T	F	T	T
F	F	F	T	T	T
		1	2	3	4

In each of the above columns, '**P#Q**' is false only under one truth-value assignment. Each of these can be represented graphically by a branching rule. For example, suppose the truth function represented by column 1 for '**P#Q**' is to be decomposed. This truth function states that '**P#Q**' is true if and only if $v(\textbf{P}) = \text{T}$ or $v(\textbf{Q}) = \text{T}$. This can be represented graphically by branching a '**P**' and '**Q**' from '**P#Q**' in the tree.

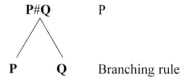

Branching '**P**' and '**Q**' graphically represents the fact that $v(\textbf{P\#Q}) = \text{T}$ if and only if $v(\textbf{P}) = \text{T}$ or $v(\textbf{Q}) = \text{T}$. We can think of the above diagram as representing the following:

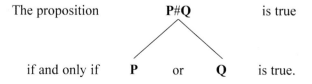

Let's turn to the truth function represented by column 2. This truth function states that $v(\mathbf{P\#Q}) = T$ if and only if $v(\mathbf{P}) = T$ or $v(\mathbf{Q}) = F$. Otherwise put, $v(\mathbf{P\#Q}) = T$ if and only if $v(\mathbf{P}) = T$ or $v(\mathbf{\neg Q}) = T$. Thus, in order to represent column 2 for '$\mathbf{P\#Q}$,' write '$\mathbf{P}$' under one branch and '$\mathbf{\neg Q}$' on the other. That is,

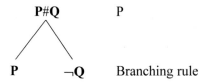

Branching rule

Again, this procedure can be used to represent the remaining truth functions in columns 3 and 4. These truth functions are left as an exercise.

Finally, *branching and stacking* occurs when the proposition is true under two truth-value assignments and false under two truth-value assignments.

Branching-and-stacking rule | A branching-and-stacking rule is a truth-tree rule where the truth conditions of a proposition '$\mathbf{P}$' are represented by branching and stacking. A branching-and-stacking rule will be used when '$\mathbf{P}$' is true under only two truth-value assignments and false under two truth-value assignments.

Below are the truth functions involving two true truth-value assignments and two false truth-value assignments.

P	**Q**	**P#Q**	**P#Q**	**P#Q**	**P#Q**	**P#Q**	**P#Q**
T	T	T	T	F	F	F	T
T	F	T	F	T	T	F	F
F	T	F	T	T	F	T	F
F	F	F	F	F	T	T	T
		1	2	3	4	5	6

The representation of these truth functions in terms of a truth tree is left as an exercise.

Exercise Set #1

A. Consider the following truth functions. Create a truth-tree decomposition rule that accurately represents columns 3 and 4.

P	**Q**	**P#Q**	**P#Q**	**P#Q**	**P#Q**
T	T	**T**	F	F	F
T	F	F	**T**	F	F
F	T	F	F	**T**	F
F	F	F	F	F	**T**
		1	2	3	4

B. Consider the following truth functions. Create a truth-tree decomposition rule that accurately represents columns 3 and 4.

P	Q	P#Q	P#Q	P#Q	P#Q
T	T	T	T	T	F
T	F	T	T	F	T
F	T	T	F	T	T
F	F	F	T	T	T
		1	2	3	4

C. A branching-and-stacking rule is a truth-tree rule where the truth conditions of a proposition 'P' are represented by branching and stacking. A branching-and-stacking rule will be used when 'P' is true under only two truth-value assignments and false under two truth-value assignments. Represent the following six truth functions with different branching-and-stacking truth-tree representations.

P	Q	P#Q	P#Q	P#Q	P#Q	P#Q	P#Q
T	T	T	T	F	F	F	T
T	F	T	F	T	T	F	F
F	T	F	T	T	F	T	F
F	F	F	F	F	T	T	T
		1	2	3	4	5	6

4.2 TRUTH-TREE DECOMPOSITION RULES

In earlier sections, we learned how to set up the truth tree for decomposition (step 1) and learned a very general approach to decomposition. In this section, we learn two of the nine decomposition rules for PL and some vocabulary for talking about and analyzing truth trees.

4.2.1 Conjunction Decomposition Rule (∧D)

Whether a stacking, branching, or branching-and-stacking rule is used depends upon the type of proposition being decomposed. Consider '(R∧¬M), R∧(W∧¬M).' Begin by setting up the tree:

$$
\begin{array}{lll}
1 & R∧¬M & P \\
2 & R∧(W∧¬M) & P
\end{array}
$$

Notice that both propositions are conjunctions of the form 'P∧Q.' From the truth table analysis of conjunctions, we know that a conjunction is true if and only if both of its conjuncts are true, and so it is true under only one truth-value assignment. Thus, in the case of line 1, $v(R∧¬M) = T$ if and only if $v(R) = T$ and $v(¬M) = T$. In order to represent that $v(R∧¬M) = T$ under one and only one truth-value assignment, a stacking rule is employed. That is, 'R' and '¬M' are stacked below the existing set of propositions.

1	R∧¬M	P
2	R ∧ (W∧¬M)	P
3	R	
4	¬M	

In the case of the example above, a stacking decomposition rule was applied to a conjunction. However, there is a more general point. Namely, since a conjunction is a type of proposition that is true under one and only truth-value assignment, a general stacking rule can be formulated that is specific to conjunctions. This is known as *conjunction decomposition* and is abbreviated as follows: '∧D.'

Conjunction Decomposition (∧D): Stacking

P∧Q	P
P	∧D
Q	∧D

We represent that a proposition has been decomposed by conjunction decomposition by writing the line number of the proposition undergoing decomposition and '∧D' in the third column. For instance, to indicate that lines 3 and 4 are the result of using conjunction decomposition on line 1, '1∧D' is written to the right of 'R' and '¬M' in lines 3 and 4, respectively:

1	R∧¬M✓	P
2	R∧(W∧¬M)	P
3	R	1∧D
4	¬M	1∧D

The above tree now indicates that 'R' at line 3 and '¬M' at line 4 came about by applying '∧D' to line 1. In order to indicate that 'R∧¬M' has had a decomposition rule applied to it, a checkmark (✓) is placed to the right of it. This indicates that the proposition has been decomposed.

A fully decomposed truth tree is a tree where all the propositions that can be decomposed have been decomposed.

Fully decomposed tree   A fully decomposed truth tree is a tree where all the propositions that can be decomposed have been decomposed.

In the tree above, the proposition at line 2 can be decomposed but has not been decomposed, so the tree is not a fully decomposed truth tree. Thus, the next step is to decompose any remaining propositions. Notice that line 2 is a conjunction. Thus, '∧D' can be applied to it. This produces the following truth tree:

1	R∧¬M✓	P
2	R∧(W∧¬M)✓	P

3	R	1∧D
4	¬M	1∧D
5	R	2∧D
6	W∧¬M	2∧D

We still do not have a fully decomposed truth tree since line 6 can be decomposed but has not been decomposed. Thus, we can apply another use of '∧D' to line 6. This produces the following:

1	R∧¬M✓	P
2	R∧(W∧¬M)✓	P
3	R	1∧D
4	¬M	1∧D
5	R	2∧D
6	W∧¬M✓	2∧D
7	W	6∧D
8	¬M	6∧D

Notice that 'R∧(W∧¬M)' had to undergo two separate decompositions: first, a decomposition of 'R∧(W∧¬M),' and then a decomposition of 'W∧¬M.' The above tree is now a fully decomposed tree since all the propositions that can be decomposed have been.

4.2.2 Disjunction Decomposition Rule (∨D)

So far, we have addressed how to set up the truth tree, three general ways propositions can be decomposed, and the specific decomposition rule for conjunctions. What remains is to consider (1) decomposition rules for other types of propositions, and (2) vocabulary for analyzing truth trees for various logical properties.

There are eight remaining decomposition rules, and each applies to one of the remaining eight basic propositional forms.

Nine Decomposable Proposition Types		Decomposition Rule
Conjunction	**P∧R**	∧D
Disjunction	**P∨R**	∨D
Conditional	**P→R**	→D
Biconditional	**P↔R**	↔D
Negated conjunction	**¬(P∧R)**	¬∧D
Negated disjunction	**¬(P∨R)**	¬∨D
Negated conditional	**¬(P→R)**	¬→D
Negated biconditional	**¬(P↔R)**	¬↔D
Double negation	**¬¬P**	¬¬D

The next decomposition rule to consider is '∨D,' which applies to disjunctions. Consider the set of propositions 'R∧W' and 'M∨¬W.' Again, start by stacking these propositions.

$$
\begin{array}{ccc}
1 & R{\wedge}W & P \\
2 & M{\vee}\neg W & P \\
\end{array}
$$

Next, decompose the 'R∧W' with conjunction decomposition by stacking 'R' and 'W' under the existing stack.

$$
\begin{array}{ccc}
1 & R{\wedge}W\checkmark & P \\
2 & M{\vee}\neg W & P \\
3 & R & 1{\wedge}D \\
4 & W & 1{\wedge}D \\
\end{array}
$$

Line 1 is decomposed, and the only decomposable proposition remaining is the disjunction at line 2. However, *conjunction decomposition* (∧D) cannot be applied at line 2 because line 2 is not a conjunction but a disjunction.

In determining how to formulate a decomposition rule for disjunctions, consider the truth table for disjunctions.

P	**Q**	**P∨Q**
T	T	T
T	F	T
F	T	T
F	F	F

Notice that $v(\mathbf{P}{\vee}\mathbf{Q}) = F$ in one and only one case, and it is true in all others. Remember that a stacking rule is applied if and only if a proposition is true in one case, and a branching rule is applied to propositions that are false only under one truth-value assignment. Thus, disjunctions will branch when decomposed. In order to determine what kind of branching rule to use, consider the four possibilities:

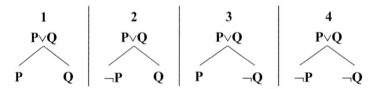

Consider (2) as a possible candidate for **P∨Q**. This tree says that $v(\mathbf{P}{\vee}\mathbf{Q}) = T$ if and only if $v(\mathbf{P}) = T$ or $v(\neg\mathbf{Q}) = T$; that is, $v(\mathbf{Q}) = F$. This does not represent the truth conditions for **P∨Q** since row 3 of the truth table says that if $v(\mathbf{P}) = F$ and $v(\mathbf{Q}) = T$, then $v(\mathbf{P∨Q}) = T$. The only acceptable candidate is (1) since it represents the fact that a disjunction is true if either of its disjuncts are true. Thus, disjunction decomposition (∨D) is the following decomposition rule:

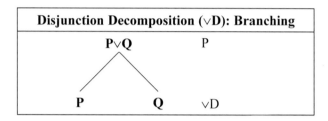

Applying '∨D' to the above tree produces the following:

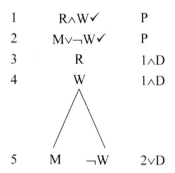

Before continuing further, it is helpful at this point to introduce some more terminology to talk about the tree and its branches.

First, all of the propositions in a branch can be catalogued by starting from the bottom of a branch and moving upward through the branch to the top of the tree.

Branch   A branch includes all the propositions obtained by starting from the bottom of the tree and reading upward through the tree.

For example, note that in the above tree, there are two branches.

Propositions in the Branch	
Branch 1	M, W, R, M∨¬W, R∧W
Branch 2	¬W, W, R, M∨¬W, R∧W

The following tree has eight branches.

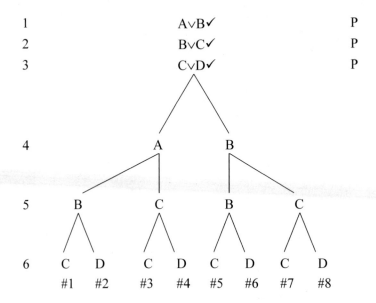

1	A∨B✓	P
2	B∨C✓	P
3	C∨D✓	P

Second, a branch of a tree is either fully or partially decomposed.

| Fully decomposed branch | A branch is fully decomposed when all propositions in the branch that can be decomposed have been decomposed. |
| Partially decomposed branch | A branch is partially decomposed when there is at least one proposition in the branch that has not been decomposed. |

For example, consider the following partially decomposed tree consisting of 'R∧W' and 'M∨¬W.'

1	R∧W✓	P
2	M∨¬W	P
3	R	1∧D
4	W	1∧D

Notice that there is only one branch in the tree above and that it contains a decomposable proposition that has not been decomposed. That is, 'M∨¬W' is decomposable but has not been decomposed.

In contrast, notice that the tree below consists of two branches, and both are fully decomposed.

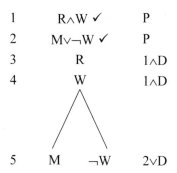

1	R∧W ✓	P
2	M∨¬W ✓	P
3	R	1∧D
4	W	1∧D
5	M ¬W	2∨D

Third, there are (1) *closed branches*, (2) *open branches*, and (3) *completed open branches*. A branch is closed (or closes) provided that branch contains a proposition and its literal negation (e.g., 'P' and '¬P').

> Closed branch   A closed branch contains a proposition '**P**' and its literal negation '¬**P**.' A closed branch is represented by an '**X**.'

For example, consider the following tree:

1	R∧W✓	P
2	¬W	P
3	R	1∧D
4	W	1∧D
	X	

Notice that there is one branch in the above tree, and that branch contains 'W' at line 4 and the literal negation of 'W' at line 2. The above branch is closed since it contains 'W' and '¬W' in the branch.

A branch is open if and only if that the branch is not closed. That is, the branch does not contain any instance of a proposition and its literal negation.

> Open branch   An open branch is a branch that is not closed, that is, a branch that does not contain a proposition '**P**' and its literal negation '¬**P**.'

For example, consider the following partially decomposed tree consisting of 'R∧W' and 'M∨¬W.'

1	R∧W✓	P
2	M∨¬W	P
3	R	1∧D
4	W	1∧D

Notice that there is one branch, and it is open since the branch does not contain a proposition and its literal negation.

However, it is important to distinguish between a *completed open branch* and an open branch more generally. In the above example, note that the branch is open but not fully decomposed. If the branch were open and fully decomposed, then the branch would be a completed open branch.

Completed open branch   A completed open branch is a fully decomposed branch that is not closed. That is, it is a fully decomposed branch that does not contain a proposition and its literal negation. An open branch is denoted by writing an '**0**' at the bottom of the tree.

To bring much of the above terminology together, consider that once line 2 in the above tree has been decomposed, there are no undecomposed propositions in the tree, and so the tree below has two fully decomposed branches.

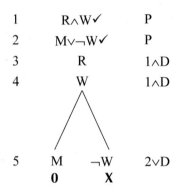

1	R∧W✓	P
2	M∨¬W✓	P
3	R	1∧D
4	W	1∧D
5	M ¬W	2∨D
	0 **X**	

In the above tree, notice that the leftmost branch (consisting of 'M,' 'W,' and 'R') is a completed open branch (indicated by '**0**'), and the rightmost branch (consisting of '¬W,' 'W,' and 'R') is a closed branch (indicated by '**X**'). The leftmost branch is a completed open branch because (1) it does not contain any instance of a proposition and its literal negation, and (2) it is fully decomposed.

The rightmost branch is a closed branch since it contains a proposition and its literal negation (i.e., '¬W' at line 5 and 'W' at line 4).

Consider another example of a tree examined earlier in this chapter:

1	R∧¬M✓	P
2	R∧(W∧¬M)✓	P
3	R	1∧D
4	¬M	1∧D
5	R	2∧D
6	W∧¬M✓	2∧D
7	W	6∧D
8	¬M	6∧D
	0	

In the above example, there is only one branch consisting of the following nonde-composable propositions: '¬M,' 'W,' and 'R.' Since the branch is fully decomposed and does not contain an instance of a proposition and its literal negation, the branch is a completed open branch.

Fourth, and finally, using the above terminology for branches, it is now possible to talk about the whole tree. A tree is classified into the following three types:

(1) An uncompleted open tree
(2) A completed open tree
(3) A closed tree

A tree is a *completed open tree* if and only if there is one completed open branch.

Completed open tree   A tree is a completed open tree if and only if it has at least one completed open branch. That is, a tree is a completed open tree if and only if it contains at least one fully decomposed branch that is not closed. A completed open tree is a tree where there is at least one branch that has an '**O**' under it.

Consider the following tree:

1	R∧¬M✓	P
2	R∧(W∧¬M)✓	P
3	R	1∧D
4	¬M	1∧D
5	R	2∧D
6	W∧¬M✓	2∧D
7	W	6∧D
8	¬M	6∧D
	O	

It was determined above that this tree has a completed open branch. Thus, the above tree is a completed open tree.

Consider a second tree:

1	R∧W✓	P
2	M∨¬W✓	P
3	R	1∧D
4	W	1∧D

```
              /\
             /  \
            /    \
5      M        ¬W      2∨D
       O         X
```

Notice that the leftmost branch is a completed open branch. Thus, the tree is a completed open tree.

A tree is *closed* if and only if all of its branches are closed. That is, if every branch in a tree has an '**X**' under it, then the tree is closed.

> Closed tree   A tree is closed when all of the tree's branches are closed. A closed tree will have an '**X**' under every branch.

Consider the following tree:

```
1        R∧W✓      P
2         ¬W       P
3          R       1∧D
4          W       1∧D
           X
```

Notice that all of the tree's branches are closed. Thus, the tree is closed. Here is another example:

```
1        R∧W✓           P

2       ¬R∨¬W✓          P

3          R            1∧D

4          W            1∧D
                /\
               /  \
              /    \
5          ¬R      ¬W      2∨D
           X        X
```

Notice that the above tree consists of two branches. Both are closed, and so the tree is closed.

Next, consider the following tree consisting of 'A→B,' '¬(A∨B),' and 'B∧C':

```
1        A→B✓           P
2       ¬(A∨B)✓         P
3        B∧C✓           P
4          B            3∧D
5          C            3∧D
6         ¬A            2¬∨D
```

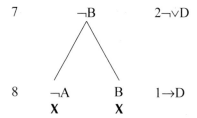

The above tree uses some decomposition rules that have not been introduced, but you should be able to determine that the tree is closed because every branch is closed. The rightmost branch is closed because it contains a proposition and its literal negation ('B' and '¬B'), but also notice that the leftmost branch is closed because 'B' (line 4) and '¬B' (line 7) are in the leftmost branch.

As one last example, consider the following tree consisting of 'P∨Q' and '¬P∧Q.'

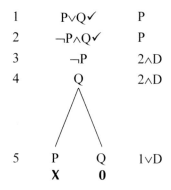

Notice that the above tree is not closed. This tree is a completed open tree because the tree has at least one completed open branch. This is the rightmost branch consisting of 'Q,' 'Q,' and '¬P.' Remember that in order for a tree to be closed, all of the branches must be closed, while in order for a tree to be a completed open tree, there needs to be at least one completed open branch.

Finally, an *uncompleted open tree* is a tree that is neither a completed open tree nor a closed tree. This is a tree that is unfinished because all of the branches have not been closed or there is not one completed open branch.

4.2.3 Decompose Propositions under Every Descending Open Branch

Before proceeding to the remaining decomposition rules, there is one more rule associated with the decomposition of truth trees. Consider the following set of propositions: 'R∧W' and 'C∨D.' First, the tree is set up by stacking all of the propositions.

1	R∧W	P
2	C∨D	P

Next, choose one of the propositions to decompose. Suppose that 'C∨D' is chosen from line 2.

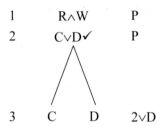

In the above tree, 'C∨D' is decomposed. However, 'R∧W' has not been decomposed. In decomposing line 1, you will decompose it under *every* open branch that descends from and contains '**P**.'

So, in the case of the above tree, there are two open branches that descend from 'R∧W,' and so 'R∧W' must be decomposed under both of these branches. The completed tree is as follows:

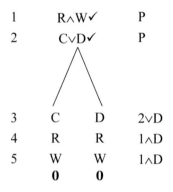

The general rule for this is called the *decomposition descending rule*. It states that when decomposing a proposition '**P**,' decompose '**P**' under every open branch that descends from '**P**.'

Decomposition descending rule	When decomposing a proposition '**P**,' decompose '**P**' under every open branch that descends from '**P**.'

Consider the following propositions: 'R∨(P∧M)' and 'C∨D.' Again, start by stacking the propositions and then decompose line 1.

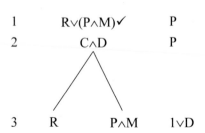

Notice that there are two decomposable propositions remaining: 'C∧D' at line 2 and 'P∧M' at line 3. According to the decomposition descending rule, 'C∧D' must be decomposed under every open branch that descends from 'C∧D.' Since the left and right branches are both open, 'C∧D' must be decomposed under both of these branches.

```
1              R∨(P∧M)✓        P
2                C∧D✓          P
                  /\
                 /  \
3          R          P∧M      1∨D
4          C           C       2∧D
5          D           D       2∧D
```

Notice that the above tree is still not fully decomposed because there is a remaining decomposable proposition in the rightmost branch. In order to complete the truth tree, 'P∧M' at line 3 must be decomposed. Since propositions are decomposed under every remaining open branch that descends from the proposition upon which a decomposition rule is applied, 'P∧M' will only be decomposed in the rightmost branch. Thus, the completed tree is as follows:

```
1              R∨(P∧M)✓        P
2                C∧D✓          P
                  /\
                 /  \
3          R          P∧M✓     1∨D
4          C           C       2∧D
5          D           D       2∧D
6          0           P       3∧D
7                      M       3∧D
                       0
```

Notice that 'P∧M' is not decomposed under the leftmost branch. This is because propositions are decomposed under open branches that descend from that proposition. Thus, decomposing 'P∧M' at line 3 in the following way would be a violation of the decomposition descending rule:

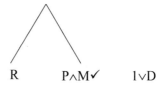

```
3                R        P∧M✓      1∨D
```

4		C	C	2∧D
5		D	D	2∧D
6	NO! →	**P**	P	3∧D
7	NO! →	**M**	M	3∧D

Finally, consider the tree consisting of 'P∧¬P' and 'W∨L.'

1	P∧¬P✓	P
2	W∨L	P
3	P	1∧D
4	¬P	1∧D
	X	

The decomposition descending rule states that when a proposition '**P**' is decomposed, '**P**' should be decomposed under every open branch that descends from '**P**.' However, in the above tree, there are no open branches since 'P' and '¬P' form a closed branch. At this point, the branch is closed, and the tree is complete and considered closed since any further decomposition will only yield more closed branches. To illustrate, consider what would happen if line 2 were decomposed:

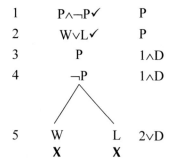

Notice that the leftmost branch closes because it contains 'P' and '¬P' and the rightmost branch closes because it contains 'P' and '¬P' at lines 3 and 4.

Exercise Set #2

A. Using the truth-tree method, decompose the following sets of propositions to determine whether the tree is an open or closed tree.

   1. * P∧(R∧S)
   2.   P∨(R∨D)
   3. * P∨Q, ¬P∧Q
   4.   P∧(R∧D), Z∧M
   5. * P∧(P∨Z), P∧¬P
   6.   W∧¬W, Z∨(D∨R)
   7. * (P∧Q)∧W, ¬W ∨ M

8.   (R∨W)∨¬P, (¬R∧¬W)∧P
9. * (R∨¬W)∨¬P, C∧D, (¬R∧R)∧M
10.   D∨¬D, Z∧[(Z∧P)∧¬R]

Solutions to Starred Exercises in Exercise Set #2

1. * P∧(R∧S); open tree.

1	P∧(R∧S)✓	P
2	P	1∧D
3	R∧S✓	1∧D
4	R	3∧D
5	S	3∧D
	0	

3. * P∨Q, ¬P∧Q; open tree.

1	P∨Q✓	P
2	¬P∧Q ✓	P
3	¬P	2∧D
4	Q	2∧D

5	P Q	1∨D
	X **0**	

5. * P∧(P∨Z), P∧¬P; closed tree.

1	P∧(P∨Z)✓	P
2	P∧¬P✓	P
3	P	2∧D
4	¬P	2∧D
	X	

7. * (P∧Q)∧W, ¬W∨M; open tree.

1	(P∧Q)∧W✓	P
2	¬W ∨ M✓	P
3	P∧Q✓	1∧D
4	W	1∧D
5	P	3∧D
6	Q	3∧D

7	¬W M	2∨D
	X **0**	

9. * (R∨¬W)∨¬P, C∧D, (¬R∧R)∧M; closed tree.

1	(R∨¬W)∨¬P	P
2	C∧D	P
3	(¬R∧R)∧M✓	P
4	¬R∧R✓	3∧D
5	M	3∧D
6	¬R	4∧D
7	R	4∧D
	X	

4.3 THE REMAINING DECOMPOSITION RULES

Thus far, we have formulated two decomposition rules (∧D and ∨D). What remains is to formulate the decomposition rules for the remaining proposition types and to learn how to read trees for logical properties (e.g., equivalence, tautology, validity, etc.). This section addresses the remaining seven decomposition rules, the next section addresses the strategic use of rules, and a later section addresses logical properties.

4.3.1 Conditional Decomposition (→D)

Consider the truth table for a conditional ('**P**→**Q**').

P	**Q**	**P→Q**
T	T	T
T	F	F
F	T	T
F	F	T

Notice that a conditional is false under one and only one truth-value assignment. That is, $v(\mathbf{P}{\to}\mathbf{Q}) = F$ if and only if $v(\mathbf{P}) = T$ and $v(\mathbf{Q}) = F$. Thus, conditionals will branch when decomposed. In order to determine what kind of branching rule to use, again consider the four possibilities:

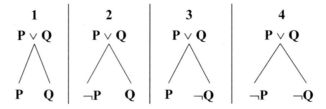

Rather than testing each of these possibilities, notice from the truth table that a conditional ('**P**→**Q**') is true if and only if either $v(\mathbf{P}) = F$ or $v(\mathbf{Q}) = T$. In other words,

$v(\mathbf{P}{\rightarrow}\mathbf{Q}) = T$ if and only if $v(\neg\mathbf{P}) = T$ or $v(\mathbf{Q}) = T$. Thus, the decomposition rule for conditionals ($\rightarrow$D) can be represented as the following branching rule:

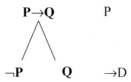

As a quick illustration, consider the following tree for 'P→(W→Z)' where '→D' is used twice:

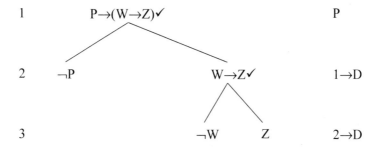

Notice that in the above tree when '→D' is used to decompose 'W→Z,' '¬W' and 'Z' are placed under 'W→Z' on the right-hand side of the branch and not the left-hand side.

4.3.2 Biconditional Decomposition (↔D)

Consider the truth table for a biconditional ('P↔Q').

P	Q	P↔Q
T	T	T
T	F	F
F	T	F
F	F	T

A biconditional is a proposition that is true in two cases and false in two cases. In order to represent this, we will use a combination of stacking and branching.

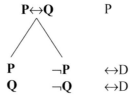

As a quick illustration, consider the following tree for 'P↔(W∧Z),' where '↔D' is used:

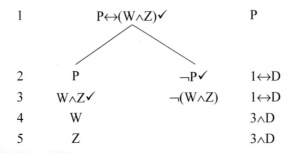

There are two things to note in the above tree. First, when ↔D is applied to line 1, on the left-hand side we have 'P' and 'W∧Z,' and the latter proposition can be decomposed using '∧D.' However, notice that on the right-hand side we have '¬P' and '¬(W∧Z).' While '¬P' cannot be decomposed further, '¬(W∧Z)' can, but we currently do not have a rule for how to decompose it.

4.3.3 Negated Conjunction Decomposition (¬∧D)

A negated conjunction is only false in one case: when both of the conjuncts are false. This means that a negated conjunction is true if either of the negated conjuncts is true. That is, '¬∧D' is a branching rule.

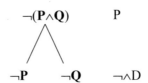

In order to illustrate '¬∧D,' we return to the following tree considered in our discussion of '↔D':

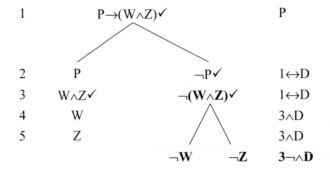

Note that an application of '¬∧D' to '¬(W∧Z)' branches where a negated proposition results on each side of the branch.

4.3.4 Negated Disjunction Decomposition (¬∨D)

A negated disjunction is only true in one case: when both of its constituents are false. That is, a negated disjunction is true if and only if both of the negated disjuncts are true. Therefore, it is a stacking rule.

$$
\begin{array}{ll}
\neg(P\lor Q) & P \\
\neg P & \neg\lor D \\
\neg Q & \neg\lor D
\end{array}
$$

As a quick illustration, consider the following tree for '¬[Z∨(B∨R)],' where '¬∨D' is used twice.

$$
\begin{array}{lll}
1 & \neg[Z\lor(B\lor R)]\checkmark & P \\
2 & \neg Z & 1\neg\lor D \\
3 & \neg(B\lor R)\checkmark & 1\neg\lor D \\
4 & \neg B & 3\neg\lor D \\
5 & \neg R & 3\neg\lor D
\end{array}
$$

4.3.5 Negated Conditional Decomposition (¬→D)

A negated conditional is true in one case: if and only if the both the antecedent is true and the consequent is false. Therefore, it is a stacking rule.

$$
\begin{array}{ll}
\neg(P\to Q) & P \\
P & \neg\to D \\
\neg Q & \neg\to D
\end{array}
$$

As a quick illustration, consider the following tree for '¬[L→(B→R)],' where '¬→D' is used twice.

$$
\begin{array}{lll}
1 & \neg[L\to(B\to R)]\checkmark & P \\
2 & L & 1\neg\to D \\
3 & \neg(B\to R)\checkmark & 1\neg\lor D \\
4 & B & 3\neg\lor D \\
5 & \neg R & 3\neg\lor D
\end{array}
$$

4.3.6 Negated Biconditional Decomposition (¬↔D)

A negated biconditional is true in two cases and false in two cases. Therefore, it both stacks and branches.

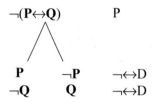

As a quick illustration, consider the following tree for '¬[P↔(W∨Z)],' where '¬↔D' is used:

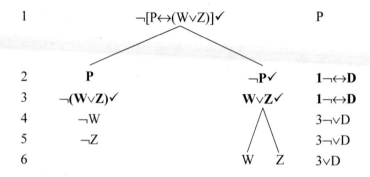

4.3.7 Double Negation Decomposition (¬¬D)

A proposition that is doubly negated is true in one case, when the proposition in un-negated form is true. Therefore, it stacks.

$$\begin{array}{cc} \neg\neg\mathbf{P} & \mathrm{P} \\ \mathbf{P} & \neg\neg\mathrm{D} \end{array}$$

As a quick illustration, consider the following tree for '¬[¬P∨¬Z)],' where there are two uses of '¬¬D' after a use of '¬∨D':

1	¬(¬P∨¬Z)✓	P
2	¬¬P✓	1¬∨D
3	¬¬Z✓	1¬∨D
4	P	2¬¬D
5	Z	3¬¬D

4.3.8 Truth-Tree Rules

Stacking	Branching	Stacking and Branching
P∧Q P ∧D Q ∧D	¬(P∧Q) ¬P ¬Q ¬∧D	P↔Q P ¬P ↔D Q ¬Q ↔D
¬(P∨Q) ¬P ¬∨D ¬Q ¬∨D	P∨Q P Q ∨D	¬(P↔Q) P ¬P ¬↔D ¬Q Q ¬↔D
¬(P→Q) P ¬→D ¬Q ¬→D	P→Q ¬P Q →D	
¬¬P P ¬¬D		

Exercise Set #3

A. Using the truth-tree method, decompose the following sets of propositions to determine whether the tree is an open or closed tree.

    1. * P→R, ¬(P→Z)
    2.   ¬(P∨L), P→Z
    3. * ¬(P∧M), P
    4.   P↔R, P→R
    5. * P∧¬L, ¬(P→L), Z∨F
    6.   ¬P→(Z∧M), ¬(P→Z)
    7. * R→(R∨L), Z↔(Q∧¬R)
    8.   ¬¬P, (P∧¬P)→(Z∨V)
    9.   ¬(Z↔L), Z→(P∧V)
  10.   ¬(¬Z∨P)↔(P∧¬V)
  11.   ¬(¬Z→P), ¬[P∨(¬V→R)]
  12.   F∨(T∨R), ¬(P∧R), P∧(¬P∧R)
  13.   M∧(¬P∧¬Z), Z∧L, F→(R∧T)
  14.   (P∧Z)→¬Z, Z∧¬(M∨V)
  15.   (M∨T)↔¬(M↔¬Z)

Solutions to Starred Exercises in Exercise Set #3

1. * P→R, ¬(P→Z); open tree.

1	P→R✓	P
2	¬(P→Z)✓	P
3	P	2¬→D
4	¬Z	2¬→D

5	¬P	R	1→D
	X	0	

3. * ¬(P∧M), P; open tree.

1	¬(P∧M)✓	P
2	P	P

3	¬P	¬M	1¬∧D
	X	0	

5. * P∧¬L, ¬(P→L), Z∨F; open tree.

1	P∧¬L✓	P
2	¬(P→L)✓	P
3	Z∨F✓	P
4	P	1∧D
5	¬L	1∧D
6	P	2¬→D
7	¬L	2¬→D

8	Z	F	3∨D
	0	0	

7. * R→(R∨L), Z↔(Q∧¬R); open tree.

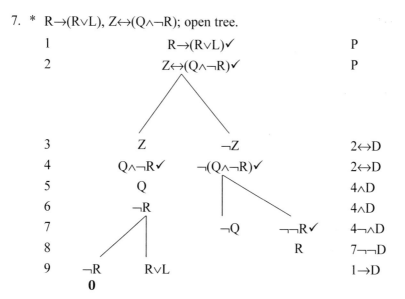

1	R→(R∨L)✔	P
2	Z↔(Q∧¬R)✔	P
3	Z ¬Z	2↔D
4	Q∧¬R✔ ¬(Q∧¬R)✔	2↔D
5	Q	4∧D
6	¬R	4∧D
7	¬Q ¬¬R✔	4¬∧D
8	R	7¬¬D
9	¬R R∨L	1→D

4.4 BASIC STRATEGIES

Before analyzing the remaining rules for truth-functional operators, it is helpful to formulate a number of strategic rules that will simplify the decomposition of truth trees.

Strategic rule 1   Use no more rules than needed.
Strategic rule 2   Use rules that close branches.
Strategic rule 3   Use stacking rules before branching rules.
Strategic rule 4   Decompose more complex propositions before simpler propositions.[1]

These rules are listed in order of decreasing priority. That is, if you have a choice between using strategic rule 1 or strategic rule 3, use 1.

4.4.1 Strategic Rule 1

Strategic rule 1 is the following:

Strategic rule 1   Use no more rules than needed.

Suppose that all you wanted to know about a particular proposition or set of propositions was whether or not it produced an open tree. It is not always necessary to produce a fully developed tree in order to make this determination since a completed open tree is a tree with at least one completed open branch. For instance, consider whether the following set of propositions is consistent: '(P∧¬W)∧¬M' and 'M∨(¬M∨P).'

```
1        (P∧¬W)∧¬M✓      P

2        ¬M∨(¬M∨P)✓      P

3          P∧¬W✓         1∧D

4            ¬M           1∧D

5             P           3∧D

6            ¬W           3∧D
                /\
               /  \
              /    \
             /      \
7        ¬M        ¬M∨P   2∨D
         0
```

Now, the above tree is not fully decomposed since there is still a complex proposition '¬M∨P' (at line 7). However, there is no reason to decompose it because a tree is a completed open tree if and only if there is at least one completed open branch.

4.4.2 Strategic Rule 2

Strategic rule 2 is the following:

Strategic rule 2   Use rules that close branches.

It is generally advisable to decompose propositions that will close branches. The reason for this is that whenever you decompose a complex proposition, it must be decomposed into every open branch below the decomposed proposition. Consider the following example:

```
1        ¬M∨¬P      P
2        M∨P        P
3        P∨Q        P
```

In the stack above, we want to construct the simplest possible tree to determine whether the tree is open or closed. Strategic rule 2 suggests that we chose rules that are likely to close branches. These are lines 1 and 2. Let us decompose 'M∨P,' then '¬M∨¬P.'

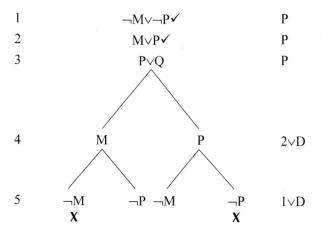

Notice two things about the above tree. First, '¬M∨¬P' had to be decomposed under every open branch below it in the tree, not only under 'P' (line 4) but also under 'M' (line 4). Second, after decomposing '¬M∨¬P,' there are two closed branches and two open branches. If 'P∨Q' were decomposed before '¬M∨¬P,' there would be four open branches. This would require '¬M∨¬P' to be decomposed under each of these four open branches. Since we made use of the strategic rule that says to use decomposition rules that close branches and decomposed '¬M∨¬P' before 'P∨Q,' we only have to decompose 'P∨Q' under the remaining two open branches.

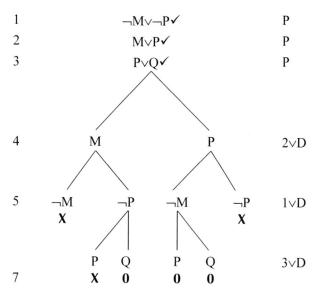

The above tree is a completed open tree since there is at least one completely decomposed open branch.

4.4.3 Strategic Rule 3

Strategic rule 3 is the following:

Strategic rule 3   Use stacking rules before branching rules.

One helpful way of remembering this rule is that you want your trees to be tall and not bushy. Whenever you decompose a complex proposition, it must be decomposed into every open branch below the decomposed proposition. Using stacking rules before branching rules has the benefit of simplifying trees. Consider the following set of propositions: '¬M∨Q' and '(R∧Q)∧P.'

$$
\begin{array}{ccc}
1 & ¬M∨Q & P \\
2 & (R∧Q)∧P & P
\end{array}
$$

In the above stack, we have the option of using a decomposition rule that stacks (∧D) or a decomposition rule that branches (∨D). Consider what happens if we branch first.

1	¬M∨Q✓		P
2	(R∧Q)∧P✓		P

3	¬M	Q	1∨D
4	R∧Q✓	R∧Q✓	2∧D
5	P	P	2∧D
6	R	R	4∧D
7	Q	Q	4∧D

The above tree is seven lines long and requires writing two separate stacks for the decomposition of '(R∧Q)∧P.' Consider what happens if we use the stacking rule (∧D) first.

1	¬M∨Q✓	P
2	(R∧Q)∧P✓	P
3	R∧Q✓	2∧D
4	P	2∧D
5	R	3∧D

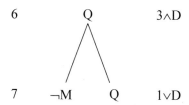

Here, the tree is again seven rows long but is less complex since it requires only one column for the decomposition of the stacking rule (∧D). Thus, stacking rules produce simpler trees than branching rules. So, stack before you branch!

4.4.4 Strategic Rule 4

Strategic rule 4 is the following:

> Strategic rule 4   Decompose more complex propositions before simpler proposi-
>   tions.

Strategic rule 4 is the weakest of the strategic rules. Compare the following two trees for '(M∨S)∨(Q∨R), R∨S.' In the first tree, the more complex proposition is decomposed:

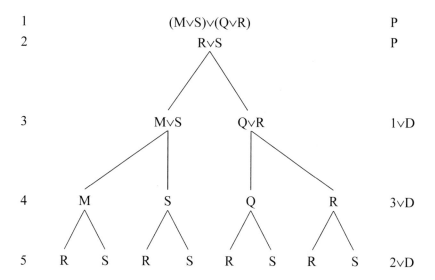

In the above tree, notice that line 5 involves four decompositions of 'R∨S' under the four open branches. In the second tree, the first proposition decomposed is the less complex proposition at line 2. In the tree below, notice that line 5 has two decompositions of 'M∨S' and two decompositions of 'Q∨R.'

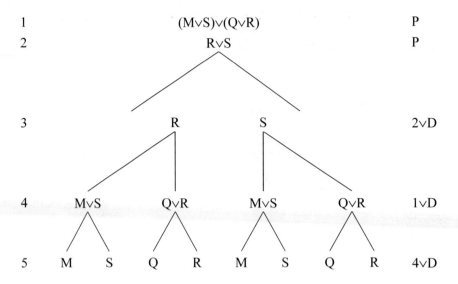

Although using strategic rule 4 may not reduce the overall size of the tree, it can help simplify the number of different applications of a rule applied.

Exercise Set #4

A. Using the truth-tree method, construct a truth tree to determine whether the following yield a completed open tree or a closed tree. Remember to use the strategic rules.

   1.   *   A∧B, B∧C, C∧D
   2.   (A∧B)∧D, (D∧¬B)∧A
   3.   *   A∧B, B∧¬C, C∧D
   4.   ¬A∨¬B, A∧¬B
   5.   *   A→ B, B→ C, C→D
   6.   ¬(A→B), ¬(A∨B)
   7.   *   A →B, ¬B→C, ¬C→D
   8.   B↔C, B→C, ¬B∨C
   9.   *   A→B, ¬(A∨B), B∧C
 10.   [P→¬(Q∨R)], ¬(Q↔¬R)
 11.   *   ¬[P→¬(Q∨R)], ¬(Q∨¬R)
 12.   (P↔¬L), P∧(¬P∧Z), L→(R→Z)
 13.   P∨(R∨D), R↔(V↔D)¬(P→P)
 14.   ¬[Z∨(¬Z∨V)], ¬(C∨P)↔(M∧¬D)
 15.   P∨¬(¬P→R), V↔(M→¬M)

Solutions to Starred Exercises in Exercise Set #4

1. *   A∧B, B∧C, C∧D; open tree.

1	A∧B✓	P
2	B∧C✓	P
3	C∧D✓	P
4	A	1∧D
5	B	1∧D
6	B	2∧D
7	C	2∧D
8	C	3∧D
9	D	3∧D
	0	

3. *   A∧B, B∧¬C, C∧D; closed tree.

1	A∧B✓	P
2	B∧¬C✓	P
3	C∧D✓	P
4	B	2∧D
5	¬C	2∧D
6	C	3∧D
7	D	3∧D
	X	

5. *   A→B, B→C, C→D; open tree.

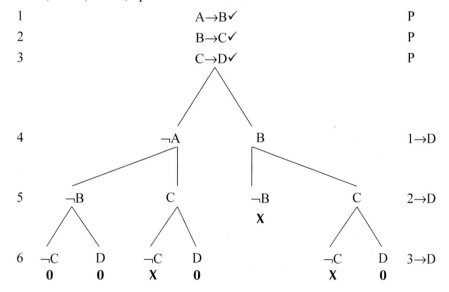

7. * A→B, ¬B→C, ¬C→D; open tree.

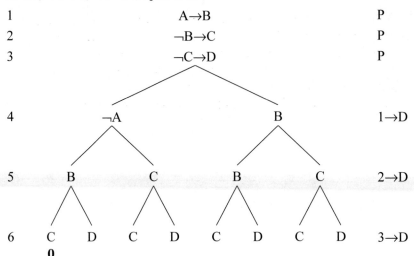

1	A→B	P
2	¬B→C	P
3	¬C→D	P

4   ¬A   B   1→D

5   B   C   B   C   2→D

6   C   D   C   D   C   D   C   D   3→D
   0

9. * A→B, ¬(A∨B), B∧C; closed tree.

1	A→B	P
2	¬(A∨B)✔	P
3	B∧C✔	P
4	B	3∧D
5	C	3∧D
6	¬A	2¬∨D
7	¬B	2¬∨D
	X	

11. * ¬[P→¬(Q∨R)], ¬(Q∨¬R); open tree.

1	¬[P→¬(Q∨R)]✔	P
2	¬(Q∨¬R)✔	P
3	¬Q	2¬∨D
4	¬¬R✔	2¬∨D
5	R	4¬¬D
6	P	1¬→D
7	¬¬(Q∨R)✔	1¬→D
8	Q∨R✔	7¬¬D

9   Q   R   8→D
     X   **0**

4.5 TRUTH-TREE WALK THROUGH

In this section, three trees are examined. The first will come completed, the second will have a step-by-step walkthrough, and the third will illustrate a simple strategic point. In each case, you should look at the original stack of propositions, try to complete the tree yourself, and check your work against the completed tree.

Consider the following set of propositions: '{M→P, ¬(P∨Q), (R∨S)∧¬P}.' First, there is the initial setup, which will simply consist of stacking the propositions:

1	M→P✓	P
2	¬(P∨Q)✓	P
3	(R∨S)∧¬P✓	P

Second, there is the decomposition:

1	M→P✓	P
2	¬(P∨Q)✓	P
3	(R∨S)∧¬P✓	P
4	R∨S✓	3∧D
5	¬P	3∧D
6	¬P	2¬∨D
7	¬Q	2¬∨D
8	¬M P	1→D
	X	
9	R S	4∨D
	0 **0**	

Third, we can analyze the tree to see whether or not it is open or closed. The above tree is complete since all propositions are decomposed, and it is open since there is at least one branch with an '**0**' under it.

For the next example, consider the following set of propositions:

¬(M→P), ¬(P↔Q), ¬(P∨M)∨¬Q, P→M

Looking at this set, there is only one proposition that stacks: '¬(M→P).' Strategic rule 2 says to use rules that stack rather than branch, so we will start decomposing the tree by applying '(¬→D)' to '¬(M→P)':

1	¬(M→P)✓	P
2	¬(P↔Q)✓	P
3	¬(P∨M)∨¬Q	P
4	P→M	P
5	M	1¬→D
6	¬P	1¬→D

Next, all of the remaining propositions branch, so we should consider using a decomposition branching rule on a proposition that will close branches. It would not be a good idea to decompose 'P→M' at line 5 since this will give us 'M' and '¬P' in a branch and would not close any branches. Another option is to decompose the negated biconditional '¬(P↔Q)' using '(¬↔D)' since this opens one branch yet closes another. Consider this choice below.

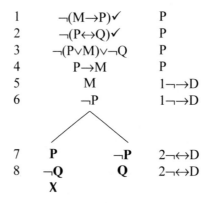

Next, decomposing 'P→M' from line 4 is still not a good idea since it will only open more branches. However, applying (∨D) upon '¬(P∨M)∨¬Q' will close one branch:

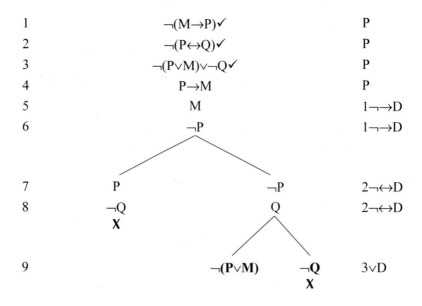

There still is not a good reason to decompose 'P→M' since it will not close any branches. Thus, try decomposing '¬(P∨M)' since it stacks rather than branches.

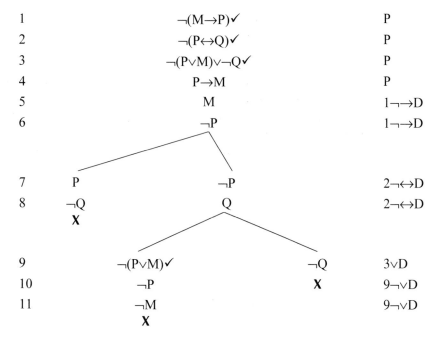

All branches are closed; therefore the tree is closed. It is important to recognize that (1) whenever a branch closes, you should not decompose any more propositions in that branch, and (2) when all branches close, the tree is finished, even if all propositions have not been fully decomposed. The above walkthrough shows that even though a tree can be fully decomposed by decomposing propositions randomly, a tactical use of the decomposition rules will reduce the complexity of the tree and your total amount of work.

Being able to identify which propositions are likely to close branches is extremely helpful. To see this more clearly, consider the following set of propositions:

P→Q, T→[(Q∨R)∨(¬S∨M)], (P∨Q)∨R, (P→M)∨[W↔(¬S∨S)], ¬(¬P∨Q)

This set of propositions is likely to yield a very complex tree if you do not proceed with a strategic use of the decomposition rules in mind. It can, however, be solved quite easily if you see that '¬(¬P∨Q)' will stack, giving '¬¬P' and '¬Q,' while 'P→Q' will branch, giving '¬P' and 'Q.' This will immediately close both branches and the tree, making the remaining decompositions irrelevant.

1	P→Q✓	P
2	T→[(Q∨R)∨(¬S∨M)]	P
3	(P∨Q)∨R	P
4	(P→M)∨[W↔(¬S∨S)]	P

5	¬(¬P∨Q)✓	P
6	¬¬P✓	5∨D
7	¬Q	5∨D
8	P	6¬¬D

| 9 | ¬P   Q | 1→D |
| | X   X | |

4.6 LOGICAL PROPERTIES OF TRUTH TREES

We have developed a procedure for determining whether a truth tree is a completed open tree or a closed tree. These two properties of trees are used to mechanically determine certain logical properties belonging to propositions, sets of propositions, and arguments. In the following sections, these properties are investigated through the use of truth trees.

4.6.1 Semantic Analysis of Truth Trees

In the previous chapter, truth tables were used to determine various properties of propositions, sets of propositions, and arguments. Truth trees allow the same determination. Recall that decomposing a proposition 'P' consists of representing the conditions under which 'P' is true by stacking, branching, or branching and stacking. In the case of a completed open branch, it is possible to recover a set of valuations of the propositional letters. To see this more clearly, consider the truth tree for the following set of propositions: '{R∧¬M, R∧(W∧¬M)}.'

1	R∧¬M✓	P
2	R∧(W∧¬M)✓	P
3	R	1∧D
4	¬M	1∧D
5	R	2∧D
6	W∧¬M✓	2∧D
7	W	6∧D
8	¬M	6∧D
	0	

Notice that there is one completed open branch of propositions consisting of '¬M,''W,''R.' We can interpret this branch to recover a set of valuations of the propositional letters in a way that 'R∧¬M' and 'R∧(W∧¬M)' are jointly true. The method by which this is done proceeds in two steps. First, for every completed open branch, identify any atomic propositions and any negated atomic propositions. In the tree above, there is one completed open branch, consisting of 'R,''W,' and '¬M.' Second, assign a value of *true* to atomic propositions and *false* to negated propositional letters. This procedure determines a *valuation set*. In the case of 'R,' 'W,' and '¬M,' we assign their truth values as follows:

$$v(R) = T$$
$$v(W) = T$$
$$v(M) = F$$

Consider a second tree consisting of '{R∧W, M∨¬W}':

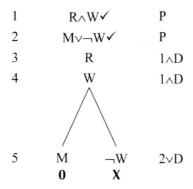

Notice that the above tree is a completed open tree where the leftmost branch is a completed open branch and the rightmost branch is closed. Focusing only on atomic propositions and their literal negations, branches 1 and 2 consist of the following:

Branch 1   M, W, R
Branch 2   ¬W, W, R

By assigning a value of true to atomic propositions and false to their literal negations, a consistent set of valuations can be assigned to propositional letters in the completed open branch but not for the closed branch. With respect to branch 1, we see that the stack of propositions is true when the following valuations are assigned to propositional letters:

$$v(M) = T$$
$$v(W) = T$$
$$v(R) = T$$

However, with respect to branch 2, a consistent set of valuations cannot be recovered. If we try to extract a valuation set for branch 2, we get the following: $v(W) = F$, $v(W) = T$, and $v(R) = T$. This type of extraction would require valuating 'W' as both true and false, which would violate the principle that a proposition is either true or false (not both and not neither). Thus, we consider the left branch closed.

	M	W	R
Branch 1	T	T	T
Branch 2	Closed branch		

In sum, valuations for propositional letters can be extracted by reading upward from the base of a completed open branch and assigning a value of true to propositional letters and false to literal negations of propositional letters.

Before turning to a discussion of how trees can be used to determine properties like consistency, contingency, validity, and so forth, we finish our discussion here with a tree that involves more than one valuation set (i.e., more than one way to assign truth values to propositional letters). Consider a tree consisting of the following propositions: 'M→P,' '¬(P∨Q),' '(R∨S)∧¬P.'

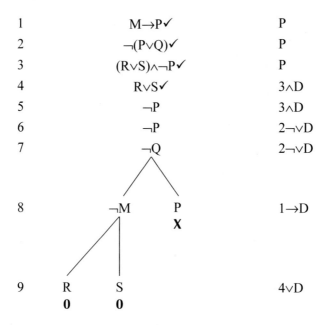

In the above tree, there are two completed open branches. Moving upward from the leftmost 'R' at the base of the completed tree, we can first assign 'R' a value of true. Next, notice that the branch consists of '¬M,' '¬P,' and '¬Q,' and so we assign a value of false to 'M,' 'P,' and 'Q.'

	R	S	M	P	Q
Valuation set 1	T	?	F	F	F

However, note that there is no 'S' in this branch. This means that the propositions in the stack can be jointly true independent of whether 's' is true or 's' is false. The propositions that compose the stack will be true independent of whether

'S' is true or false. Let's call the valuation set where $v(S) = T$ valuation set 1, and the one where $v(S) = F$ valuation set 2.

	R	S	M	P	Q
Valuation set 1	T	T	F	F	F
Valuation set 2	T	F	F	F	F

Turning to the second completed open branch, we begin from the base of the tree and move upward from 'S,' assigning it true as a value, then assigning truth values to the other propositional letters: $v(M) = F$, $v(P) = F$, and $v(Q) = F$. However, note that in this branch, 'R' is not present and so can be either $v(R) = F$ (valuation set 3) or $v(R) = T$ (valuation set 4).

	R	S	M	P	Q
Valuation set 1	T	T	F	F	F
Valuation set 2	T	F	F	F	F
Valuation set 3	F	T	F	F	F
Valuation set 4	T	T	F	F	F

Notice, however, that valuation set 4 and valuation set 1 are identical, and so one of these is superfluous.

	R	S	M	P	Q
Valuation set 1	T	T	F	F	F
Valuation set 2	T	F	F	F	F
Valuation set 3	F	T	F	F	F

Thus, using the truth-tree method, we have been able to determine that ''M→P,' '¬(P∨Q),' and '(R∨S)∧¬P' are jointly true under three different possible valuations.

Exercise Set #5

A. Decompose the following truth trees. For any completed open branch, give the truth-value assignments (valuations) to the propositional letters that compose the branch.
   1. * ¬(P→¬Q), P∧Q
   2.   ¬P∨¬Z, P→¬(Z∨R)
   3. * ¬[P→¬(M∧¬Z)], ¬(¬P∨Z)
   4.   ¬(P↔Z), ¬P→¬Z
   5. * (P↔¬Z), ¬(P→Z)
   6.   P∨(R→¬Z), ¬P∨¬(R→¬Z)
   7.   P∧(R↔¬Z), ¬P∧¬(R↔¬Z)
   8.   P∨(Q∨¬M)→R, ¬R∧Q

Solutions to Starred Exercises in Exercise Set #5

1. * ¬(P→¬Q), P∧Q; completed open tree: $v(P) = T$, $v(Q) = T$.

1	¬(P→¬Q)✓	P
2	P∧Q✓	P
3	P	2∧D
4	Q	2∧D
5	P	1¬→D
6	¬¬Q✓	1¬→D
7	Q	6¬¬D
	0	

3. * ¬[P→¬(M∧¬Z)], ¬(¬P∨Z); completed open tree: $v(P) = T$, $v(M) = T$, $v(Z) = F$.

1	¬[P→¬(M∧¬Z)]✓	P
2	¬(¬P∨Z)✓	P
3	P	1¬→D
4	¬¬(M∧¬Z)✓	1¬→D
5	M∧¬Z✓	4¬¬D
6	M	5∧D
7	¬Z	5∧D
8	¬¬P✓	2¬∨D
9	¬Z	2¬∨D
10	P	8¬¬D
	0	

5. * (P↔¬Z), ¬(P→Z); completed open tree: $v(P) = T$, $v(Z) = F$.

1	P↔¬Z✓		P
2	¬(P→Z)✓		P
3	P		2¬→D
4	¬Z		2¬→D

5	P	¬P	1↔D
6	¬Z	¬¬Z	1↔D
7		Z	6¬¬D
	0	**X**	

4.6.2 Consistency

Now that you are familiar with how to decompose truth trees, how to determine whether the tree is open or closed, and how to retrieve a set of valuations from a tree with a completed open branch, the next step is to learn how to analyze trees for various properties. In this section, we learn how to determine whether a set of propositions is consistent (or inconsistent).

In the previous chapter, we saw that a truth table can be used to determine whether a set of propositions is consistent by determining whether or not there is some row (truth-value assignment) where every proposition is true. For instance, in considering 'P→Q,' 'Q∨P,' and 'P↔Q,' we construct a truth table and identify the row where all of the propositions are true.

P	Q	(P	→	Q)	(Q	∨	P)	(P	↔	Q)
T	T	T	**T**	T	T	**T**	T	T	**T**	T
T	F	T	F	F	F	T	T	T	F	F
F	T	F	T	T	T	T	F	F	F	T
F	F	F	T	F	F	F	F	F	T	F

From the above, 'P→Q,' 'Q∨P,' and 'P↔Q' are consistent when $v(P)$ = T and $v(Q)$ = T. We are now in a position to define consistency and inconsistency for truth trees.

Consistent   A set of propositions '{**P, Q, R**, . . ., **Z**}' is consistent if and only if there is at least one valuation where '**P**,' '**Q**,' '**R**,' . . ., '**Z**' are true. A truth tree shows that '{**P, Q, R**, . . ., **Z**}' is consistent if and only if a complete tree of the stack of '**P**,' '**Q**,' '**R**,' . . ., '**Z**' determines a completed open tree, that is, if there is at least one completed open branch.

Inconsistent   A set of propositions '{**P, Q, R**, . . ., **Z**}' is logically inconsistent if and only if there is no valuation where '**P**,' '**Q**,' '**R**,' . . ., '**Z**' are jointly true. A truth tree shows that '{**P, Q, R**, . . ., **Z**}' is inconsistent if and only if a tree of the stack of '**P**,' '**Q**,' '**R**,' . . ., '**Z**' is a closed tree, that is, if all branches close.

The method for determining whether the set of propositions '{**P, Q, R**, . . ., **Z**}' is consistent (or inconsistent) begins by putting each of the propositions in the set on its own line in a stack. For example, in the case of '{R∧¬M, R∧(W∧¬M)},' the propositions are first stacked:

1   R∧¬M   P
2   R∧(W∧¬M)   P

Second, the tree is decomposed until the tree either closes or there is a completed open branch:

1   R∧¬M✓   P
2   R∧(W∧¬M)✓   P
3   R   1∧D
4   ¬M   1∧D
5   R   2∧D
6   W∧¬M✓   2∧D
7   W   6∧D
8   ¬M   6∧D
        O

Finally, the tree is analyzed. Since the above tree contains at least one completed open branch, there is at least one valuation set that would make the propositions in the set jointly true. The definition of consistency states that if the tree has a completed open branch, the set of propositions '{R∧¬M, R∧(W∧¬M)}' is consistent.

As a second example, consider the set '{R∧W, M∨¬W}':

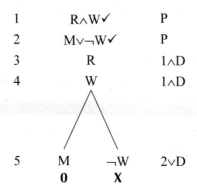

1	R∧W✓	P
2	M∨¬W✓	P
3	R	1∧D
4	W	1∧D
5	M ¬W	2∨D
	0 **X**	

The above tree is a completed open tree (since it contains at least one completed open branch), and so the set of propositions '{R∧W, M∨¬W}' that formed the stack is consistent.

Consider a final example:

1	P→Q✓	P
2	T→[(Q∨R)∨(¬S∨M)]	P
3	(P∨Q)∨R	P
4	(P→M)∨[W↔(¬S∨S)]	P
5	¬(¬P∨Q)✓	P
6	¬¬P✓	5∨D
7	¬Q	5∨D
8	P	6¬¬D
9	¬P Q	1→D
	X **X**	

The above tree is complete and closed. It is thus inconsistent because there is not a completed open branch, for all of the branches are closed.

Property	Number of Trees	Propositions in the Tree	Has the Property. . .
Consistency	At Least 1	For '{P, Q, R, . . ., Z},' a truth tree for 'P,' 'Q,' 'R,' . . ., 'Z'	Iff 'P,' 'Q,' 'R,' . . ., 'Z' determines a completed open tree.
Inconsistency	At Least 1	For 'P,' a truth tree for 'P,' 'Q,' 'R,' . . ., 'Z'	Iff 'P,' 'Q,' 'R,' . . ., 'Z' determines a closed tree.

Before moving on to the remaining logical properties, notice that a benefit of the truth-tree method is that it is more efficient. For instance, a complete truth table of the tree considered above would require constructing an eight-row table, involving 104 'Ts' and 'Fs.' The above truth tree is a much more economical method for testing for the same property since it tested whether 'R∧¬M' and 'R∧(W∧¬M)' are logically consistent in eight lines.

4.6.3 Tautology, Contradiction, and Contingency

In this section, truth trees are used to determine whether a proposition is a tautology, contradiction, or contingency.

Tautology   A proposition 'P' is a tautology if and only if 'P' is true under every valuation. A truth tree shows that 'P' is a tautology if and only if a tree of the stack of '¬P' determines a closed tree.

Contradiction   A proposition 'P' is a contradiction if and only if 'P' is false under every valuation. A truth tree shows that 'P' is a contradiction if and only if a tree of the stack of 'P' determines a closed tree.

Contingency   A proposition 'P' is a contingency if and only if 'P' is neither always false under every valuation nor always true under every valuation. A truth tree shows that 'P' is a contingency if and only if a tree of 'P' does not determine a closed tree and a tree of '¬P' does not determine a closed tree.

In testing a proposition for whether it is a tautology, contradiction, or contingency, the initial tree construction is important. For instance, when testing a proposition 'P' to see if it is a tautology, you begin by writing '¬P' as the first line of the stack. For example, consider the proposition 'P∨¬P.' This proposition appears to be a tautology, but to test to see whether it is, you must begin by placing '¬(P∨¬P)' at the first line of the tree and not 'P∨¬P.'

```
1        ¬(P∨¬P)✓        P
2           ¬P           1∧D
3          ¬¬P✓          1∧D
4            P           3¬¬D
             X
```

Notice that the above tree is closed. A closed tree for a stack consisting of '{¬(P∨¬P)}' means that there is no valuation set that makes '¬(P∨¬P)' true. If there is no valuation set that makes '¬(P∨¬P)' true, then '¬(P∨¬P)' is a contradiction. However, the proposition we want to know about is 'P∨¬P.' If '¬(P∨¬P)' is always false, then '(P∨¬P)' is always true. And if '(P∨¬P)' is always true, then '(P∨¬P)' is a tautology.

Next, consider 'P→(Q∧¬P).' Let's begin by testing this proposition to see if it is a tautology by placing the literal negation of 'P→(Q∧¬P)' as the first line of the stack:

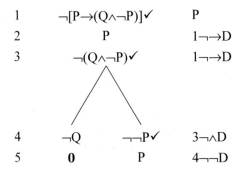

```
1        ¬[P→(Q∧¬P)]✓        P
2             P               1¬→D
3          ¬(Q∧¬P)✓           1¬→D

4        ¬Q        ¬¬P✓        3¬∧D
5         0          P         4¬¬D
```

Notice that the above tree does not determine a closed tree, and so 'P→(Q∧¬P)' is not a tautology. From the above tree, we know that there is at least one valuation set that makes '¬[P→(Q∧¬P)]' true. In other words, there is at least one valuation set that makes 'P→(Q∧¬P)' false. We do not know, however, whether every way of valuating the propositional letters in 'P→(Q∧¬P)' would make it false, in which case 'P→(Q∧¬P)' would be a contradiction, or if some ways of valuating 'P→(Q∧¬P)' make it true and some make it false, making 'P→(Q∧¬P)' a contingency. In order to find this out, we need to make use of another test.

Before considering the other tests for 'P→(Q∧¬P),' let's briefly turn to the truth-tree test for contradiction. Consider 'P∧¬P,' which is obviously a contradiction. The test for contradiction begins by simply writing 'P∧¬P' on the first line, decomposing the proposition, and then checking to see whether the tree is open or closed.

```
1          P∧¬P✓          P
2            P            1∧D
3           ¬P            1∧D
             X
```

Notice that the above tree is closed, which means that there is no way of assigning truth values to the propositional letters in 'P∧¬P' so as to make it true. In other words, 'P∧¬P' is a contradiction.

Let's return to 'P→(Q∧¬P)' and see whether it is a contradiction. Begin the tree de-composition by writing 'P→(Q∧¬P)' at line 1 and then decomposing the proposition.

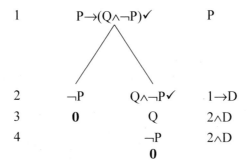

1	P→(Q∧¬P)✓	P
2	¬P Q∧¬P✓	1→D
3	**O** Q	2∧D
4	¬P	2∧D
	O	

Notice that the above tree does not close, and so 'P→(Q∧¬P)' is not a contradic-tion. An earlier tree (above) showed that 'P→(Q∧¬P)' is also not a tautology. This leaves one option for 'P→(Q∧¬P),' namely, that it is a contingency.

To summarize, the truth-tree method can be used to determine whether a proposi-tion '**P**' is a tautology, contradiction, or contingency. In testing '**P**' to see if it is a tautology, begin the tree with '¬**P**.' If the tree closes, you know that it is a tautology. If the tree is open, then '**P**' is either a contradiction or a contingency. Similarly, in testing '**P**' to see if it is a contradiction, begin the tree with '**P**.' If the tree closes, you know that it is a contradiction. If the tree is open, then '**P**' is either a tautology or a contingency. Lastly, if the truth-tree test shows that 'P' is neither a contradiction nor a tautology, then '**P**' is a contingency.

Property	Number of Trees	Propositions in the Tree	Has the Property . . .
Tautology	At Least 1	For '**P**,' a truth tree for '¬**P**'	Iff '¬**P**' determines a closed tree.
Contradiction	At Least 1	For '**P**,' a truth tree for '**P**'	Iff '**P**' determines a closed tree.
Contingency	At Least 2	For '**P**,' truth trees for '**P**' and '¬**P**'	Iff neither '**P**' nor '¬**P**' determines a closed tree.

4.6.4 Logical Equivalence

Logical equivalence concerns a set consisting of two propositions, or a pair of propositions.

Equivalence   A pair of propositions '**P**,' '**Q**' is equivalent if and only if '**P**' and '**Q**' have identical truth values under every valuation. A truth tree shows that '**P**' and '**Q**' are equivalent if and only if a tree of '¬**(P↔Q)**' determines a closed tree.

Provided that you are able to do a truth tree that checks a proposition for a tautology, a truth tree checking for logical equivalence requires little more knowledge. Two
propositions 'P' and 'Q' are logically equivalent if and only if 'P' and 'Q' never have
different truth values. It follows that if 'P' and 'Q' can never have different truth values, then 'P↔Q' is a tautology. Thus, in order to determine whether 'P' and 'Q' are
logically equivalent, we only need to determine whether 'P↔Q' is a tautology. This
is done by considering whether '¬(P↔Q)'determines a closed tree.

Consider the following two propositions: 'P∨¬P' and '¬(P∧¬P).' In order to
determine whether or not they are equivalent, we combine them in the form of a
biconditional, giving us '(P∨¬P)↔¬(P∧¬P)'; we then negate the biconditional,
which yields'¬[(P∨¬P)↔¬(P∧¬P)],' and test to see whether the tree closes.

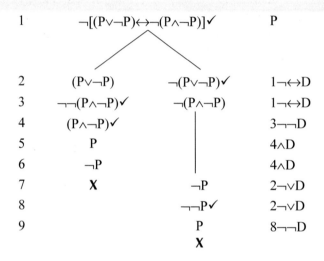

Since both branches close, the tree is closed. Therefore, 'P∨¬P' and '¬(P∧¬P)' are
logically equivalent.

Property	Number of Trees	Propositions in the Tree	Has the Property . . .
Equivalence	At least 1	For 'P,' 'Q,' a truth tree for '¬(P↔Q)'	Iff '¬(P↔Q)' determines a closed tree.

4.6.5 Validity

In this section, we use the truth-tree method to determine whether an argument is valid
or invalid. An argument is valid if and only if it is impossible for the premises to be
true and the conclusion to be false. In chapter 1, the negative test was used to determine whether or not arguments were valid. This test asks you to imagine whether it is
possible for the premises to be true and the conclusion to be false. In chapter 3, it was

argued that the truth table method provides a better way to test for validity because it does not rely on an individual's ability to imagine whether it is possible for the premises to be true and conclusion false. At the beginning of this chapter, it was pointed out that the truth table method becomes increasingly unmanageable when arguments involve a large number of propositional letters.

The truth-tree method circumvents these problems because the complexity of a truth tree is not a function of the number of propositional letters in the argument. In the case of a truth tree, an argument 'P, Q, R, . . ., Y ⊢ Z' is valid if and only if the stack 'P,''Q,''R,'. . ., 'Y,''¬Z' determines a closed tree.

> Validity   An argument 'P, Q, . . ., Y ⊢ Z' is valid in PL if and only if it is impossible for the premises to be true and the conclusion false. A truth tree shows that an argument 'P, Q, . . ., Y ⊢ Z' is valid in PL if and only if 'P,' 'Q,' 'R,' . . ., 'Y,' '¬Z' determine a closed tree.

Consider the following example:

$$P \rightarrow Q, P \vdash Q.$$

A corresponding truth tree can be created by stacking the premises and the negation of the conclusion, then testing for consistency, that is, testing 'P→Q,''P,' and '¬Q.'

1	P→Q✓	P
2	P	P
3	¬Q	P

$$\begin{array}{ccc} & \diagup & \diagdown & \\ 4 & \neg P & Q & 1\rightarrow D \\ & X & X & \end{array}$$

Since the stack 'P→Q,' 'P,' and '¬Q' closes, it is inconsistent; therefore, 'P→Q, P ⊢ Q' is a valid argument.

Consider a more complex example:

$$P \rightarrow \neg(Q \rightarrow \neg W), (Q \rightarrow \neg W) \vee (P \leftrightarrow S), P \vdash S \rightarrow P$$

In order to test this argument for validity, we test the following stack:

$$P \rightarrow \neg(Q \rightarrow \neg W), (Q \rightarrow \neg W) \vee (P \leftrightarrow S), P, \neg(S \rightarrow P)$$

Notice that nothing about the premises is changed. The only difference is that the conclusion is negated.

1	P→¬(Q→¬W)	P
2	(Q→¬W)∨(P↔S)	P
3	P	P
4	¬(S→P)✓	P
5	S	4¬→D
6	¬P	4¬→D
	X	

The tree immediately closes because there is an inconsistency in the trunk or main branch of the tree. Therefore, 'P→¬(Q→¬W), (Q→¬W)∨(P↔S), P ⊢ S→P' is valid.

Property	Number of Trees	Propositions in the Tree	Has the Property . . .
Validity	At Least 1	For '**P**, . . ., **Y** ⊢ **Z**,' a truth tree for '**P**,' . . ., '**Y**,' '¬**Z**'	Iff '**P**,' . . ., '**Y**,' '¬**Z**' determines a closed tree.

END-OF-CHAPTER EXERCISES

A. Using the truth-tree method, determine whether the following sets of propositions are logically consistent or inconsistent.
   1. * P∧Q, P→Q
   2.   ¬P∧¬Q, ¬(P∨Q)
   3. * P, P→Q, ¬Q
   4.   P→Q, P∨Q, ¬P∨Q
   5. * ¬(P∧Q), P→¬Q, ¬(P→Q)
   6.   (R→¬S)→¬M, R↔M
   7.   R↔M, M↔R, R→¬M
   8.   ¬(R↔¬R), (R→R)
   9.   (P∨Q), P→Q, ¬Q∧P
   10. W∧P, S↔Q, ¬(P↔W), W→Z, ¬P

B. Determine which of the following propositions are contradictions, tautologies, or contingencies by using the truth-tree method.
   1. * [(P∨¬W)∧P]∧W
   2.   (P→Q)∧P
   3. * (P∨¬P)∨Q
   4.   (¬P∨Q)∧¬(P→Q)
   5. * (P↔Q)∧¬[(P→Q)∧(Q→P)]
   6.   [(R→¬S)→¬M]→(¬P∧P)
   7.   ¬(¬P→¬P)
   8.   ¬(R↔¬R)∧(R↔R)

C. Determine which of the following sets of propositions are logically equivalent by using the truth-tree method.
   1. * P→Q, ¬P∨Q
   2.   ¬P→¬Q, Q→P
   3. * P↔Q, (P→Q)∧(Q→P)
   4   T∨¬¬S, ¬S∧(T∨W)
   5. * ¬(¬P∨¬Q), ¬¬P∧¬¬Q
   6.   ¬(P↔Q), ¬[(P→Q)∧(Q→P)]

D. Determine which of the following arguments are valid by using the truth-tree method.
   1. * P→Q, ¬Q ⊢ ¬P
   2.   P∧¬P ⊢ (W∨M)∨T
   3. * P∨(Q∧R) ⊢ (P∨Q)∧(P∨R)
   4.   R ⊢ R∨[M∨(T∨P)]
   5. * P→Q, Q→R ⊢ ¬R→(¬P∧¬Q)
   6.   ¬(P∧Q), ¬(P∨Q) ⊢ ¬P∧¬Q
   7. * ¬P∨¬Q ⊢ ¬(P∨Q)
   8.   (S→Q)∨[(R∨M)∨P], ¬(S→Q), ¬P, ¬M ⊢ R
   9. * ⊢ P→P
   10.   ⊢ P∨¬P

Solutions to Starred Exercises in End-of-Chapter Exercises

A.

   1. * P∧Q, P→Q; consistent.

1	P∧Q✓	P
2	P→Q✓	P
3	P	1∧D
4	Q	1∧D

| 5 | ¬P   Q | 2→D |
| | X   0 | |

   3. * P, P→Q, ¬Q; inconsistent.

1	P	P
2	P→Q✓	P
3	¬Q	P

| 4 | ¬P   Q | 2→D |
| | X   X | |

5. * ¬(P∧Q), P→¬Q, ¬(P→Q); consistent.

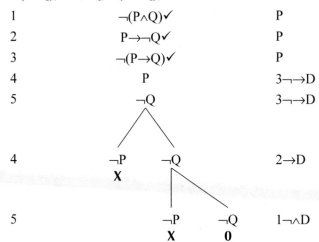

1	¬(P∧Q)✓	P
2	P→¬Q✓	P
3	¬(P→Q)✓	P
4	P	3¬→D
5	¬Q	3¬→D
4	¬P ¬Q	2→D
	X	
5	¬P ¬Q	1¬∧D
	X 0	

B.

  1. * [(P∨¬W)∧P]∧W; contingent.

Tree #1: Not a Contradiction

1	[(P∨¬W)∧P]∧W✓	P
2	(P∨¬W)∧P✓	1∧D
3	W	1∧D
4	(P∨¬W) ✓	2∧D
5	P	2∧D
6	¬W P	4∨D
	X 0	

Tree #2: Not a Tautology

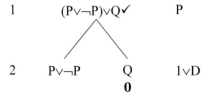

1	¬{[(P∨¬W)∧P]∧W}✓	P
2	¬[(P∨¬W)∧P] ¬W	¬∧D
	0	

  3. * (P∨¬P)∨Q; tautology.

1	(P∨¬P)∨Q✓	P
2	P∨¬P Q	1∨D
	0	

```
1        ¬[(P∨¬P)∨Q]✔        P
2          ¬(P∨¬P)✔          1¬∨D
3             ¬Q             1¬∨D
4             ¬P             2¬∨D
5             P              2¬∨D
              X
```

5. * (P↔Q)∧¬[(P→Q)∧(Q→P)]; contradiction.

```
1     (P↔¬Q)∧¬[(P→Q)∧(Q→P)]✔        P
2            (P↔¬Q)✔               1∧D
3       ¬[(P→Q)∧(Q→P)]✔            1∧D


4      ¬(P→Q)✔        ¬(Q→P)✔       3¬∧D
5         P              Q          4¬→D
6        ¬Q             ¬P          4¬→D


7     P     ¬P      P      ¬P       2↔D
8     Q     ¬Q      Q      ¬Q       2↔D
      X      X      X       X
```

C.

1. * P→Q, ¬P∨Q; equivalent.

```
1           ¬[(P→Q)↔(¬P∨Q)]✔                        P
            ¬[(P↔Q)↔(¬P∨Q)]✔

2      P→Q✔                    ¬(P→Q)✔              ¬↔D
3     ¬(¬P∨Q)✔                  ¬P∨Q✔               ¬↔D
4      ¬¬P✔                                         3¬D
5       ¬Q                                          3¬D
6        P                                          4¬¬D
7                                 P                 2¬→D
8                                ¬Q                 2¬→D
9    ¬P      Q                                      2→D
     X       X
10                            ¬P      Q             3∨D
                              X       X
```

3. * P↔Q, (P→Q)∧(Q→P); equivalent.

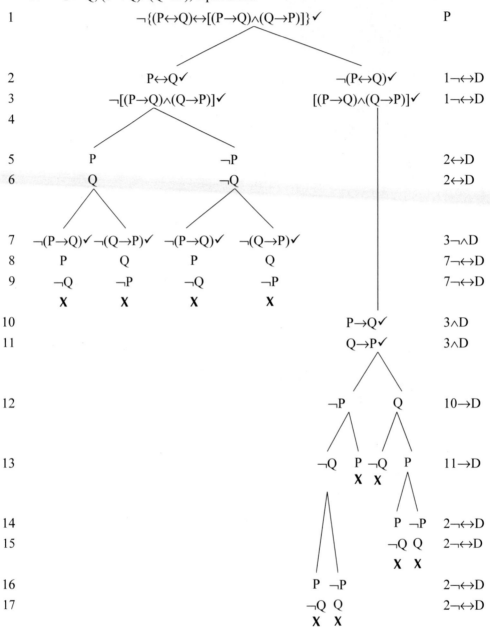

1	¬{(P↔Q)↔[(P→Q)∧(Q→P)]} ✓	P
2	P↔Q ✓ ¬(P↔Q) ✓	1¬↔D
3	¬[(P→Q)∧(Q→P)] ✓ [(P→Q)∧(Q→P)] ✓	1¬↔D
4		
5	P ¬P	2↔D
6	Q ¬Q	2↔D
7	¬(P→Q)✓ ¬(Q→P)✓ ¬(P→Q)✓ ¬(Q→P)✓	3¬∧D
8	P Q P Q	7¬↔D
9	¬Q ¬P ¬Q ¬P	7¬↔D
	X X X X	
10	P→Q ✓	3∧D
11	Q→P ✓	3∧D
12	¬P Q	10→D
13	¬Q P ¬Q P	11→D
	X X	
14	P ¬P	2¬↔D
15	¬Q Q	2¬↔D
	X X	
16	P ¬P	2¬↔D
17	¬Q Q	2¬↔D
	X X	

5. *   ¬(¬P∨¬Q), ¬¬P∧¬¬Q; equivalent.

1   ¬[¬(¬P∨¬Q)↔(¬¬P∧¬¬Q)]✓   P

2   ¬(¬P∨¬Q)✓   ¬¬(¬P∨¬Q)✓   1¬↔D
3   ¬(¬¬P∧¬¬Q)   ¬¬P∧¬¬Q✓   1¬↔D
4   ¬¬P✓   2¬∨D
5   ¬¬Q✓   2¬∨D
6   P   4¬¬D
7   Q   5¬¬D
8   ¬¬P✓   3∧D
9   ¬¬Q✓   3∧D
10   P   8¬¬D
11   Q   9¬¬D
12   ¬¬¬P✓   ¬¬¬Q✓   ¬P∨¬Q✓   2¬∧D + ¬¬D
13   ¬P   ¬Q   12¬¬D
          X   X
14   ¬P   ¬Q   12¬∨D
                                        X   X

D.

1. *   P→Q, ¬Q ⊢ ¬P; valid.

1   P→Q✓   P
2   ¬Q   P
3   ¬¬P✓   P
4   P   3¬¬D

5   ¬P   Q   1→D
           X   X

3. * P∨(Q∧R) ⊢ (P∨Q)∧(P∨R); valid.

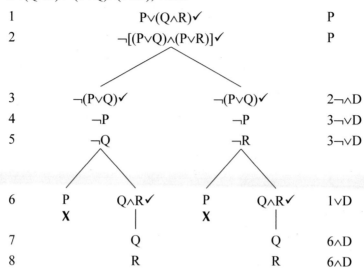

1	P∨(Q∧R)✔	P
2	¬[(P∨Q)∧(P∨R)]✔	P
3	¬(P∨Q)✔ ¬(P∨Q)✔	2¬∧D
4	¬P ¬P	3¬∨D
5	¬Q ¬R	3¬∨D
6	P Q∧R✔ P Q∧R✔	1∨D
	X X	
7	Q Q	6∧D
8	R R	6∧D
	X X	

5. * P→Q, Q→R ⊢ ¬R→(¬P∧¬Q); valid.

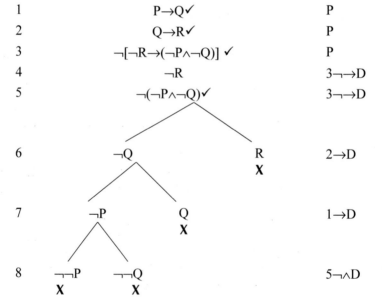

1	P→Q✔	P
2	Q→R✔	P
3	¬[¬R→(¬P∧¬Q)] ✔	P
4	¬R	3¬→D
5	¬(¬P∧¬Q)✔	3¬→D
6	¬Q R	2→D
	X	
7	¬P Q	1→D
	X	
8	¬¬P ¬¬Q	5¬∧D
	X X	

7. * ¬P∨¬Q⊢¬(P∨Q); invalid.

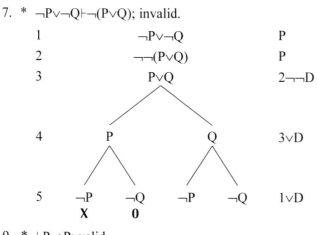

1	¬P∨¬Q	P
2	¬¬(P∨Q)	P
3	P∨Q	2¬¬D
4	P Q	3∨D
5	¬P ¬Q ¬P ¬Q	1∨D
	X 0	

9. * ⊢P→P; valid.

1	¬(P→P)✓	P
2	P	1¬→D
3	¬P	1¬→D
	X	

DEFINITIONS

Consistent   A set of propositions '{**P, Q, R**, . . ., **Z**}' is consistent if and only if there is at least one valuation where '**P**,' '**Q**,' '**R**,' . . .,'**Z**' are true. A truth tree shows that '{**P, Q, R**, . . ., **Z**}' is consistent if and only if a complete tree of the stack of '**P**,' '**Q**,' '**R**,' . . .,'**Z**' determines a completed open tree, that is, if there is at least one completed open branch.

Inconsistent   A set of propositions '{**P, Q, R**, . . .,**Z**}' is logically inconsistent if and only if there is no valuation where '**P**,' '**Q**,' '**R**,' . . .,'**Z**' are jointly true. A truth tree shows that '{**P, Q, R**, . . ., **Z**}' is inconsistent if and only if a tree of the stack of '**P**,' '**Q**,' '**R**,' . . .,'**Z**' is a closed tree. That is, if all branches close.

Tautology   A proposition '**P**' is a tautology if and only if '**P**' is true under every valuation. A truth tree shows that '**P**' is a tautology if and only if a tree of the stack of '¬**P**' determines a closed tree.

Contradiction   A proposition '**P**' is a contradiction if and only if '**P**' is false under every valuation. A truth tree shows that '**P**' is a contradiction if and only if a tree of the stack of '**P**' determines a closed tree.

Contingency   A proposition '**P**' is a contingency if and only if '**P**' is neither always false under every valuation nor always true under every valuation. A truth tree shows that '**P**' is a contingency if and only if a tree of '**P**' does not determine a closed tree and a tree of '¬**P**' does not determine a closed tree.

| Equivalence | A pair of propositions 'P,' 'Q' is equivalent if and only if 'P' and 'Q' have identical truth values under every valuation. A truth tree shows that 'P' and 'Q' are equivalent if and only if a tree of '¬(P↔Q)' determines a closed tree. |
| Validity | An argument 'P, Q, . . ., Y ⊢ Z' is valid in PL if and only if it is impossible for the premises to be true and the conclusion false. A truth tree shows that an argument 'P, Q, . . ., Y ⊢ Z' is valid in PL if and only if 'P,' 'Q,' 'R,' . . .,'Y,' '¬ Z' determines a closed tree. |

TRUTH-TREE VOCABULARY

Branch	A branch contains all the propositions obtained by starting from the bottom of the tree and reading upward through the tree.
Branching rule	A branching rule is a truth-tree rule where the condition under which a proposition 'P' is true is represented by branching. A branching rule is applied to propositions that are false only under one truth-value assignment.
Branching-and-stacking rule	A branching-and-stacking rule is a truth-tree rule where the condition under which a proposition 'P' is true is represented by branching and stacking. A branching-and-stacking rule will be used when 'P' is true under only two truth-value assignments and false under two truth-value assignments.
Closed branch	A closed branch contains a proposition and its literal negation (e.g., 'P' and '¬P'). A closed branch is represented by an 'X.'
Closed tree	A tree is closed when all of the tree's branches are closed. A closed tree will have an 'X' under every branch.
Completed open branch	A completed open branch is a fully decomposed branch that is not closed. That is, it is a fully decomposed branch that does not contain a proposition and its literal negation. An open branch is denoted by an '0' at the bottom of the branch.
Completed open tree	A tree is a completed open tree if and only if it has at least one completed open branch. That is, a tree is a completed open tree if and only if it contains at least one fully decomposed branch that is not closed. A completed open tree is a tree where at least one branch has an '0' under it.
Decomposition descending rule	When decomposing a proposition 'P,' decompose 'P' under every open branch that descends from 'P.'
Fully decomposed branch	A branch is fully decomposed when all propositions in the branch that can be decomposed have been decomposed.
Fully decomposed tree	A fully decomposed truth tree is a tree where all the propositions that can be decomposed have been decomposed.

Open branch   An open branch is a branch that is not closed, that is, a branch that does not contain a proposition and its literal negation.

Stacking rule   A stacking rule is a truth-tree rule where the condition under which a proposition 'P' is true is represented by stacking. A stacking rule is applied to propositions that are true under one truth-value assignment.

Truth tree   A truth tree is a schematic decision procedure typically used for the purpose of testing propositions, pairs of propositions, and arguments for logical properties.

KEY FOR TRUTH-TREE ANALYSIS

Property	Number of Trees	Propositions in the Tree	Has the Property . . .
Consistency	At Least 1	For '{P, Q, R, . . ., Z},' a truth tree for 'P,' 'Q,' 'R,' . . ., 'Z'	Iff 'P,' 'Q,' 'R,' . . ., 'Z' determines a completed open tree.
Inconsistency	At Least 1	For 'P,' a truth tree for 'P,' 'Q,' 'R,' . . ., 'Z'	Iff 'P,' 'Q,' 'R,' . . ., 'Z' determines a closed tree.
Tautology	At Least 1	For 'P,' a truth tree for '¬P'	Iff '¬P' determines a closed tree.
Contradiction	At Least 1	For 'P,' a truth tree for 'P'	Iff 'P' determines a closed tree.
Contingency	At Least 2	For 'P,' truth trees for 'P' and '¬P'	Iff neither 'P' nor '¬P' determines a closed tree.
Equivalence	At Least 1	For 'P' and 'Q,' a truth tree for '¬(P↔Q)'	Iff '¬(P↔Q)' determines a closed tree.
Validity	At Least 1	For 'P, . . ., Y ⊢ Z,' a truth tree for 'P,' . . ., 'Y,' '¬Z'	Iff 'P,' . . ., 'Y,' '¬Z' determines a closed tree.

DECOMPOSABLE PROPOSITIONS

Nine Decomposable Proposition Types		Decomposition Rule
Conjunction	**P∧R**	∧D
Disjunction	**P∨R**	∨D
Conditional	**P→R**	→D
Biconditional	**P↔R**	↔D
Negated conjunction	**¬(P∧R)**	¬∧D
Negated disjunction	**¬(P∨R)**	¬∨D
Negated conditional	**¬(P→R)**	¬→D
Negated biconditional	**¬(P↔R)**	¬↔D
Double negation	**¬¬P**	¬¬D

TRUTH-TREE DECOMPOSITION RULES

Stacking	Branching	Stacking∧Branching
P∧Q 　**P**　∧D 　**Q**　∧D	**¬(P∧Q)** 　／＼ **¬P**　**¬Q**　¬∧D	**P↔Q** 　／＼ **P**　**¬P**　↔D **Q**　**¬Q**　↔D
¬(P∨Q) 　**¬P**　¬∨D 　**¬Q**　¬∨D	**P∨Q** 　／＼ **P**　**Q**　∨D	**¬(P↔Q)** 　／＼ **P**　**¬P**　¬↔D **¬Q**　**Q**　¬↔D
¬(P→Q) 　**P**　¬→D 　**¬Q**　¬→D	**P→Q** 　／＼ **¬P**　**Q**　→D	
¬¬P 　**P**　¬¬D		

NOTE

1. See Merrie Bergmann, James Moore, and Jack Nelson, *The Logic Book*, 5th edition (Boston: McGraw Hill Education, 2009), pp. 137–40.

Chapter Five

Propositional Logic Derivations

In the previous chapters, we used truth tables and truth trees to test whether individual propositions, sets of sentences, and arguments had a given semantic property. For example, with respect to a set of propositions, a truth tree could be devised to test whether the propositions in the set were consistent. These tests do not, however, correspond to the reasoning that takes place in daily life. The goal of this chapter is to introduce a system of natural deduction. A natural deduction system is a set of derivational rules (general steps of reasoning) that mirror everyday reasoning in certain noteworthy ways. The particular system will be called a system of *propositional derivations*, or PD for short. Unlike truth tables and truth trees, PD is a system of syntactic rules insofar as they are formulated on the basis of the structure of everyday reasoning. Once the basics of PD have been mastered, we turn to a more advanced system of reasoning and a set of reasoning strategies that, while deviating from everyday reasoning, makes reasoning more efficient.

5.1 PROOF CONSTRUCTION

In this chapter, our principal concern will be learning how to solve a variety of proofs in an efficient manner. A *proof* is a finite sequence of well-formed formulas (wffs), or propositions, each of which is either a premise, an assumption, or the result of preceding formulas and a derivation rule.

> Proof   A proof is a finite sequence of well-formed formulas (or propositions), each of which is either a premise, an assumption, or the result of preceding formulas and a derivation rule.

Let's unpack this definition. First, in calling a proof a finite sequence of well-formed formulas, we are simply saying that no proof will be infinitely long, and no proof will contain propositions that are not well-formed formulas. Let's call the end point of any proof the *conclusion*. Thus, every proof will have an ending point, and every proposition in the proof should obey the formation rules laid down in chapter 2. Second, every line or proposition in the proof will be one of three types: (1) a premise, (2) an assumption, or (3) the result of some combination of (1) or (2) plus a derivation rule. A *derivation rule* is an explicitly stated rule of PD that allows for moving forward in a proof. For example, if there is a derivation rule that says whenever you have 'P' and 'Q,' you can move a step forward in the proof to the proposition 'P∧Q,' then this derivation rule would justify (or legitimate) 'P∧Q' in a proof involving 'P' and 'Q.'

A special symbol is introduced in proofs. This symbol is the syntactic (or single) turnstile (⊢). In previous chapters, we used '⊢' to represent arguments. In this chapter, '⊢' takes on a more precise meaning, namely, that of *syntactic entailment*. Thus,

$$P∧R ⊢ R$$

means that there is a proof of '**R**' from '**P∧R**,' or '**R**' is a syntactic consequence of '**P∧R**.' In the above example, '**R**' is the conclusion of the proof, whereas '**P∧R**' are its premises. In addition,

$$⊢R$$

means that there is a proof of '**R**,' or that '**R**' is a theorem. In the above example, '**R**' is the conclusion of a proof with no premises.

The simple idea of a proof then is one that begins with a set of premises and in which each subsequent step in the proof is justified by a specified rule. To see how this might look, let's examine the following argument: 'R∨S, ¬S ⊢ R.' Setting up a proof is relatively simple. There are three components. The first is a numerical ordering of the lines of the proof. The second is a listing of the set of propositions that are the premises of the proof. The third is the labeling of each of these premises with a 'P' for premise. Thus, setting up the proof for 'R∨S, ¬S ⊢ R' looks as follows:

Line Number	Proposition	Premises
1	R∨S	P
2	¬S	P

As the proof proceeds, derivation rules will be used to derive propositions. These propositions are listed under the premises and justified by citing any propositions used in the argument form and the abbreviation for a the derivation rule. Thus,

Line Number	Proposition	Premises/Justification
1	R∨S	P
2	¬S	P
3	R	1,2 + derivation rule abbreviation

Once the conclusion is reached in a proof, the proof is finished. Thus, in the case of 'R∨S, ¬S ⊢ R,' since 'R' is the conclusion, the proof is completed at line 3.

5.2 PREMISES AND THE GOAL PROPOSITION

Certain propositions are premises. The justification for premises is symbolized by the letter 'P.' So, if we were asked to prove 'A→B, B→C ⊢ C→D,' the proof would be set up as follows:

$$
\begin{array}{lll}
1 & A{\rightarrow}B & \mathbf{P} \\
2 & B{\rightarrow}C & \mathbf{P}
\end{array}
$$

Sometimes it is helpful to write the goal proposition or conclusion of the proof to the right of the last 'P' in the justification column. The *goal proposition* is the proposition you are trying to solve at a given state in a proof. Writing this proposition can serve as a helpful reminder in the course of a long proof. In the case of the above proof, the conclusion 'C→D' is written in the third column. That is,

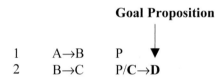

Goal Proposition

$$
\begin{array}{lll}
1 & A{\rightarrow}B & \mathbf{P} \\
2 & B{\rightarrow}C & \mathbf{P/C{\rightarrow}D}
\end{array}
$$

5.3 INTELIM DERIVATION RULES

In what follows, the derivation rules for PD are introduced. PD is an *intelim system.* That is, for every propositional operator (¬, ∧, ∨, →, ↔), there are two derivation rules: an *introduction rule* and an *elimination rule.* An introduction rule for a particular operator is a derivation rule that introduces a proposition with that operator into the proof. An elimination rule for a particular rule for a particular operator is a derivation rule that crucially begins (or uses) a proposition with that operator in the proof.

5.3.1 Conjunction Introduction (∧I)

Conjunction introduction (∧I) is a derivation rule whereby a conjunction is derived by conjoining two propositions (atomic or complex) already occurring in the proof. Conjunction introduction states that from '**P**' and '**Q**,' the complex conjunction '**P∧Q**' or '**Q∧P**' can be validly inferred.

| 1 | **Conjunction Introduction (∧I)**
From '**P**' and '**Q**,' we can derive '**P∧Q**.'
Also, from '**P**' and '**Q**,' we can derive '**Q∧P**.' | | **P**
Q
P∧Q
Q∧P |

∧I
∧I |

Consider the following argument:

$$W, Q, R \vdash W \land R$$

Begin by writing the premises, labeling them with 'P,' and indicating that the goal of the proof is 'W∧R.'

1	W	P
2	Q	P
3	R	P/W∧R

The goal of the proof is to derive 'W∧R,' which can be derived by using '∧I' on lines 1 and 3.

1	W	P
2	Q	P
3	R	P/W∧R
4	W∧R	1,3∧I

Notice that line 4 is justified by using the conjunction introduction derivation rule, and it is applied to lines 1 and 3.

It is important to remember that '**P**' and '**Q**' in the above form are metavariables for propositions. Since '**P**' and '**Q**' are metavariables, '∧I' can be used on both atomic and complex propositions. Consider the following argument:

$$A \rightarrow B, D \vdash (A \rightarrow B) \land D$$

1	A→B	P
2	D	P/(A→B)∧D
3	(A→B)∧D	1,2∧I

If conjunction introduction were formulated only by using propositional letters instead of metalinguistic variables, then it would only be acceptable if the premises were 'P' and 'Q.' But since derivation rules are formulated using metalinguistic variables, the argument above involves a legitimate use of conjunction introduction. For '**P**' picks out 'A→B,' and '**Q**' picks out 'D,' and so conjunction introduction allows for deriving '**P**∧**Q**,' that is,'(A→B)∧D.'

Finally, consider a more complex example:

$$(R \leftrightarrow S), W \land \neg T, Z \vdash [(R \leftrightarrow S) \land (W \land \neg T)] \land Z$$

1	(R↔S)	P
2	W∧¬T	P
3	Z	P/[(R↔S)∧(W∧¬T)]∧Z

Note that this proof will require multiple uses of '∧I' in a particular order.

1	(R↔S)	P
2	W∧¬T	P
3	Z	P/[(R↔S)∧(W∧¬T)]∧Z
4	**(R↔S)∧(W∧¬T)**	**1,2∧I**
5	**[(R↔S)∧(W∧¬T)]∧Z**	**3,4∧I**

In the introduction, it was noted that a natural deduction system is a set of derivational rules (general steps of reasoning) that mirror everyday reasoning in certain noteworthy ways. Conjunction introduction seems to do just this. For consider the following argument:

1	John is angry.	P
2	Liz is angry.	P
3	John is angry, and Liz is angry.	Conjunction introduction

In the above argument, *John is angry, and Liz is angry* follows from *John is angry* and *Liz is angry*. More generally, it seems whenever we have two propositions, it is legitimate to derive a second proposition that is the conjunction of these two propositions.

Finally, it is important to note that conjunction introduction, as a derivation form, is a statement about *syntactic entailment*. That is, the derivation rule is formulated independently of the truth or falsity of the premises. However, while the rules are formulated purely in terms of their structure, the rules we include in PD are guided by *semantic concerns*. That is, we choose rules that are *deductively valid*. Given this consideration, we can check whether or not we should include a given derivation rule into PD using a truth table or truth tree.

In the case of '∧I', consider the following truth table:

P	**Q**	⊢	**P**	∧	**Q**
T	T		T	T	T
T	F		T	F	F
F	T		F	F	T
F	F		F	F	F

The above table shows that for '**P**, **Q**⊢**P**∧**Q**,' in no truth-value assignment is it possible for the premises '**P**' and '**Q**' to be true and the conclusion '**P**∧**Q**' false.

Alternatively, consider a truth tree for the same argument:

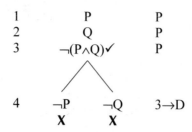

1	P	P
2	Q	P
3	¬(P∧Q)✓	P
4	¬P ¬Q	3→D
	X X	

The above tree shows that the premises and the negation of the conclusion produces a closed tree and is therefore valid.

5.3.2 Conjunction Elimination (∧E)

Conjunction elimination (∧E) is a derivation rule that allows for deriving either conjunct of a conjunction. In other words, it is a derivation rule that eliminates a conjunction by inferring either of the conjuncts.

2	**Conjunction Elimination (∧E)**	**P∧Q**	
	From '**P∧Q**,' we can derive '**P**.'	**P**	∧E
	Also, from '**P∧Q**,' we can derive '**Q**.'	**Q**	∧E

Note that conjunction elimination is an elimination rule, and as such, it is a derivation rule that can only be applied to conjunctions. Remember that a proposition is a conjunction only if its main truth-functional operator is the symbol for conjunction (i.e.,'∧'). Thus, considering '(P∧Q)∧W' and '(P∧Q)∨R,' the derivation rule '∧E' cannot be applied to '(P∧Q)∨R' since the main operator in that proposition is '∨.' However, '∧E' can be applied to '(P∧Q)∧W' since it has '∧' as its main operator. Consider the following application of the rule to '(P∧Q)∧W':

| 1 | (P∧Q)∧W | P |
| 2 | P∧Q | 1∧E |

Notice that '∧E' was applied to line 1 to infer 'P∧Q.' When using '∧E' either of the conjuncts of the conjunction can be inferred. Thus, another application of '∧E' allows for inferring the right conjunct 'W':

1	(P∧Q)∧W	P
2	P∧Q	1∧E
3	W	1∧E

Notice that line 2 is also a conjunction. This means that '∧E' can also be applied to 'P∧Q,' allowing for an inference to either of its conjuncts.

1	(P∧Q)∧W	P
2	P∧Q	1∧E
3	W	1∧E
4	P	2∧E
5	Q	2∧E

Remember that '∧E' is a derivation rule that is formulated using metalinguistic variables, and so '∧E' can be applied to a variety of different propositions that are conjunctions. For example, consider the following proof:

$$(A→B)∧(C∧D) \vdash D$$

1	(A→B)∧(C∧D)	P/D
2	C∧D	1∧E
3	D	2∧E

Notice that the proposition in line 1 is a conjunction. Thus, '∧E' can be applied to it in order to derive either of the conjuncts. Since the goal of the proof is to derive 'D' rather than 'A→B,' '∧E' is used to derive 'C∧D' at line 2. At this point, another use of '∧E' allows for deriving 'D' and finishes the proof.

Again, conjunction elimination closely resembles the way everyday reasoning occurs. For instance,

John is in the park, and Mary is in the subway. Therefore, John is in the park.

In addition, the choice to include conjunction elimination in PD is guided by semantic concerns. Thus, a truth table or truth tree is capable of showing that '∧E' is deductively valid.

1	P∧Q✓	P
2	¬P	P
3	P	1∧D
4	Q	1∧D
	X	

Finally, it is important to recognize that everyday argumentation is rarely restricted to the use of a single derivation rule. Instead we find individuals who use a variety of different derivation rules. At this point, our system of derivations (PD) is limited to two derivation rules, but we can consider an argument that involves the use of both. That is, consider the following argument:

$$R∧B, D \vdash (B∧D)∧R$$

1	R∧B	P
2	D	P/(B∧D)∧R

3	R	1∧E
4	B	1∧E
5	B∧D	2,4∧I
6	(B∧D)∧R	3,5∧I

In the above proof, it was necessary to make use of conjunction elimination to break the complex propositions down into their component parts and to use conjunction introduction on key atomic propositions to build up the desired complex proposition. Keep this strategy of breaking down and building up in mind as you practice the exercises below.

Exercise Set #1

A. Prove the following
  1. * (P∧Q)∧W ⊢ P
  2.   P∨Q, W, R ⊢ (W∧R)
  3. * M, F, R ⊢ (M∧F)∧R
  4.   P→Q, S ⊢ (P→Q)∧S
  5. * P→Q, S∧M ⊢ (P→Q)∧M
  6.   W∧R, M∧P ⊢ (W∧M)∧(P∧R)
  7. * (M∧N)∧W, (P→Q)∧Y ⊢ (P→Q)∧W
  8.   F∧[G∧(P∨Q)], (L∧M) ⊢ L∧(P∨Q)

Solutions to Starred Exercises in Exercise Set #1

1. * (P∧Q)∧W ⊢ P

1	(P∧Q)∧W	P
2	P∧Q	1∧E
3	P	2∧E

3. * M, F, R ⊢ (M∧F)∧R

1	M	P/(M∧F)∧R
2	F	P
3	R	P
4	M∧F	1,2∧I
5	(M∧F)∧R	3,4∧I

5. * P→Q, S∧M ⊢ (P→Q)∧M

1	P→Q	P/(P→Q)∧M
2	S∧M	P
3	M	2∧E
4	(P→Q)∧M	1,3∧I

7. * (M∧N)∧W, (P→Q)∧Y ⊢ (P→Q)∧W

1	(M∧N)∧W	P
2	(P→Q)∧T	P/(P→Q)∧W
3	W	1∧E
4	P→Q	2∧E
5	(P→Q)∧W	3,4∧I

5.3.3 Assumptions (A) and Subproofs

An *assumption* (A) is a proposition taken—or assumed—to be true for the purpose of proof. Before we consider how to represent assumptions in PD, let's consider two examples where assumptions are made in an everyday argument. Consider, for example, a case where you and a friend disagree about whether or not God exists. You think that God does not exist, and your friend thinks that God does exist. In order to persuade your friend, you might reason as follows:

> Let's assume that God does exist. If God is as great as you say he is, then there should be no poverty or suffering in the world. But there is suffering in the world. This is inconsistent! Therefore, God does not exist.

The above argument does not start by simply asserting, *God exists*. It instead begins by assuming that God exists and then, given this proposition and others, involves a line of reasoning on the basis of this assumption. One way to look at the argument above is that it involves two proofs. There is the *main proof*, which aims at the conclusion *God does not exist*, and there is a *subproof*, which aims to show that anyone who assumes (or believes) that *God does exist* is forced into believing something that cannot be the case, namely, something inconsistent (i.e., 'There is both suffering and no suffering in the world').

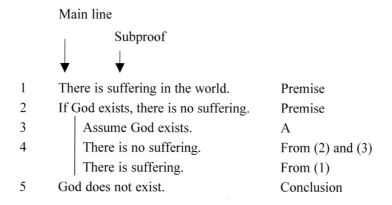

	Main line / Subproof	
1	There is suffering in the world.	Premise
2	If God exists, there is no suffering.	Premise
3	Assume God exists.	A
4	There is no suffering.	From (2) and (3)
	There is suffering.	From (1)
5	God does not exist.	Conclusion

Let's turn to your friend's counterargument:

Let's assume that God does not exist. If God does not exist, then the universe just magically came into existence out of nothing. But the universe did not come into existence from nothing. Therefore, you are contradicting yourself in saying that God does not exist and the world exists. Therefore, God does exist.

In the above case, your friend starts with an assumption that she does not believe, namely, that *God does not exist*. Using this assumption, your friend reasons—within a subproof—that if *God does not exist* is true, then *The world both exists and does not exist* is true, which is inconsistent. Thus, your friend ultimately concludes that God does exist.

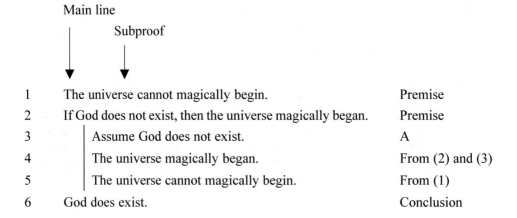

1	The universe cannot magically begin.	Premise
2	If God does not exist, then the universe magically began.	Premise
3	Assume God does not exist.	A
4	The universe magically began.	From (2) and (3)
5	The universe cannot magically begin.	From (1)
6	God does exist.	Conclusion

Whenever an assumption is made in a proof, we indicate that the proposition is an assumption with an 'A' in the justification column. In addition, we also want to separate the reasoning done in a subproof from the reasoning done in the main part of the proof. In order to indicate the presence of a subproof, we indent from the main part of the proof and begin the subproof by drawing a descending line from where the subproof starts to the point where the subproof ends. Here is an example:

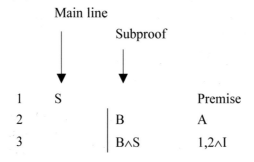

1	S		Premise
2		B	A
3		B∧S	1,2∧I

Notice that at line 2, 'B' is assumed, and so the proof enters a subproof. In the above case, the subproof is nested in (or built on) the main line of the proof.

Within a subproof, it is possible to make use of derivation rules and even to make further assumptions. An example of a legitimate use of reasoning is '∧I' at line 3 above. In addition, additional assumptions can be made, producing a subproof within a subproof. For example,

1	Q			P
2		S		A
3			W	A

In the example above, the proof starts with 'Q' in the main line of the proof. Next, at line 2, 'S' is assumed, which begins the first subproof (call this subproof₁). Note that subproof₁ is not simply an independent proof but is part of a proof that is built on the main line of the proof. Building one part of the proof on another in this way is called *nesting*. In the above example, the main line of the proof nests subproof₁ (or, alternatively, subproof₁ is in the nest of the main line) since subproof₁ is built on the main line. In addition, subproof₁ is more deeply nested than the main line since the main line nests (or contains) subproof₁.

Next, at line 3, 'W' is assumed, which begins a second subproof (call this subproof₂). Again, notice that the assumption that begins subproof₂ is not independent of subproof₁. Instead, subproof₂ is built upon subproof₁ and the main line of the proof. This means that the main line nests subproof₁,and subproof₁ nests subproof₂. In addition, subproof₂ is more deeply nested than both subproof₁ and the main line of the proof.

Here is a graphical representation:

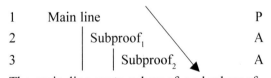

1	Main line			P
2		Subproof₁		A
3			Subproof₂	A

The main line nests subproof₁ and subproof₂, while subproof₁ nests subproof₂.

One way to think about the above argument is in terms of a conversation. That is, the above example is similar to someone uttering the following:

We agree that 'Q' is true.
Given that 'Q' is the case, assume 'S.'
Given that 'Q' is true and our assumption 'S,' assume 'W.'

In each case, an assumption is built on propositions already occurring in the proof.

However, it is not necessary that the proof structure develop in this way. It is often the case that once one assumption has been made, it is desirable to make a separate assumption that does not depend upon the first assumption. Consider the following example:

1	A		P
2		B	A
3		A∧B	1,2∧I
4		C	A
5		A∧C	1,4∧I

In the above case, there are two subproofs, but neither subproof nests the other. That is, the subproof beginning with 'C' at line 4 is not built upon the subproof beginning with 'B' at line 2. It is a separate subproof. In plain English, it is as though the following conversation is taking place:

> We agree that 'A' is true.
> Given the truth of 'A,' assume 'B.'
> Given the truth of 'A,' assume 'C.'

Finally, another important feature of reasoning within subproofs is that propositions outside the subproof can be used inside the subproof, but propositions inside the subproof cannot be used outside the subproof. In the language of nests, a proposition less nested can be used in a more deeply nested part, but a proposition more deeply nested cannot be used in a less deeply nested part.

It is perhaps easiest to see this graphically:

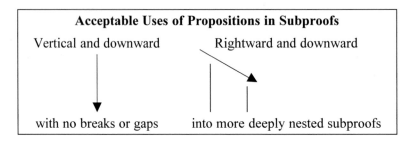

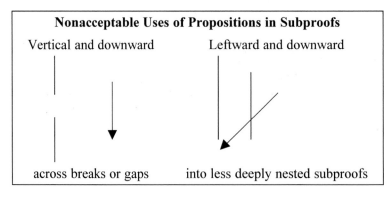

For some concrete illustrations, consider a few examples of nonacceptable forms of reasoning within subproofs:

1	A		P
2		B	A
3		A∧B	1,2∧I
4		C	A
5		C∧B	4,2∧I—**NO!**

Notice that the above example involves transferring a proposition from one sub-proof into another subproof. That is, 'B' is assumed and then used in a different subproof at line 5.

Consider another example:

1	A		P
2		B∧C	A
3		B	2∧E
4	A∧(B∧C)		1,2∧I—**NO!**
5	A∧B		1,3∧I—**NO!**

In the above case, lines 4 and 5 are not acceptable because propositions within the subproof are taken outside the subproof. It would be as if you said, 'Assume that I am the greatest person in the world; therefore, I am the greatest person in the world.'

However, propositions from outside the subproof can be taken into a subproof that it contains. For example,

1	A		P
2		B∧C	A
3		A∧(B∧C)	1,2∧I

Given these restrictions on reasoning within subproofs, it may seem as though reasoning in a subproof is pointless since there is no way to move a proposition from within a subproof to the main line of the proof. Everything is, as it were, trapped in a subproof. This is not the case since there are specific rules that allow for inferring propositions "outside" a subproof. The first rule that allows for such a procedure is conditional introduction (→I).

5.3.4 Conditional Introduction (→I)

Conditional introduction (→I) is a derivation rule that begins with an assumption in a subproof and allows for deriving a conditional outside the subproof. The derived conditional consists of the assumed proposition as the *antecedent* and the derived conclusion in the subproof as the *consequent*.

3	**Conditional Introduction (→I)** From a derivation of '**Q**' within a subproof involving an assumption '**P**,' we can derive '**P→Q**' out of the subproof.	P . . . Q	A
		P→Q	→I

Here is a simple example:

$$Q \vdash P \rightarrow Q$$

1	Q		P/Q
2		P	A
3		P∧Q	1,2∧I
4		Q	3∧E
5	P→Q		2–4→I

Notice that conditional introduction begins by assuming '**P**' (the antecedent of the conclusion); then '**Q**' (the consequent of the conclusion) is derived in the subproof; finally, a conditional '**P→Q**' is derived outside the subproof. When a subproof is completed, we say that the assumption of the subproof has been *discharged*. That is, the force or positive charge of the assumption, as well the corresponding subproof in which it occurs, is no longer in effect. In addition, we call a subproof with an undischarged assumption an *open subproof* and a subproof with a discharged assumption a *closed subproof.*

Let's consider the following proof in a more step-by-step manner:

$$S \rightarrow D \vdash S \rightarrow (S \rightarrow D)$$

First, start by writing down the premises and indicating the goal proposition of the proof:

| 1 | S→D | P/S→(S→D) |

Next, notice that our goal conclusion is a conditional: 'S→(S→D).' We might be able to derive this conditional by using '→I,' where we assume the antecedent of the conclusion in a subproof and derive the consequent of the conclusion in that same subproof. Thus, start by assuming 'S' and making 'S→D' our goal for the subproof:

```
1      S→D           P/S→(S→D)
2          | S       A/S→D
```

With 'S' assumed, the next step is to derive 'S→D' in the subproof.

```
1      S→D                  P/S→(S→D)
2          | S              A/A→D
3          | S∧(S→D)        1,2∧I
4          | S→D            3∧E
```

With 'S→D' derived in the subproof, we can discharge the assumption (close the subproof) by using '→I':

```
1      S→D                  P/S→(S→D)
2          | S              A/A→D
3          | S∧(S→D)        1,2∧I
4          | S→D            3∧E
5      S→(S→D)              2–4→I
```

The proof is now complete! When using '→I,' you exit the subproof with a conditional consisting of the assumption of the subproof as the antecedent and the derived conclusion in the subproof as the consequent. In the case above, 'S' is the assumption in the subproof, and 'S→D' is the derived proposition in the subproof. Conditional introduction allows for deriving the proposition 'S→(S→D)' out of the subproof in which 'S' and 'S→D' are found. Once '→I' is used at line 5, the assumption 'S' is discharged and the subproof is closed (i.e., not open).

When working with multiple subproofs, it is important to realize that the use of conditional introduction only allows for deriving a conditional out of the subproof containing the assumption and the derived proposition. In short, you can only use conditional introduction to derive a conditional out of one subproof. Consider the following proofs:

```
1      P                        P
2          | R                  A
3          |    | Z             A
4          |    | Z∧R           2,3∧I
5          |    | R             4∧E
6          | Z→R                3–5→I—OK!
```

```
1        P                          P
2            | R                    A
3            |      | Z             A
4            |      | Z∧R           2,3∧I
5            |      | R             4∧E
6        Z→R                        3–5→I—NO!
```

In the first proof, 'Z→R' is properly derived out of the subproof where 'Z' is the assumption. In the second proof, 'Z→R' is improperly derived not only out of the subproof where 'Z' is the assumption but also the subproof involving 'R.'

Here is a final example involving the use of conditional introduction. Prove the following:

$$A \vdash B \to [B \land (C \to A)]$$

```
1        A                          P/B→[B∧(C→A)]
2            | B                    A/B∧(C→A)
3            |      | C             A/A
4            |      | C∧A           1,3∧I
5            |      | A             4∧E
6            | C→A                  3–5→I
7            | B∧(C→A)              2,6∧I
8        B→[B∧(C→A)]                2–7→I
```

Notice that the above proof involves an auxiliary assumption 'C,' and while propositions can be brought into the subproof, any proposition involving 'C' cannot be used outside the subproof until it has been discharged with '→I.'

5.3.5 Conditional Elimination (→E)

Conditional elimination (→E), more commonly known as *modus ponens*, is a derivation rule that allows for inferring the consequent of a conditional by affirming the antecedent of that conditional.

4	**Conditional Elimination (→E)** From 'P→Q' and P, we can derive 'Q.'	**P→Q** **P** **Q**	→E

Consider the proof for '(A∨B)→C, A, A∨B ⊢ C.'

1	(A∨B)→C	P
2	A	P
3	A∨B	P/C
4	C	1,3→E

In the above example, notice that '→E' is used on the propositions occurring in lines 1 and 3. It is not permissible to use '→E' on lines 1 and 2 because in order to use '→E,' you need a proposition that is *the entire antecedent* of the conditional. One easy way to remember '→E' is that it is a rule that affirms the antecedent, and since only conditionals have antecedents, this rule will only apply to conditionals.

Let's consider a more complicated use of '→E' in the following proof:

$$A→B, A, B→C, C→D ⊢ D.$$

1	A→B	P
2	A	P
3	B→C	P
4	C→D	P/D
5	B	1,2→E
6	C	3,5→E
7	D	4,6→E

5.3.6 Reiteration (R)

Reiteration (R) is one of the most intuitive derivation rules. It allows for deriving any proposition that already occurs as a premise or as a derived proposition in the proof.

5	**Reiteration (R)** Any proposition '**P**' that occurs in a proof or subproof may be rewritten at a level of the proof that is equal to '**P**' or more deeply nested than '**P**.'	**P** . . . **P**	 R

Consider a very simple example:

1	A→B	P
2	A	P
3	A	2R

Line 3 involves a use of reiteration from line 2. Next, turn to a slightly more complex example:

1	A→B	P
2	A	P

3	¬(B∨C)	P
4	A→B	1R
5	A	2R
6	¬(B∨C)	3R

When using reiteration, keep in mind restrictions on reasoning in and deriving propositions out of subproofs. With respect to reiteration, it is not acceptable to

(1) reiterate a proposition from a more deeply nested part of the proof into a less deeply nested part of the proof;
(2) reiterate a proposition from one part of the proof into another part that is not within its nest.

Another way of putting (1) is that it is not acceptable to reiterate a proposition *out of* a subproof, but it is acceptable to reiterate a proposition *into* a subproof. For example,

1	P→Q			P
2		S		A
3			W	A
4			P→Q	1R
5			S	2R
6		W		3R—**NO!**
7	S			2R—**NO!**

Notice that while the use of 'R' at lines 4 and 5 is acceptable, the use of 'R' at lines 6 and 7 reiterates propositions from a more nested part of the proof into a less nested part of the proof, which is not acceptable.

To see why this use of reiteration is invalid, consider the following argument, which does not obey the above restriction.

1	Assume that I am the richest person.	A
2	Therefore, it follows that I am the richest person.	1R—**NO!**

This argument is clearly invalid since from the assumption that I am the richest person in the world, it does not follow that I am the richest person in the world.

Consider the second restriction on the use of 'R'; namely, it is not acceptable to

(2) reiterate a proposition from one part of the proof into another part that is not within its nest.

Consider the following example:

```
1      P                    P
2           | S            A
3           | P∧S          1,2∧I


4           | T            A
```

Notice that line 4 begins a subproof that is not nested (or contained) in the previous subproof beginning at line 2. That is, it begins a subproof that, while contained in the main line of the proof, is independent of the subproof that begins at line 2.

```
1      P                    P
2           | S            A
3           | P∧S          1,2∧I


4           | T            A
5           | S            2R—NO!
```

Notice that line 5 violates the second restriction since 'S' is reiterated from one subproof into another that is not within its nest.

Finally, it is also important to note one distinguishing feature of reiteration, namely, that it is a derived rule. This means that any use of reiteration is somewhat superfluous since the inference that it achieves can be achieved using the existing set of derivation rules. To see this more clearly, consider the proof of the valid argument 'R ⊢ R.'

```
1      R                    P/R
2           | R            A/R→R
3           | R∧R          1,2∧I
4           | R            3∧E
5      R→R                 2–4→I
6      R                   1,5→E
```

Although the introduction of reiteration into our set of derivation rules is not essential, it is extremely convenient since the proof above can be simplified into the following proof.

```
1      R      P/R
2      R      1R
```

Exercise Set #2

A. Prove the following:
1. *   (A∧B)→C, A, B ⊢ C
2.   R→(M→P), R, M ⊢ P
3. *   P, Q ⊢ P→Q
4.   A, B, C, [(A∧B)∧C]→D ⊢ D
5. *   A, B ⊢ C→(A∧B)
6.   A∨B, B∨C ⊢ (C∨D)→(A∨B)
7. *   P ⊢ P, without using reiteration
8.   ⊢ P→P
9. *   ⊢ A→(A∧A)
10.   ⊢ ¬A→(¬A→¬A)

Solutions to Starred Exercises in Exercise Set #2

1. *   (A∧B)→C, A, B ⊢ C

1	(A∧B)→C	P/C
2	A	P
3	B	P
4	A∧B	2,3∧I
5	C	1,4→E

3. *   P, Q ⊢ P→Q

1	P		P
2	Q		P/P→Q
3		P	A/Q
4		Q	2R
5	P→Q		3–4→I

5. *   A, B ⊢ C→(A∧B)

1	A		P
2	B		P/C→(A∧B)
3		C	A/A∧B
4		A∧B	1,2∧I
5	C→(A∧B)		3–4→I

7. *   P ⊢ P, without using reiteration

1	P		P/P
2		P	A/P→P
3		P∧P	1,2∧I
4		P	3∧E
5	P→P		2–4→I
6	P		1,5→E

9. * ⊢A→(A∧A)

1		A	A/A∧A
2		A	1R
3		A∧A	1,2∧I
4	A→(A∧A)		1–3→I

5.3.7 Negation Introduction (¬I) and Elimination (¬E)

Two additional derivation rules are negation introduction (¬I) and negation elimination (¬E). Proofs involving these two forms are sometimes called *indirect proofs* or *proofs by contradiction*.

6	**Negation Introduction (¬I)** From a derivation of a proposition 'Q' and its literal negation '¬Q' within a subproof involving an assumption 'P,' we can derive '¬P' out of the subproof.		P . . . ¬Q Q	A
		¬P		¬I

Negation introduction is a derivation rule where 'P' is assumed, and in the course of a subproof, an inconsistency of the form 'Q' and '¬Q' is derived. Once an inconsistency is shown to follow from 'P,' negation introduction (¬I) allows for deriving '¬P' out of the subproof.

Negation elimination follows a similar procedure, except the initial assumption is a negated proposition '¬P' and the proposition derived is the unnegated form (i.e., 'P').

7	**Negation Elimination (¬E)** From a derivation of a proposition 'Q' and its literal negation '¬Q' within a subproof involving an assumption '¬P,' we can derive 'P' out of the subproof.		¬P . . . ¬Q Q	A
		P		¬E

Notice that '¬P' is assumed, a contradiction is derived, and 'P' is discharged. The idea again is that if '¬P' leads to a contradiction, then 'P' must be the case.

The basic idea in using '¬I' and '¬E' is to (1) assume a proposition in a subproof, (2) derive a contradiction, and (3a) derive the literal negation of the assumed proposition outside the subproof, or (3b) derive the unnegated form of the assumed proposition outside the subproof. Here is an example:

$$A→D, ¬D ⊢¬A$$

Start by assuming 'A' in the subproof.

$$
\begin{array}{llll}
1 & A{\rightarrow}D & & P \\
2 & \neg D & & P \\
3 & & A & A/\neg I \\
4 & & D & 1,3{\rightarrow}E \\
5 & & \neg D & 2R \\
6 & \neg A & & 3\text{-}5\neg I
\end{array}
$$

Thus, since the assumption of 'A' leads to an explicit inconsistency, it must be the case that '¬A.' Let us consider a case of negation elimination, that is, a use of '¬E.' For example, '(A→D)∧A, ¬D ⊢C.'

$$
\begin{array}{llll}
1 & (A{\rightarrow}D){\wedge}A & & P \\
2 & \neg D & & P \\
3 & & \neg C & A/\neg E \\
4 & & A{\rightarrow}D & 1{\wedge}E \\
5 & & A & 1{\wedge}E \\
6 & & D & 4,5{\rightarrow}E \\
7 & & \neg D & 2R \\
8 & C & & 3\text{-}7\neg E
\end{array}
$$

Exercise Set #3

A. Prove the following:
  1. * A ⊢B→A
  2.   A∧B, B→C ⊢C
  3. * (P→Q)→W, Q ⊢W
  4.   C∧¬C ⊢S
  5. * (A∧B)∧C, R∧(W∧¬C) ⊢S
  6.   ¬A→(C→D), ¬A→(C→¬D), ¬A∧C ⊢R
  7. * (A∨B)→M, M→¬(A∨B) ⊢¬(A∨B)
  8.   ⊢A→A
  9. * A, B, B→¬A ⊢C→D
 10.   (R∨S)→¬B, ¬B→¬S, ¬S→¬(R∨S) ⊢¬(R∨S)

Solutions to Starred Exercises in Exercise Set #3

  1. * A ⊢B→A

$$
\begin{array}{llll}
1 & A & & P/B{\rightarrow}A \\
2 & & B & A/A \\
3 & & A & 1R \\
4 & B{\rightarrow}A & & 2\text{-}3{\rightarrow}I
\end{array}
$$

3. * (P→Q)→W, Q ⊢W

1	(P→Q)→W	P
2	Q	P/W
3	P	A/Q
4	Q	2R
5	P→Q	3–4→I
6	W	1,5→E

5. * (A∧B)∧C, R∧(W∧¬C) ⊢S

1	(A∧B)∧C	P
2	R∧(W∧¬C)	P/S
3	C	1∧E
4	W∧¬C	2∧E
5	¬C	4∧E
6	¬S	A/P, ¬P
7	C	3R
8	¬C	5R
9	S	6–8¬E

7. * (A∨B)→M, M→¬(A∨B) ⊢¬(A∨B)

1	(A∨B)→M	P
2	M→¬(A∨B)	P/¬(A∨B)
3	(A∨B)	A/P, ¬P
4	M	1,3→E
5	¬(A∨B)	2,4→E
6	(A∨B)	3R
7	¬(A∨B)	3–6¬I

9. * A, B, B→¬A ⊢C→D

1	A	P
2	B	P
3	B→¬A	P/C→D
4	¬A	2,3→E
5	¬(C→D)	A/P, ¬P
6	A	1R
7	¬A	4R
8	C→D	5–7¬E

5.3.8 Disjunction Introduction (∨I)

Disjunction introduction (∨I) is a derivation rule whereby a disjunction is derived from a proposition (atomic or complex) already occurring in the proof. Disjunction introduction states that from a proposition 'P,' a disjunction 'P∨Q' or 'Q∨P' can be derived. Here is the derivation rule for disjunction introduction.

8	**Disjunction Introduction (∨I)** From 'P,' we can validly infer 'P∨Q' or 'Q∨P.'	**P** **P∨Q** **Q∨P**	∨I ∨I

Consider the following example:

$$P \vdash (P∨W)∧(Z∨P)$$

1	P	P/(P∨W)∧(Z∨P)
2	P∨W	1∨I
3	Z∨P	1∨I
4	(P∨W)∧(Z∨P)	2,3∧I

In the above proof, there are two different uses of '∨I' on 'P' in line 1. Notice that the use of '∨I' only applies to a single proposition and allows for deriving a disjunction.

Consider a more complex example. Prove the following:

$$[P∨(Q∨R)]→W, R \vdash W$$

1	[P∨(Q∨R)]→W	P
2	R	P/W
3	Q∨R	2∨I
4	P∨(Q∨R)	3∨I
5	W	1,4→E

In the above example, the desired conclusion is 'W.' Notice that 'W' could be derived if 'P∨(Q∨R)' were in the proof. Using multiple instances of '∨I' on 'R' in line 2 allows us to obtain this proposition. Notice that '∨I' can be applied to complex propositions, even propositions that are already disjunctions.

One more proof:

$$P, (P∨¬W)→R \vdash R∨¬(Z∨Q)$$

1	P	P
2	(P∨¬W)→R	P/R∨¬(Z∨Q)
3	P∨¬W	1∨I
4	R	2,3→E
5	R∨¬(Z∨Q)	4∨I

This proof requires two uses of '∨I.' The first use is similar to the one in the previous proof. '∨I' is applied to 'P' to derive '∨¬W' in order to derive 'R' from line 2. Once 'R' is inferred at line 4, '∨I' is used again to infer the disjunction.

5.3.9 Disjunction Elimination (∨E)

Disjunction elimination (∨E) is a derivation rule such that from a disjunction '**P∨Q**,' a proposition '**R**' can be derived, provided that the proposition can be derived from two separate subproofs where each subproof begins with one of the disjunction's disjuncts (i.e., one with '**P**' and one with '**Q**'). Here is the basic form for disjunction elimination.

9	**Disjunction Elimination (∨E)**
	From '**P∨Q**' and two derivations of '**R**'—one involving '**P**' as an assumption in a subproof, the other involving '**Q**' as an assumption in a subproof—we can derive '**R**' out of the subproof.
		P ∨ Q
		  P A
		  .
		  .
		  .
		  R
		
		    Q A
		    .
		    .
		    .
		    R
		R ∨E

In the above argument form, each of the disjuncts from '**P∨Q**' is separately assumed. From within both of these subproofs, '**R**' is derived by some unspecified form of valid reasoning. If '**R**' can be derived in both subproofs, then '**R**' can be derived from '**P∨Q**.'

Here is an example involving '∨E':

1	P∨Q	P
2	P→R	P
3	Q→R	P/R
4	⎸P	A
5	⎸R	2,4→E
6	⎸Q	A
7	⎸R	3,6→E
8	R	1,4–5,6–7∨E

Notice that since 'P' implies 'R' in a subproof, and 'Q' implies 'R' in a subproof, 'R' can be discharged from the subproof. Also, notice that in the justification column, the use of '∨E' requires citing the original disjunction and all propositions in the two subproofs. Here is another example:

1	P∨(Q∧T)	P
2	P→T	P/T
3	P	A
4	T	2,4→E
5	Q∧T	A
6	T	5∧E
7	T	1,3–4,5–6∨E

Again, the use of '∨E' involves two separate subproofs. First, the proof begins by assuming one of the disjuncts and deriving 'T'; then there is a separate assumption (involving the other disjunct), and the same proposition (i.e., 'T') is derived. Once 'T' is derived in both subproofs, then 'T' can be derived outside the subproof.

Here is a more complex proof involving '∨E.' Prove the following:

$$S \to W, M \to W, P \land (R \lor T), R \to (S \lor M), T \to (S \lor M) \vdash W$$

1	S→W	P
2	M→W	P
3	P∧(R∨T)	P
4	R→(S∨M)	P
5	T→(S∨M)	P
6	R∨T	3∧E
7	R	A
8	S∨M	4,7→E
9	T	A
10	S∨M	5,9→E
11	S∨M	6,7–8,9–10∨E
12	S	A
13	W	1,12→E
14	M	A
15	W	2,14→E
16	W	11,12–13,14–15∨E

Disjunction elimination is perhaps the most complicated derivation rule to master, and how it relates to everyday reasoning is perhaps somewhat unclear. To illustrate consider the following argument:

> Suppose that you wanted to show that *John will have a great evening* follows from *John will either go to the party or stay home.* In order to show this, you need to show both that if John goes to the party, he will have a great time and that if John stays home, he will also have a great time. The reason that it needs to follow from both disjuncts is because if John will have a great evening *only if* he goes to the party and not if he stays home (and vice versa), then it is possible for the premise *John will either go to the party or stay home* to be true, and the conclusion *John will have a great evening* to be false.

Thus, if we want to reason by beginning with a disjunction, it is necessary to show that a proposition follows from both of the disjuncts. This is done by separately assuming both disjuncts and showing how the same propositions follow in each subproof.

5.3.10 Biconditional Elimination and Introduction (↔E and ↔I)

The derivation rules applicable to biconditionals are '↔E' and '↔I.' For biconditional introduction (↔I), from a derivation of 'Q' within a subproof involving an assumption 'P,' and from a derivation of 'P' within a separate subproof involving an assumption 'Q,' we can derive 'P↔Q'out of the subproofs. In other words, '↔I' introduces a biconditional from two separate subproofs.

10	**Biconditional Introduction (↔I)** From a derivation of 'Q' within a subproof involving an assumption 'P' and from a derivation of 'P' within a separate subproof involving an assumption 'Q,' we can derive 'P↔Q' out of the subproof.		**P** . . . **Q**	A
			Q . . . **P**	A
		P↔Q		↔I

Here is an example of a proof involving biconditional introduction:

$$P \rightarrow Q, Q \rightarrow P \vdash P \leftrightarrow Q.$$

1	P→Q		P
2	Q→P		P/P↔Q
3		P	A
4		Q	1,3→E
5		Q	A
6		P	2,5→E
7	P↔Q		3–4,5–6↔I

Biconditional elimination (↔E), in contrast, allows for deriving one side of a biconditional, provided the proof contains a biconditional and the other side of the biconditional. That is, from '**P↔Q**' and '**P**,' we can derive '**Q**.' And from '**P↔Q**' and '**Q**,' we can derive '**P**.'

11	**Biconditional Elimination (↔E)** From '**P↔Q**' and '**P**,' we can derive '**Q**.' And from '**P↔Q**' and '**Q**,' we can derive '**P**.'	P↔Q P Q	↔E
		P↔Q Q P	↔E

Let's look at two examples (simple and complex) that use '↔E' and then look at a proof involving both '↔I' and '↔E.' Prove the following:

$$(P \leftrightarrow Q) \leftrightarrow (R \rightarrow T), R \rightarrow T \vdash P \leftrightarrow Q$$

1	(P↔Q)↔(R↔T)	P
2	R→T	P/P↔Q
3	P↔Q	1,2↔E

In order to infer one side of the biconditional, it is necessary to have the other side at some line in the proof. Since the right-hand side of the biconditional is at line 2, the left-hand side can be derived using '↔E.' Here is another example:

$$P \leftrightarrow Q, P, (Q \wedge P) \leftrightarrow W \vdash W.$$

1	P↔Q	P
2	P	P
3	(Q∧P)↔W	P/W

4	Q	1,2↔E
5	Q∧P	2,4∧I
6	W	3,5↔E

The above proof involves two uses of '↔E.' The first use at line 4 is straightforward. However, notice that in the second use, in order to derive 'W' at line 6, 'Q∧P' is needed.

Finally, consider an example that combines both '↔I' and '↔E.' Prove the following:

$$P↔Q, Q ⊢, (P∨¬Z)↔(¬Z∨P)$$

1	P↔Q		P
2	Q		P/(P∨¬Z)↔(¬Z∨P)
3	P		1,2↔E
4		P∨¬Z	A/(¬Z∨P)
5		¬Z∨P	3∨I
6		¬Z∨P	A
7		P∨¬Z	3∨I
8	(P∨¬Z)↔(¬Z∨P)		4–5,6–7↔I

Exercise Set #4

A. Prove the following:

1. * P→Q, P⊢ Q
2.   P⊢ P∨Q
3. * P, (P∨Q)→R⊢ R
4.   P, Q, (P∧Q)→R⊢ R
5. * P, P→Q⊢ Q∨M
6.   A∨B, A→D,B→D⊢ D
7. * P→Q, Q→R, P⊢ R
8.   (P∨M)→Q, Q→R, P⊢ R∨W
9. * (A∨B)∨C, (A∨B)→D, C→D⊢ D∨M
10.   [(A∨B)∨D]→R, A⊢ R
11. * P, Q⊢ P↔Q
12.   A→B, ¬B, A⊢ ¬W
13. * A→B, ¬B, A⊢ W
14.   ⊢ A→A
15.   ⊢ (A→B)→(A→A)

16.   ⊢ ¬(A∧¬A)

17.   P→Q, P⊢ Q→Q

18.   A∨B, A→R, B→R⊢ W→R

19.   P⊢ P→P

20.   P→Q, Q→P⊢ P↔Q

21.   (P∨W)→¬Q, P↔¬Q, W⊢ P

22.   P, Q, R, S⊢ (P↔Q)↔(R↔S)

23.   P↔(Q∨R), P, ¬Q⊢ R

B. Translate the following English arguments into propositional logic and prove these arguments using the derivation rules in PD.

1.   If John is happy, then Mary is sad. John is happy. Therefore, Mary is sad.

2.   If John is happy, then Mary is sad. Mary is not sad. Therefore, John is not happy.

3.   John is happy, and Mary is happy. If John is happy, then John loves Mary. If Mary is happy, then Mary loves John. Therefore, John loves Mary, and Mary loves John.

4.   God is good, and God is great. It follows that if God is good or God is not good, then God is great.

5.   If God is all-knowing and all-powerful and all-loving, then there is no evil in the world. There is evil in the world. Therefore, it is not the case that God is all-knowing and all-powerful and all-loving.

6.   John will run from the law if and only if (iff) the police are after him. John will run from the law. Thus, the police are after John.

7.   John is a criminal. If John is a criminal, then the police are after him. If John is a criminal, then John will run from the law. It follows that John will run from the law if and only if the police are after him.

8.   If the murder weapon is John's or a witness saw John commit the crime, then John is not innocent. A witness did see John commit the crime. Thus, it follows that John is not innocent.

9.   John is the murderer, or he isn't. If John is the murderer, then there is strong evidence showing that he is guilty. There is strong evidence showing John is guilty if and only if the prosecution can show he pulled the trigger. But it is not the case that the prosecution can show John pulled the trigger. Thus, it follows that John is not the murderer.

10.   Taxes will go up, or they won't. If taxes go up, then people will lose their jobs, and the price of housing will decline. If taxes don't go up, then people will lose their jobs, and the price of housing will decline. It follows that people will lose their jobs.

11.   If Mr. Z wins the election or gets control of the military, then Mr. Z will either dissolve the federal government or save the country. Mr. Z has gotten control of the military if and only if he has convinced the generals that the current president is inept. Mr. Z has convinced the generals that the current president is inept and will not save the country. It follows that Mr. Z will dissolve the federal government.

12.   If John runs every day, then he has a healthy heart. If John has a healthy heart, he won't have a heart attack. It follows that if John runs every day, he won't have a heart attack.

13.   If John runs every day and eats properly, then he will live a long life. John will not live a long life. It follows that John does not both run every day and eat properly.

14.   If John is innocent, then the bloody glove found at the scene of the crime and the murder weapon are not John's, and John was not seen at the scene of the crime. But the bloody glove is John's, the murder weapon is John's, and John was seen at the scene of the crime. It follows that John is not innocent.

Solutions to Starred Exercises in Exercise Set #4

1. * $P{\rightarrow}Q,\ P \vdash Q$

1	$P{\rightarrow}Q$	P
2	P	P/Q
3	Q	$1,2{\rightarrow}E$

3. * $P,\ (P{\vee}Q){\rightarrow}R \vdash R$

1	P	P
2	$(P{\vee}Q){\rightarrow}R$	P/R
3	$P{\vee}Q$	$1{\vee}I$
4	R	$2,3{\rightarrow}E$

5. * $P,\ P{\rightarrow}Q \vdash Q{\vee}M$

1	P	P
2	$P{\rightarrow}Q$	$P/Q{\vee}M$
3	Q	$1,2{\rightarrow}E$
4	$Q{\vee}M$	$3{\vee}I$

7. * $P{\rightarrow}Q,\ Q{\rightarrow}R,\ P \vdash R$

1	$P{\rightarrow}Q$	P
2	$Q{\rightarrow}R$	P
3	P	P/R
4	Q	$1,3{\rightarrow}E$
5	R	$2,4{\rightarrow}E$

9. * $(A{\vee}B){\vee}C,\ (A{\vee}B){\rightarrow}D,\ C{\rightarrow}D \vdash D{\vee}M$

1	$(A{\vee}B){\vee}C$	P
2	$(A{\vee}B){\rightarrow}D$	P
3	$C{\rightarrow}D$	$P/D{\vee}M$
4	$A{\vee}B$	A
5	D	$2,4{\rightarrow}D$

6		C	A
7		D	6,3→D
8	D		1,4–5,6–7∨E
9	D∨M		8∨I

11. * P, Q ⊢P↔Q

1	P		P
2	Q		P/P↔Q
3		P	A
4		Q	2R

5		Q	A
6		P	1R
7	P↔Q		3–4,5–6↔I

13. * A→B, ¬B, A ⊢W

1	A→B		P
2	¬B		P
3	A		P/W
4	B		1,3→E
5		¬W	A/P, ¬P
6		¬B	2R
7		B	4R
8	W		5–7¬E

Eleven Derivation Rules for PD					
1	**Conjunction Introduction (∧I)** From 'P' and 'Q,' we can derive 'P∧Q.' Also, from 'P' and 'Q,' we can derive 'Q∧P.'			**P** **Q** **P∧Q** **Q∧P**	∧I ∧I
2	**Conjunction Elimination (∧E)** From 'P∧Q,' we can derive 'P.' Also, from 'P∧Q,' we can derive 'Q.'			**P∧Q** **P** **Q**	∧E .∧E
3	**Conditional Introduction (→I)** From a derivation of 'Q' within a subproof involving an assumption 'P,' we can derive 'P→Q' out of the subproof. **P→Q**		**P** . . . **Q**		A →I

4	**Conditional Elimination (→E)** From 'P→Q' and 'P,' we can derive 'Q.'		P→Q P Q		→E
5	**Reiteration (R)** Any proposition 'P' that occurs in a proof or subproof may be rewritten at a level of the proof that is equal to 'P' or more deeply nested than '**P**.'		P . . . P		R
6	**Negation Introduction (¬I)** From a derivation of a proposition '**Q**' and its literal negation '¬**Q**' within a subproof involving an assumption '**P**,' we can derive '¬**P**' out of the subproof.			P . . . ¬Q Q	A
			¬P		¬I
7	**Negation Elimination (¬E)** From a derivation of a proposition '**Q**' and its literal negation '¬**Q**' within a subproof involving an assumption '¬**P**,' we can derive 'P' out of the subproof.			¬P . . . ¬Q Q	A
			P		¬E
8	**Disjunction Introduction (∨I)** From '**P**,' we can validly infer '**P**∨**Q**' or '**Q**∨**P**.'		P P∨Q Q∨P		 ∨I ∨I
9	**Disjunction Elimination (∨E)** From '**P**∨**Q**' and two derivations of '**R**'—one involving 'P' as an assumption in a subproof, the other involving 'Q' as an assumption in a subproof—we can derive '**R**' out of the subproof.		P∨Q	P . . . R Q . . . R	A A
			R		∨E

10	**Biconditional Introduction (↔I)** From a derivation of '**Q**' within a subproof involving an assumption '**P**' and from a derivation of '**P**' within a separate subproof involving an assumption '**Q**,' we can derive '**P↔Q**' out of the subproof.		**P** . . . **Q**	A
			Q . . . **P**	A
			P↔ Q	↔I
11	**Biconditional Elimination (↔E)** From '**P↔Q**' and '**P**,' we can derive '**Q**.' And from '**P↔Q**' and '**Q**,' we can derive '**P**.'		**P↔Q** **P** **Q**	↔E
			P↔Q **Q** **P**	↔E

5.4 STRATEGIES FOR PROOFS

If you struggled on a number of the exercises in the preceding sections, you are not alone. Learning to do proofs quickly and accurately requires practice and a familiarity with a basic set of *proof strategies*. In this section, two different kinds of strategies for solving proofs are formulated: (1) strategies aimed at the direct manipulation of propositions in the proof, and (2) strategies aimed at the deliberate and tactical use of assumptions. In the next section, we supplement our existing set of derivation rules with some additional derivation rules, and we then finish things off by refining our strategic rules.

5.4.1 Strategies Involving Premises

First, we begin with two strategies that do not involve assumptions. We will call these *strategic proof rules*, or SP rules.

SP#1(E)   First, eliminate any conjunctions with '∧E,' disjunctions with '∨E,' conditionals with '→E,' and biconditionals with '↔E.' Then, if necessary, use any necessary introduction rules to reach the desired conclusion.

SP#2(B)   First, work backward from the conclusion using introduction rules (e.g., '∧I,' '∨I,'' →I,' '↔I'). Then, use SP#1(E).

Beginning with SP#1(E), the basic idea behind this strategic rule is to start a proof by simplifying or breaking down any available premises. Consider the following:

$$P{\rightarrow}(R{\wedge}M), (P{\wedge}S){\wedge}Z \vdash R$$

1	P→(R∧M)	P
2	(P∧S)∧Z	P/R

SP#1(E) suggests that you should use elimination rules to break down any complex propositions into simpler propositions. Since line 2 is a complex conjunction, conjunction elimination (∧E) can be used to derived a number of atomic propositions:

1	P→(R∧M)	P
2	(P∧S)∧Z	P/R
3	**P∧S**	**2∧E**
4	**Z**	**2∧E**
5	**P**	**3∧E**
6	**S**	**3∧E**

At this point, we can follow SP#1(E) further and apply additional elimination rules. Line 1 is a conditional, and since the antecedent of this conditional 'P' occurs at line 5, conditional elimination allows for deriving 'R∧M.'

1	P→(R∧M)	P
2	(P∧S)∧Z	P/R
3	P∧S	2∧E
4	Z	2∧E
5	P	3∧E
6	S	3∧E
7	**R∧M**	**1,5→E**

Following SP#1(E) still further, notice that 'R∧M' is a conjunction, and so we can apply conjunction elimination to finish the proof.

1	P→(R∧M)	P
2	(P∧S)∧Z	P/R
3	P∧S	2∧E
4	Z	2∧E
5	P	3∧E
6	S	3∧E
7	R∧M	1,5→E
8	**R**	**7∧E**

Consider a variant of the above proof:

$$P{\rightarrow}R,\ (P{\wedge}S){\wedge}Z \vdash R{\wedge}S$$

This proof will require not only the initial use of elimination rules to break propositions into simpler propositions but also the use of introduction rules to derive the desired conclusion.

| 1 | P→R | P |
| 2 | (P∧S)∧Z | P/R∧S |

SP#1(E) suggests that we should begin by using as many elimination rules as possible.

1	P→R	P
2	(P∧S)∧Z	P/R∧S
3	P∧S	2∧E
4	Z	2∧E
5	P	3∧E
6	S	3∧E
7	R	1,5→E

Now that elimination rules have been applied, SP#1(E) suggests trying to reach the conclusion by applying any introduction rules that would lead to the conclusion. In the case of the above proof, since the goal of the proof is 'R∧S,' and 'R∧S' is a conjunction, we can apply conjunction introduction. Thus,

1	P→R	P
2	(P∧S)∧Z	P/R∧S
3	P∧S	2∧E
4	Z	2∧E
5	P	3∧E
6	S	3∧E
7	R	1,5→E
8	R∧S	6,7∧I

Moving to SP#2(B), the basic idea behind this strategy is this: rather than moving forward (downward) in a proof from the premises or assumptions to a conclusion, work backward (upward) from the conclusion to the premises or assumptions.

SP#2(B)   First, work backward from the conclusion using introduction rules (e.g., '∧I,' '∨I,' '→I,' '↔I'). Then, use SP#1(E).

Consider the following proof:

$$P \rightarrow R, \ Z \rightarrow W, \ P \ Z \vdash R \lor W$$

1	P→R	P
2	Z→W	P
3	P	P/R∨W

Rather than trying to use elimination rules, we might start the proof by skipping a few lines in the proof and writing the conclusion at the bottom. That is,

1	P→R	P
2	Z→W	P
3	P	P/R∨W
	.	
	.	
	.	
	.	
#	R∨W	?

Next, consider what derivation rule would have allowed us to reach 'R∨W.' Since 'R∨W' is a disjunction, we can speculate that it could be derived by the use of disjunction introduction from either 'W' or 'R.' We will call propositions that are obtained as a result of the working-backward method *intermediate conclusions*. Thus, working backward we obtain the intermediate conclusions 'R' and 'W':

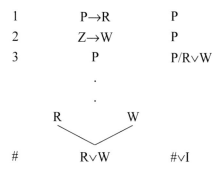

Now that we have worked backward a line, the next step will be to try to use the premises and reach either of the intermediate conclusions. Using elimination rules, we can derive the following:

1	P→R	P
2	Z→W	P
3	P	P/R∨W

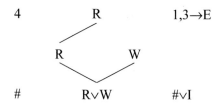

In the above, we have created two paths. The first path links the premises to one of the intermediate conclusions. The second path links the intermediate conclusion to the conclusion of the proof. With both of these paths, we can finish the proof as follows:

1	P→R	P
2	Z→W	P
3	P	P/R∨W
4	R	1,3→E
5	R∨W	4∨I

Thus, the general idea behind SP#2(B) is to start with the conclusion and work backward, using any introduction rules that would yield an intermediate conclusion. Once you have worked backward to a sufficient degree, try to use any elimination rules that would lead you to an intermediate conclusion.

Exercise Set #5

A. Solve the following proofs using strategic rules SP#1(E) and SP#2(B). However, start each proof by using SP#2(B).
1. * Z∧(B∧F), (M∧T)∧(L→P), Q∧(R∧P) ⊢ ¬R∨(S∨T)
2.   (S∧W)∧(T∧X), (P∧W)∧F, F→R ⊢ (P∧R)∨(S∧L)
3. * (Z∧Q)∧(F∧L), R∧P, W∧B ⊢ (Z∨T)∨(M→R)
4.   (L∧F)→S, W∧(F∧X), W→L ⊢ (S∨R)∨P
5.   M∧(R∧¬Z), S∧(P∧W), Q ⊢ (S↔Q)∨[M∧(R∧Z)]
6.   [(P∧Q)∧(W∧L)]∧[R∧(S∧T)], Z∧[(W∧R)∧(T∧Z)], (F→P)↔W ⊢ A∨Z

Solutions to Starred Exercises in Exercise Set #5

1. * Z∧(B∧F), (M∧T)∧(L→P), Q∧(R∧P)⊢¬R∨(S∨T)

1	Z∧(B∧F)	P
2	(M∧T)∧(L→P)	P/¬P
3	Q∧(R∧P)	P/¬R∨(S∨T)
4	M∧T	2∧E
5	T	4∧E
6	S∨T	5∨I
7	¬R∨(S∨T)	6∨I

3. * (Z∧Q)∧(F∧L), R∧P, W∧B⊢(Z∨T)∨(M→R)

1	(Z∧Q)∧(F∧L)	P
2	R∧P	P
3	W∧B	P/(Z∨T)∨(M→R)
4	Z∧Q	1∧E
5	Z	4∧E
6	Z∨T	5∨I
7	(Z∨T)∨(M→R)	6∨I

5.4.2 Strategies Involving Assumptions

Probably the most difficult part of solving any proof is the first few steps. The previous section formulated two strategic rules for solving proofs involving premises. In this section, we offer some advice with respect to proofs that either do not involve premises or that require the use of assumptions.

There are roughly four strategic rules for making assumptions, each classified by the main operator of the goal proposition. That is, there is a strategic rule for atomic propositions and negated propositions ('P,''¬Q'), one for conditionals (→), one for disjunctions (∨), and one for conjunctions (∧).

SA#1(P,¬Q)   If the conclusion is an atomic proposition (or a negated proposition), assume the negation of the proposition (or the non-negated form of the negated proposition), derive a contradiction, and then use '¬I' or '¬E.'

SA#2(→)   If the conclusion is a conditional, assume the antecedent, derive the consequent, and use '→I.'

SA#3(∧)   If the conclusion is a conjunction, you will need two steps. First, assume the negation of one of the conjuncts, derive a contradiction, and then use '¬I' or '¬E.' Second, in a separate subproof, assume the negation of the other conjunct, derive a contradiction, and then use '¬I' or '¬E.' From this point, a use of '∧I' will solve the proof.

SA#4(∨)   If the conclusion is a disjunction, assume the negation of the whole disjunction, derive a contradiction, and then use '¬I' or '¬E.'

Consider the following:

P→Q, ¬Q⊢¬P.

1	P→Q	P
2	¬Q	P/¬P

First, notice that our strategic rules involving premises do not seem to offer any help, for we cannot apply '→E' to lines 1 and 2, and there is no obvious way to work backward with introduction rules.

However, consider that the goal proposition '¬P' is a negated proposition, and so there is a strategic rule SA#1(P,¬Q) for atomic propositions and negated propositions:

SA#1(P,¬Q)   If the conclusion is an atomic proposition (or a negated proposition), assume the negation of the proposition (or the non-negated form of the negated proposition), derive a contradiction, and then use '¬I' or '¬E.'

Since '¬P' is a negated proposition, SA#1(P,¬Q) says to start by assuming the non-negated form of '¬P':

1	P→Q	P
2	¬Q	P/¬P
3	**│ P**	**A/contra**

Each strategic rule involving assumptions will offer advice on what the goal of the subproof will be. In the case of SA#1(P,¬Q), it says that the goal of the subproof will be to derive a proposition and its literal negation.

1	P→Q	P
2	¬Q	P/¬P
3	│ P	A/contra
4	│ Q	1,3→E
5	│ ¬Q	2R

Once 'Q' and '¬Q' are derived, SA#1(P,¬Q) offers advice on how to close the subproof. In the case of SA#1(P,¬Q), close the subproof using either '¬I' or '¬E.' In our case, since 'P' is assumed and '¬P' is the goal, we will use '¬I':

1	P→Q	P
2	¬Q	P/¬P
3	│ P	A/contra
4	│ Q	1,3→E
5	│ ¬Q	2R
6	¬P	3–5¬I

As a second example, consider the following:

$$P \vdash \neg\neg P$$

1	P	P/¬¬P
2	│ ¬P	A/P∧¬P
3	│ P	1R
4	¬¬P	2–3¬I

The goal proposition of this proof is a negated proposition. SA#1(P, ¬Q) says to start by assuming the opposite of our desired goal. In this case, '¬P' is assumed. Next, SA#1(P,¬Q) says to derive a contradiction. This is done at lines 2 and 3. Finally, SA#1(P,¬Q) says to close the subproof with '¬I.' This is done at line 4.

Consider one more example:

$$\vdash \neg(P \wedge \neg P)$$

This proof is a zero-premise derivation, so it will require starting the proof by making an assumption.

1	P∧¬P	A
2	P	1∧E
3	¬P	1∧E
4	¬(P∧¬P)	1–3¬I

The goal of the first proof is a negated proposition. SA#1(P,¬Q) says to start by assuming the opposite of our desired goal. Since '¬(P∧P)' is the goal, 'P∧¬P' is assumed. Next, SA#1(P,¬Q) says to derive a proposition and its literal negation within the subproof. This is done at lines 2 and 3. Finally, SA#1(P,¬Q) says to exit the subproof with '¬I.' This is done at line 4.

Next, we turn to the second strategic rule involving assumptions. Consider the following:

$$R \vdash P \rightarrow R$$

Notice that the conclusion of this argument is the conditional 'P→R.' In order to prove this, first consider a basic strategy for solving for proofs whose conclusions are conditionals.

SA#2(→)   If the conclusion is a conditional, assume the antecedent, derive the consequent, and use '→I.'

The strategic rule for conditionals, SA#2(→), says to start by assuming the antecedent of the conditional 'P→R' in the subproof and work toward deriving 'R':

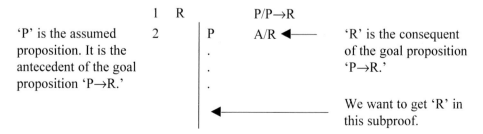

Once the antecedent is assumed, SA#2(→) says to derive the consequent of the goal proposition. This is 'R,' which can be derived by using reiteration.

```
1        R                    P/P→R
2              | P            A/R
3              | R            1R
```

Once the consequent is derived, SA#2(→) says to use '→I,' which completes the proof

```
1        R                    P/P→R
2              | P            A/P
3              | R            1R
4        P→R                  2–3→I
```

Consider a second illustration of the strategic rule for assumptions, where the conclusion is a conditional:

$$R \vdash (P \lor R) \rightarrow P$$

```
1        R        P/(P∨R)→P
```

When making an assumption, first look at the main operator of the conclusion. Since the main operator of '(P∨R)→R' is the arrow, SA#2(→) says to assume the antecedent of that proposition. Thus,

```
1        R                    P/(P∨R)→R
2              | P∨R          A/R
```

Next, SA#2(→) says to derive the consequent of the conditional.

```
1        R                    P/(P∨R)→R
2              | P∨R          A/R
3              | R            1R
```

With the consequent derived, SA#2(→) says to use '→I.' This will take us out of the subproof and complete the proof.

```
1        R                    P/(P∨R)→R
2              | P∨R          A/R
3              | R            1R
4        (P∨R)→R              2–3→I
```

Finally, let's consider an example involving zero premises:

$$\vdash P \rightarrow (Q \rightarrow P)$$

Again, when making an assumption, first look at the main operator of the goal proposition. Since the main operator of 'P→(Q→P)' is the arrow, SA#2(→) says to assume the antecedent of that proposition and derive the consequent. Thus,

$$1 \quad | \;\; P \qquad A/Q \rightarrow P$$

The next step will be to derive 'Q→P' in the subproof. However, there is no immediately obvious way to do this. At this point, it might be helpful to make another assumption. It is important to recognize that since our goal conclusion is 'Q→P,' what we assume will be guided by this proposition. Since 'Q→P' is a conditional, SA#2(→) says to assume the antecedent of that proposition. Thus,

1	P		A/Q→P
2		**Q**	**A/P**

Now that we have assumed 'Q,' the goal proposition is 'P,' which we can easily derive:

1	P		A/Q→P
2		Q	A/P
3		**P**	**1R**

Next, close the most deeply nested subproof with '→I,' and then close the remaining open subproof with another use of '→I':

1	P		A/Q→P
2		Q	A/P
3		P	1R
4	**Q→P**		**2–3→I**
5	**P→(Q→P)**		**1–4→I**

Next, we turn to our third strategy involving assumptions:

SA#3(∧)   If the conclusion is a conjunction, you will need two steps. First, assume the negation of one of the conjuncts, derive a contradiction, and then use '¬I' or '¬E.' Second, in a separate subproof, assume the negation of the other conjunct, derive a contradiction, and then use '¬I' or '¬E.' From this point, a use of '∧I' will solve the proof.

Consider the following proof:

$$\neg(P \lor Q) \vdash \neg P \land \neg Q$$

In order to solve this, an assumption is needed. In proceeding, notice that the main operator of the conclusion is the operator for the conditional (i.e.,'$\land$'). According to SA#3($\land$), solving a proof of this sort will require two separate assumptions, one for each conjunct. Let's begin by focusing on the left conjunct,'$\neg P$':

1	$\neg(P \lor Q)$		$P/\neg P \land \neg Q$
2		P	$A/P \land \neg P$
3		$P \lor Q$	$2 \lor I$
4		$\neg(P \lor Q)$	$1R$
5	$\neg P$		$2\text{–}4 \neg I$

Now that the left conjunct has been derived, it is time to derive the right conjunct using a very similar procedure:

1	$\neg(P \lor Q)$		$P/\neg P \land \neg Q$
2		P	$A/P \land \neg P$
3		$P \lor Q$	$2 \lor I$
4		$\neg(P \lor Q)$	$1R$
5	$\neg P$		$2\text{–}4 \neg I$
6		Q	$A/Q \land \neg Q$
7		$P \lor Q$	$6 \lor I$
8		$\neg(P \lor Q)$	$1R$
9	$\neg Q$		$6\text{–}8 \neg I$
10	$\neg P \land \neg Q$		$5,9 \land I$

Looking at the above proof, notice that two assumptions were made. First, 'P' was assumed, a contradiction was derived, and then a use of '$\neg I$' introduced '$\neg P$' to the main line of the proof. Second, 'Q' was assumed, a contradiction was derived, and then a use of '$\neg I$' introduced '$\neg Q$' to the main line of the proof. Finally, both '$\neg P$' and '$\neg Q$' were conjoined with '$\land I$.'

Finally, let's consider the fourth strategic rule involving assumptions:

SA#4($\lor$)   If the conclusion is a disjunction, assume the negation of the whole disjunction, derive a contradiction, and then use '$\neg I$' or '$\neg E$.'

Consider this strategic rule with respect to the following proof:

$$\neg(\neg P \land \neg Q) \vdash P \lor Q.$$

1   ¬(¬P¬∧¬Q)   P/P∨Q

Since there are no elimination rules involving negated conjunctions, an assumption is necessary. SA#4(∨) suggests that we begin by assuming the negation of our goal proposition and work toward a contradiction:

1 ¬(¬P∧¬Q)   P/P∨Q
2 | ¬(P∨Q)   A/contra

Obtaining the contradiction in the subproof is no easy task. At this point, however, we might try to work backward toward a proposition that would generate a contradiction. That is, our goal is two different intermediate conclusions:

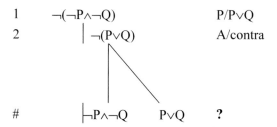

With either of the above goal propositions, we could derive a contradiction and close the subproof with our desired conclusion. Let's choose '¬P∧¬Q' as our desired conclusion. Since '¬P∧¬Q' is a conjunction, we will use SA#3(∧) as our strategy. Thus, begin by assuming the non-negated form of the left conjunct and work toward a contradiction:

1   ¬(¬P∧¬Q)   P/P∨Q
2 | ¬(P∨Q)   A/¬P∧¬Q
3 | | P   A/contra
4 | | P∨Q   3∨I
5 | | ¬(P∨Q)   2R
6 | ¬P   3–5¬I

Next, assume the non-negated form of the right conjunct and derive a contradiction:

1   ¬(¬P∧¬Q)   P/P∨Q
2 | ¬(P∨Q)   A/¬P∧¬Q
3 | | P   A/contra
4 | | P∨Q   3∨I
5 | | ¬(P∨Q)   2R

6	¬P	3–5¬I
7	Q	A / **contra**
8	P∨Q	7∨**I**
9	¬(P∨Q)	2R
10	¬Q	7–9¬**I**

Now, we can generate our desired contradiction and finish the proof:

1	¬(¬P∧¬Q)	P/P∨Q
2	¬(P∨Q)	A/¬P∧¬Q
3	P	A/contra
4	P∨Q	3∨I
5	¬(P∨Q)	2R
6	¬P	3–5¬I
7	Q	A/contra
8	P∨Q	7∨I
9	¬(P∨Q)	2R
10	¬Q	7–9¬I
11	¬P∧¬Q	6,10∧**I**
12	¬(¬P∧¬Q)	1R
13	**P∨Q**	2–12¬**E**

Let's take stock. We are working with two different kinds of strategies for solving proofs. First, there are the strategic proof rules that involve manipulating propositions that are available in the proof by either breaking down complex propositions with elimination rules or working backward with introduction rules. Second, there are strategic rules involving assumptions. Whenever you need to make an assumption, the proposition you assume should be guided by the main operator in the proposition you are trying to derive; that is, if it is a conditional, use SA#2(→).

Exercise Set #6

A. Identify the strategic assumption rule that you would use and the proposition you would assume (if necessary) if you were to solve the proof.

1. * ⊢ P→(R→R)
2.   ⊢ P→(R∨¬R)
3. * ⊢ P∨¬P
4.   ⊢ P∨¬P
5. * ⊢ ¬(P∧¬P)
6.   ⊢ (P∨S)∨¬(P∨S)
7. * ⊢ (R∨S)→(P→P)

8.　　¬(P∨Q) ⊢ ¬P∧¬Q
9.　*　¬(P→Q) ⊢ P∧¬Q

B.　Using the strategic assumption rules and strategic proof rules (if needed), solve the proofs below.

　　1.　*　⊢ P→(R→R)
　　2.　　⊢ P→(R∨¬R)
　　3.　*　⊢ P∨¬P
　　4.　　⊢ P∨¬P
　　5.　*　⊢ ¬(P∧¬P)
　　6.　　⊢ (P∨S)∨¬(P∨S)
　　7.　*　⊢ (R∨S)→(P→P)
　　8.　　¬(P∨Q) ⊢ ¬P∧¬Q
　　9.　*　¬(P→Q) ⊢ P∧¬Q

Solutions to Starred Exercises in Exercise Set #6

A.

　　1.　*　⊢ P→(R→R). Assume 'P' and use SA#2(→).
　　3.　*　⊢ P∨¬P. Assume '¬(P∨¬P)' and use SA#4(∨).
　　5.　*　⊢ ¬(P∧¬P). Assume 'P∧¬P' and use SA#1(P, ¬Q).
　　7.　*　⊢ (R∨S)→(P→P). Assume 'R∨S' and use SA#2(→).
　　9.　*　¬(P→Q) ⊢ P∧¬Q. Assume '¬P' and derive a contradiction. Assume 'Q' and derive a contradiction. Use SA#3(∧).

B.

　　1.　*　⊢ P→(R→R)

1		P	A / R→R
2			R	A / R
3			R	2R
4		R→R	2-3→I
5	P→(R→R)		1-4→I

　　3.　*　⊢ P∨¬P

1		¬(P∨¬P)	A / P, ¬P
2			P	A / P, ¬P
3			P∨¬P	2∨I
4			¬(P∨¬P)	1R
5		¬P	2-4¬I
6		P∨¬P	5∨I
7	P∨¬P		1-6¬E

5. *   ⊢ ¬(P∧¬P)

1		P∧¬P	A / P, ¬P
2		P	1∧E
3		¬P	1∧E
4	¬(P∧¬P)		1-3¬I

7. *   ⊢ (R∨S)→(P→P)

1		R∨S	A / P→P
2			P A / P
3			P 2R
4		P→P	2-3→I
5	(R∨S)→(P→P)		1-4→I

9. *   ¬(P→Q) ⊢ P∧¬Q

1	¬(P→Q)		P / P∧¬Q
2		Q	A / P, ¬P
3			P A / Q
4			Q 2R
5		P→Q	3-4→I
6		¬(P→Q)	1R
7	¬Q		2-6¬I
8		¬P	A / P, ¬P
9			P A / Q
10			¬P∨Q 8∨I
11			¬P A / Q
12			¬Q A / P, ¬P
13			P 9R
14			¬P 8R
15			Q 12-14¬E
16			Q A
17			Q 16R
18		Q	10, 11-15, 16-17∨E
19		P→Q	9-18→I
20		¬(P→Q)	1R
21	P		8-20¬E
22	P∧¬Q		7,21∧I

5.5. ADDITIONAL DERIVATION RULES (PD+)

While the various proof and assumption strategies provide some initial guidance on how to solve proofs, you may notice that some proofs cannot be solved in a quick and efficient manner. In order to further simplify proofs, we introduce six additional derivation forms into our derivation system. The addition of these six derivation rules to PD gives us PD+.

PD	PD+
∧I, ∧E, ∨I, ∨E, →I, →E, ↔I, ↔E, ¬I, ¬E, R	DS, MT, HS, DN, DEM, IMP

5.5.1 Disjunctive Syllogism (DS)

Consider the derivation commonly known as *disjunctive syllogism* (DS), which says that from a disjunction 'P∨Q' and the negation of one of the disjuncts (e.g.,'¬Q'), we can derive the other disjunct (i.e., 'P').

Disjunctive Syllogism (DS)		
From 'P∨Q' and '¬Q,' we can derive 'P.'	P∨Q	
From 'P∨Q' and '¬P,' we can derive 'Q.'	¬Q	
	P	DS
	P∨Q	
	¬P	
	Q	DS

To see why we might want to introduce DS into the existing set of derivation rules, consider the following proof of 'P∨Q, ¬Q ⊢ P' without the use of DS:

1	P∨Q			P
2	¬Q			P
3		P		A/P
4		P		3R
5		Q		A/P
6			¬P	A/Q∧¬Q
7			Q	5R
8			¬Q	2R
9		P		6–8¬E
10	P			1,3–4,5–9∨E

This is a lengthy proof to infer something so obvious. With DS, the other disjunct can be derived in one step. That is, the above proof simplifies to the following:

1	P∨Q	P
2	¬Q	P
3	P	1,2DS

This rule is extremely helpful for considering the following:

$$P \leftrightarrow (Q \lor R), P, \neg Q \vdash R$$

1	P↔(Q∨R)			P
2	P			P
3	¬Q			P/R
4	Q∨R			1,2↔E
5		R		A/R
6		R		5R
7		Q		A/R
8			¬R	A/Q∧¬Q
9			Q	7R
10			¬Q	3R
11		R		8–10¬E
12	R			4,5–6,7–11∨E

Again, the above is difficult to solve without the use of DS. The proof is now much simpler:

1	P↔(Q∨R)	P
2	P	P
3	¬Q	P/R
4	Q∨R	1,2↔E
5	R	3,4DS

While DS simplifies reasoning with disjunctions, be careful not to confuse DS with the following invalid form of reasoning:

1	P∨Q	P
2	P	P
3	Q	DS—**NO!**

To see why this is invalid, remember that a disjunction is true provided either of the disjuncts is true. From the fact that one of the disjuncts is true, the other cannot be validly inferred since it is possible for $v(P \lor Q) = T$ and $v(P) = T$, yet $v(Q) = F$. Consider a more concrete example.

1	Sally will go to the party, or John will.	P
2	Sally will go to the party.	P
3	John will go to the party.	DS—**NO!**

5.5.2 Modus Tollens (MT)

Consider the derivation commonly known as *modus tollens*, which says that from a conditional 'P→Q' and the negation of the consequent of that conditional, '¬Q,' we can derive the negation of the antecedent of that conditional, '¬P.'

12	**Modus Tollens (MT)** From 'P→Q' and '¬Q,' we can derive '¬P.'	**P→Q** **¬Q** **¬P**	 MT

Again, the introduction of MT also allows for simplifying a number of proofs. Consider the proof of 'P→Q, ¬Q ⊢¬P':

1	P→Q		P
2	¬Q		P/¬P
3		P	A/contra
4		Q	1,3→E
5		¬Q	2R
6	¬P		3–5¬I

So, while it is not necessary to introduce MT in order to solve 'P→Q, ¬Q⊢¬P,' its introduction serves to expedite the proof process.

Consider a slightly more complicated use of MT:

$$(P∧Z)→(Q∨Z), ¬(Q∨Z)⊢¬(P∧Z)$$

1	(P∧Z)→(Q∨Z)	P
2	¬(Q∨Z)	P
3	¬(P∧Z)	1,2MT

5.5.3 Hypothetical Syllogism (HS)

Consider the derivation commonly known as *hypothetical syllogism* (HS), which says that from two conditionals 'P→Q' and 'Q→R,' we can derive a conditional 'P→R.'

14	**Hypothetical Syllogism (HS)** From 'P→Q' and 'Q→R,' we can derive 'P→R.'	**P→Q** **Q→R** **P→R**	 HS

Again, HS simplifies proofs involving multiple conditionals (e.g., P→Q, Q→R ⊢ P→R).

1	P→Q	P
2	Q→R	P/P→R
3	⎸P	A/R
4	⎸Q	1,3→E
5	⎸R	2,4→E
6	P→R	3–5→I

5.5.4 Double Negation (DN)

Consider the derivation rule commonly known as *double negation* (DN), which says that from a proposition '**P**,' we can derive a doubly negated proposition '¬¬**P**,' and vice versa.

15	**Double Negation (DN)** From '**P**,' we can derive '¬¬**P**.' From '¬¬**P**,' we can derive '**P**.'	**P**⊣⊢¬¬**P**	DN

Unlike previous derivation rules, DN is an *equivalence rule*. An equivalence rule allows for substituting a proposition of one particular form for a proposition of another form that is logically equivalent to it. Sometimes equivalence rules are called *rules of replacement* because they allow for replacing equivalent propositions. In the case of DN, any proposition '**P**' can be replaced with '¬¬**P**,' and vice versa.

One way of expressing this equivalence rule is as follows:

$$\textbf{P} ⊣⊢ ¬¬\textbf{P}$$

This is an abbreviated form of the following two derivation rules:

$$\textbf{P} ⊢ ¬¬\textbf{P}$$
$$¬¬\textbf{P} ⊢ \textbf{P}$$

The above rule states that from a proposition '**P**,' the proposition '¬¬**P**' can be derived, and from a proposition '¬¬**P**,' the proposition '**P**' can be derived.

Consider the uses of DN below:

1	P∧Q	P
2	¬¬(P∧Q)	1DN
3	¬¬(¬¬P∧Q)	2DN

4	¬¬(¬¬P∧¬¬Q)	3DN
5	¬¬P∧¬¬Q	4DN
6	P∧Q	5DNx2

Notice that DN can apply to the complex proposition 'P∧Q' at line 1 or to any constituent proposition in 'P∧Q.' That is, DN can apply to subformulas. Consider some invalid examples below:

1	¬(¬P∧¬Q)	P
2	P∧¬Q	1DN—**NO!**
3	¬P∧¬Q	2DN—**NO!**
4	P∧Q	1DN—**NO!**
5	¬(¬P∧Q)	4DN—**NO!**

Notice that the use of DN at line 1 to remove two negations does not remove two negations from a single proposition but removes one negation from the complex proposition '(¬P∧¬Q)' and one from 'P.' This is an incorrect use of the equivalence rule because the rule demands that a single proposition can be replaced with a proposition that is doubly negated or a single doubly negated proposition can be replaced by removing its double negation.

5.5.5 De Morgan's Laws (DeM)

While there are elimination rules for conjunctions (∧E), disjunctions (∨E) and DS, conditionals (→E) and MT, and biconditionals (↔E), there are no elimination rules for negated conjunctions ('¬[**P∧Q**]'), negated disjunctions('¬[**P∨Q**]'), negated conditionals ('¬[**P→Q**]'), and negated biconditionals ('¬[**P↔Q**]'). One way of dealing with negated conjunctions and negated disjunctions is by introducing an equivalence rule that allows for expressing every negated conjunction in terms of a disjunction and every negated disjunction in terms of a conjunction. These equivalence rules are known as *De Morgan's Laws* (DeM).[1]

| 16 | **De Morgan's Laws (DeM)** From '¬(P∨Q),'we can derive '¬P∧¬Q.' From '¬P∧¬Q,' we can derive '¬(P∨Q).' From '¬(P∧Q),'we can derive '¬P∨¬Q.' From '¬P∨¬Q,' we can derive '¬(P∧Q).' | ¬(P∨Q)⊣⊢¬P∧¬Q ¬(P∧Q)⊣⊢¬P∨¬Q | DeM DeM |

First, it is important to note that DeM is an equivalence rule, and so it allows for substituting a formula of one type for a formula of another type. For example, De Morgan's Laws allow for deriving '¬P∧¬Q' from '¬(P∨Q),' and vice versa.

De Morgan's Laws will make many proofs much easier for two reasons. First, DeM will considerably shorten many proofs. For example, consider the following proof for '¬(P∧Q) ⊢ ¬P∨¬Q':

1	¬(P∧Q)		P
2		¬(¬P∨¬Q)	A/P∧¬P
3		¬P	A/P∧¬P
4		¬P∨¬Q	3∨I
5		¬(¬P∨¬Q)	2R
6		P	3–5¬E
7		¬Q	A/P∧¬P
8		¬P∨¬Q	7∨I
9		¬(¬P∨¬Q)	2R
10		Q	7–9¬E
11		P∧Q	6,10∧I
12		¬(P∧Q)	1R
13	¬P∨¬Q		2–12¬E

Notice that the proof is thirteen lines long and involves a number of nested assumptions. By adding DeM to our set of derivation rules, the above proof can be simplified to the following:

1	¬(P∧Q)	P
2	¬P∨¬Q	2DeM

Another benefit of adding DeM to our existing set of derivation rules is that certain versions of its use will offer a way to apply elimination rules. To see this more clearly, consider the following:

$$¬(P∨Q) \vdash ¬P$$

1	¬(P∨Q)	P/¬P
2	¬P∧¬Q	1DeM
3	¬P	2∧E

Notice that none of our elimination rules apply to line 1, but once DeM is applied to line 1, '∧E' can be applied to line 2.

5.5.6 Implication (IMP)

In the previous section, the introduction of DeM allowed for simplifying proofs that involve negated disjunctions and negated conjunctions. What about negated conditionals ('¬[**P→Q**]') and negated biconditionals ('¬[**P↔Q**]')? Currently, there is no rule in the derivation rule set that pertains to this type of proposition. One helpful equivalence rule for dealing with the former is *implication* (IMP):

17	**Implication (IMP)** From 'P→Q,' we can derive '¬P∨Q.' From '¬P∨Q,' we can derive 'P→Q.'	**P→Q⊣⊢¬P∨Q**	IMP

IMP is an equivalence rule that greatly increases the elegance of our proofs. For example, consider the following:

$$P{\rightarrow}Q \vdash \neg P{\lor}Q$$

Notice that the conclusion '¬P∨Q' is a disjunction, so in making an assumption, use SA#4(∨):

1	P→Q	P
2	¬(¬P∨Q)	A/contra
3	¬¬P∧¬Q	2DEM
4	¬¬P	1∧E
5	P	4DN
6	Q	1,5→E
7	¬Q	3∧E
8	¬P∨Q	2–7¬E

With the addition of IMP, 'P→Q ⊢ ¬P∨Q' can be solved in one line:

1	P→Q	P
2	¬P∨Q	1IMP

Now consider the following:

$$\neg P{\lor}Q \vdash P{\rightarrow}Q$$

Since the conclusion is a conditional, use SA#2(→):

1	¬P∨Q	P
2	P	A/Q
3	¬¬P	2DN
4	Q	1,3DS
5	P→Q	2–4→I

Again, with the addition of IMP, '¬P∨Q ⊢ P→Q' can be solved in one line:

1	¬P∨Q	P
2	P→Q	1IMP

The above proofs show that while proofs involving IMP can be derived without the use of IMP, the addition of this rule greatly shortens proofs. A second benefit of introducing IMP is that it simplifies proofs involving negated conditionals. Consider the following proof:

$$\neg(P{\rightarrow}Q) \vdash \neg Q$$

1	$\neg(P{\rightarrow}Q)$	P/$\neg$Q
2	$\neg(\neg P{\vee}Q)$	1IMP

Although we cannot apply any elimination rules to line 2, we can apply DeM and then apply elimination rules to solve the proof.

1	$\neg(P{\rightarrow}Q)$	P/$\neg$Q
2	$\neg(\neg P{\vee}Q)$	1 IMP
3	$\neg\neg\mathbf{P}{\wedge}\neg\mathbf{Q}$	**2DeM**
4	$\neg\mathbf{Q}$	**3$\wedge$E**

5.6 ADDITIONAL DERIVATION STRATEGIES

In this section, we conclude our discussion of proofs by refining our proof strategies given the introduction of our new derivation rules.

SP#1(E+)   First, eliminate any conjunctions with '$\wedge$E,' disjunctions with DS or '$\vee$E,' conditionals with '$\rightarrow$E' or MT, and biconditionals with '$\leftrightarrow$E.' Then, if necessary, use any introduction rules to reach the desired conclusion.

SP#2(B)   First, work backward from the conclusion using introduction rules (e.g., '$\wedge$I,' '$\vee$I,' '$\rightarrow$I,' '$\leftrightarrow$I'). Then, use SP#1(E).

SP#3(EQ+)   Use DeM on any negated disjunctions or negated conjunctions, and then use SP#1(E). Use IMP on negated conditionals, then use DeM, and then use SP#1(E).

Consider the following:

$$P{\rightarrow}Q, \neg Q, P{\vee}R, R{\rightarrow}W \vdash W$$

1	$P{\rightarrow}Q$	P
2	$\neg$Q	P
3	$P{\vee}R$	P
4	$R{\rightarrow}W$	P/W

According to SP#1(E+), we should start this proof by using as many elimination rules as possible. These now include MT for conditionals and DS for disjunctions:

1	P→Q	P
2	¬Q	P
3	P∨R	P
4	R→W	P/W
5	**¬P**	**1,2MT**
6	**R**	**3,5DS**
7	W	4,6→E

Again, the underlying idea behind this strategic proof rule is to start a proof by simplifying or breaking down any available propositions. Consider the following:

$$(P∧M)∧(¬Q∨R), P→L, P↔T, ¬R∧W⊢(L∧T)∧¬Q$$

1	(P∧M)∧(¬Q∨R)	P
2	P→L	P
3	P↔T	P
4	¬R∧W	P/(L∧T)∧¬Q

At first glance, it may not be immediately obvious how to derive '(L∧T)∧¬Q.' SP#1(E+) suggests beginning the proof by applying elimination rules. This strategic proof rule says to make use of '∧E,' '→E,' MT, DS, and '↔E' wherever possible. Start by applying '∧E' to any available conjunctions.

1	(P∧M)∧(¬Q∨R)	P
2	P→L	P
3	P↔T	P
4	¬R∧W	P/(L∧T)∧¬Q
5	**P∧M**	**1∧E**
6	**¬Q∨R**	**1∧E**
7	**¬R**	**4∧E**
8	**W**	**4∧E**
9	**P**	**5∧E**
10	**M**	**5∧E**

Notice that '∧E' is used on every available conjunction in the proof. However, the strategic rule suggests using '→E,' MT, DS, and '↔E' if possible. The remaining elimination rules give the following:

1	(P∧M)∧(¬Q∨R)	P
2	P→L	P
3	P↔T	P
4	¬R∧W	P/(L∧T)∧¬Q
5	P∧M	1∧E
6	¬Q∨R	1∧E
7	¬R	4∧E

8	W	4∧E
9	P	5∧E
10	M	5∧E
11	**L**	**2,9→E**
12	**T**	**3,9↔E**
13	**¬Q**	**6,7DS**

No more elimination rules can be used. Since the conclusion is '(L∧T)∧¬Q,' using '∧I' allows for deriving the conclusion. Thus, the proof ends as follows.

1	(P∧M)∧(¬Q∨R)	P
2	P→L	P
3	P↔T	P
4	¬R∧W	P/(L∧T)∧¬Q
5	P∧M	1∧E
6	¬Q∨R	1∧E
7	¬R	4∧E
8	W	4∧E
9	P	5∧E
10	M	5∧E
11	L	2,9→E
12	T	3,9↔E
13	¬Q	6,7DS
14	**(L∧T)**	**11,12∧I**
15	**(L∧T)∧¬Q**	**13,14∧I**

Finally, consider the addition of our third strategic proof rule:

SP#3(EQ+)   Use DeM on any negated disjunctions or negated conjunctions, and then use SP#1(E). Use IMP on negated conditionals, then use DeM, and then use SP#1(E).

Consider the proof of the following argument:

$$¬[P∨(R∨M)], ¬M→T ⊢ T$$

1	¬[P∨(R∨M)]	P
2	¬M→T	P/T

Notice that SP#1(E+) cannot be used because there is no proposition to which '∧E,' '→E,' MT, DS, or '↔E' can be applied. However, notice the negated disjunction at line 1. SP#3(EQ+) suggests using DeM. Thus,

1	¬[P∨(R∨M)]	P
2	¬M→T	P/T
3	**¬P∧¬(R∨M)**	**1DeM**

After using DeM on the negated disjunction, '∧E' can be applied.

1	¬[P∨(R∨M)]	P
2	¬M→T	P/T
3	¬P∧¬(R∨M)	1DeM
4	**¬P**	**3∧E**
5	**¬(R∨M)**	**3∧E**

Again, no more elimination rules can be used, but since there is a negated disjunction, SP#3(EQ+) suggests using DeM again. Thus,

1	¬[P∨(R∨M)]	P
2	¬M→T	P/T
3	¬P∧¬(R∨M)	1DeM
4	¬P	3∧E
5	¬(R∨M)	3∧E
6	**¬R∧¬M**	**5DeM**

This use of De Morgan's Laws again allows for the use of elimination rules.

1	¬[P∨(R∨M)]	P
2	¬M→T	P/T
3	¬P∧¬(R∨M)	1DeM
4	¬P	3∧E
5	¬(R∨M)	3∧E
6	¬R	5DeM
7	**¬M**	**5DeM**
8	**T**	**2,7→E**

5.6.1 Sample Problem

It may be instructive to consider a complicated example involving multiple uses of strategic rules. In many case, you will find that you will need to make more than one assumption. Try to use the above strategies iteratively.

Consider the following proof:

$$⊢[P→(Q→R)]→[(¬Q→¬P)→(P→R)]$$

Since the valid argument form above does not involve any premises, the first step will be to make an assumption.

First, notice that the main operator of the conclusion is '→.' Since the main operator is '→,' we will use SA#2(→). It reads,

SA#2(→)   If the conclusion is a conditional, assume the antecedent, derive the consequent, and use '→I.'

Thus, start by assuming the antecedent, which is '[P→(Q→R)].'

$$1 \quad \Big|\; P→(Q→R) \qquad A/(¬Q→¬P)→(P→R)$$

In making this assumption, our goal is to derive the consequent '(¬Q→¬P)→(P→R)' and then use '→I.' Since we cannot apply any elimination rules to line 1, one option is to make a second assumption. The second assumption is guided by the goal proposition of the first assumption. That is, in assuming 'P→(Q→R),' our goal is '(¬Q→¬P)→(P→R),' so the next assumption is based on this main operator of '(¬Q→¬P)→(P→R).'

The main operator of '(¬Q→¬P)→(P→R)' is '→,' so assume the antecedent '(¬Q→¬P)' of this proposition and try to derive the consequent '(P→R).'

$$
\begin{array}{lll}
1 & P→(Q→R) & A/(¬Q→¬P)→(P→R) \\
2 & \quad ¬Q→¬P & A/P→R
\end{array}
$$

Again, no elimination rules apply, so make a third assumption, using 'P→R' as the goal.

The main operator of 'P→R' is '→,' so assume the antecedent 'P' of this proposition, and the goal will be to derive the consequent 'R.'

$$
\begin{array}{lll}
1 & P→(Q→R) & A/(¬Q→¬P)→(P→R) \\
2 & \quad ¬Q→¬P & A/P→R \\
3 & \quad\quad P & A/R
\end{array}
$$

Now, we can apply some elimination rules. One way to solve this proof is now simply to use '→E' and MT. That is,

$$
\begin{array}{lll}
1 & P→(Q→R) & A/(¬Q→¬P)→(P→R) \\
2 & \quad ¬Q→¬P & A/P→R \\
3 & \quad\quad P & A/R \\
4 & \quad\quad Q→R & 1,3→E \\
5 & \quad\quad ¬¬P & 3DN \\
6 & \quad\quad ¬¬Q & 2,5MT \\
7 & \quad\quad Q & 6DN \\
8 & \quad\quad R & 4,7→E
\end{array}
$$

Now that we have derived 'R' at line 8, SA#2(→) says to use '→I.'

$$
\begin{array}{lll}
1 & P→(Q→R) & A/(¬Q→¬P)→(P→R) \\
2 & \quad ¬Q→¬P & A/P→R
\end{array}
$$

3		P	A/R
4		Q→R	1,3→E
5		¬¬P	3DN
6		¬¬Q	2,5MT
7		Q	6DN
8		R	4,7→E
9		P→R	3–8→I

In assuming '¬Q→¬P' at line 2, our goal was to derive 'P→R' in the same sub-proof. Since we derived 'P→R,' SA#2(→) says to use '→I.'

1	P→(Q→R)	A/(¬Q→¬P)→(P→R)
2	¬Q→¬P	A/P→R
3	P	A/R
4	Q→R	1,3→E
5	¬¬P	3DN
6	¬¬Q	2,5MT
7	Q	6DN
8	R	4,7→E
9	P→R	3–8→I
10	(¬Q→¬P)→(P→R)	2–9→I

In assuming 'P→(Q→R)' at line 1, our goal was to derive '(¬Q→¬P)→(P→R)' in the same subproof. Since we derived '(¬Q→¬P)→(P→R),' SA#2(→) says to use '→I.' This completes the proof.

1	P→(Q→R)	A/(¬Q→¬P)→(P→R)
2	¬Q→¬P	A/P→R
3	P	A/R
4	Q→R	1,3→E
5	¬¬P	3DN
6	¬¬Q	2,5MT
7	Q	6DN
8	R	4,7→E
9	P→R	3–8→I
10	(¬Q→¬P)→(P→R)	2–9→I
11	[P→(Q→R)]→[(¬Q→¬P)→(P→R)]	1–10→I

It is important to see that strategic rules can be used iteratively and that what to assume is almost always guided by the conclusion (or goal proposition) of the proof (or subproofs).

Let's look at another way to solve the same proof that starts in a similar manner.

1	P→(Q→R)		A/(¬Q→¬P)→(P→R)
2		¬Q→¬P	A/P→R
3			P	A/R

In the previous example, we used the elimination rules and then used '→I' to complete the proof. Another way to solve the same proof would be to see that in assuming 'P' at line 3, the goal of that proof is 'R.' Since 'R' is an atomic proposition, we can use the strategic rule SA#1(P,¬Q). This rule says to assume the negation of 'R' and try to derive a contradiction.

1	P→(Q→R)		A/(¬Q→¬P)→(P→R)
2		¬Q→¬P	A/P→R
3			P	A/R
4				¬R	A/R∧¬R

Our goal is to produce a contradiction in the proof. From here we will use elimination rules to do this and then '¬E' to exit the subproof; then we will finish the proof with repeated uses of '→I.'

1	P→(Q→R)	A/(¬Q→¬P)→(P→R)
2	¬Q→¬P	A/P→R
3	P	A/R
4	¬R	A/R∧¬R
5	Q→R	1,3→E
6	¬Q	4,5MT
7	¬P	2,6→E
8	P	3R
9	R	4–8¬E
10	P→R	3–9→I
11	(¬Q→¬P)→(P→R)	2–10→I
12	[P→(Q→R)]→[(¬Q→¬P)→(P→R)]	1–11→I

END-OF-CHAPTER EXERCISES

A. There are 125 valid arguments below. They are broken into four sections: easy, medium, hard, and zero-premise deductions. In mastering the derivation rules associated with proofs, start with the easy proofs and work your way up to the hard ones. Once you've mastered the hard proofs, try to prove some theorems (zero-premise deductions) of PL.

Easy Proofs

1. * $P \land \neg Q, T \lor Q \vdash T$
2. $P \rightarrow Q, Q \rightarrow R, \neg R \vdash \neg P$
3. * $A \rightarrow C, A \land D \vdash C$
4. $[(A \land B) \land C] \land D \vdash A$
5. * $A \vdash B \rightarrow (B \land A)$
6. $\neg(A \lor B) \vdash \neg A \land \neg B$
7. * $\neg(\neg A \lor B) \vdash \neg \neg A \land \neg B$
8. $(P \lor Q) \lor R, (T \lor W) \rightarrow \neg R, T \land \neg P \vdash Q$
9. * $A \lor (\neg B \rightarrow D), \neg(A \lor B) \vdash D$
10. $\neg A \lor \neg B \vdash \neg(A \land B)$
11. * $A \lor (\neg B \rightarrow D), \neg(A \lor D) \vdash B$
12. $(\neg B \land \neg D), \neg(B \lor D) \rightarrow S \vdash S$
13. * $(B \leftrightarrow D) \lor S, \neg S, B \vdash D \lor W$
14. $\neg A \rightarrow Q, (A \lor D) \rightarrow S, \neg S \vdash Q$
15. * $R \lor R \vdash R$
16. $\neg(\neg R \lor R) \vdash W$
17. * $A \rightarrow B, B \rightarrow C, \neg C \vdash \neg A$
18. $A \rightarrow B, B \rightarrow C, A \vdash C$
19. * $A \rightarrow B, A \lor C, \neg C \vdash B$
20. $A \rightarrow (B \rightarrow C), A, C \rightarrow D \vdash B \rightarrow D$
21. * $A \rightarrow B, C \rightarrow D, B \rightarrow C \vdash A \rightarrow D$
22. $P \vdash P \lor \{[(M \lor W) \land (F \land Z)] \rightarrow (C \leftrightarrow D)\}$
23. * $A \rightarrow (B \rightarrow C), A, \neg C \vdash \neg B \lor (W \rightarrow S)$
24. $M \lor (Q \land D), M \rightarrow F, (Q \land D) \rightarrow F \vdash F \lor S$
25. * $P \vdash (\neg Q \lor P) \lor \neg W$
26. $(C \land D) \rightarrow (E \land Q), C, D \vdash Q$
27. * $P \rightarrow (Q \land \neg S), F \rightarrow S, R \rightarrow P, R, \neg F \rightarrow \neg M \vdash \neg F \land \neg M$
28. $P, Q \land F, (P \land F) \rightarrow (W \lor R), \neg R, Q \rightarrow S \vdash W \land S$
29. * $P \rightarrow Q, Q \rightarrow W, P \vdash W$
30. $P \rightarrow \neg(Q \lor \neg R), P \vdash \neg Q$

31. *   ¬(¬P∧¬Q), ¬Q ⊢ P
32.   ¬[W∨(S∨M)] ⊢ ¬W∧¬M
33.   ¬[P∨(S∧M)], M ⊢ ¬P∧¬S

Solutions to Starred Easy Proofs

1. *   P∧¬Q, T∨Q ⊢ T

1	P∧¬Q	P
2	T∨Q	P/T
3	P	1∧E
4	¬Q	1∧E
5	T	2,4DS

3. *   A→C, A∧D ⊢ C

1	A→C	P
2	A∧D	P
3	A	2∧E
4	C	1,3→E

5. *   A ⊢ B→(B∧A)

1	A		P
2		B	A/B∧A
3		A	1R
4		B∧A	2,3∧I
5	B→(B∧A)		2–4→I

7. *   ¬(¬A∨B) ⊢ ¬¬A∧¬B

1	¬(¬A∨B)	A/¬¬A∧¬B
2	¬¬A∧¬B	1DeM

9. *   A∨(¬B→D), ¬(A∨B) ⊢ D

1	A∨(¬B→D)	P
2	¬(A∨B)	P/D
3	¬A∧¬B	2DeM
4	¬A	3∧E
5	¬B	3∧E
6	¬B→D	1,4DS
7	D	5,6→E

11. *   A∨(¬B→D), ¬(A∨D) ⊢ B

1	A∨(¬B→D)	P
2	¬(A∨D)	P/B
3	¬A∧¬D	2DeM

4	¬A	3∧E
5	¬D	3∧E
6	¬B→D	1,4DS
7	¬¬B	5,6MT
8	B	7DN

13. * (B↔D)∨S, ¬S, B ⊢ D∨W

1	(B↔D)∨S	P
2	¬S	P
3	B	P
4	B↔D	1,3DS
5	D	3,4↔E
6	D∨W	5∨I

15. * R∨R ⊢ R

1	R∨R		P
2		¬R	A
3		R	1,2DS
4	R		2–3¬E

17. * A→B, B→C, ¬C ⊢ ¬A

1	A→B	P
2	B→C	P
3	¬C	P/¬A
4	¬B	2,3MT
5	¬A	1,4MT

19. * A→B, A∨C, ¬C ⊢ B

1	A→B	P
2	A∨C	P
3	¬C	P/B
4	A	2,3DS
5	B	1,4→E

21. * A→B, C→D, B→C ⊢ A→D

1	A→B	P
2	C→D	P
3	B→C	P/A→D
4	A→C	1,3HS
5	A→D	4,2HS

23. * A→(B→C), A, ¬C ⊢ ¬B∨(W→S)

1	A→(B→C)	P
2	A	P
3	¬C	P

4	B→C	1,2→E
5	¬B	3,4MT
6	¬B∨(W→S)	5∨I

25. * P ⊢ (¬Q∨P)∨¬W

1	P	P/(¬Q∨P)∨¬W
2	¬Q∨P	1∨I
3	(¬Q∨P)∨¬W	2∨I

27. * P→(Q∧¬S), F→S, R→P, R, ¬F→¬M ⊢ ¬F∧¬M

1	P→(Q∧¬S)	P
2	F→S	P
3	R→P	P
4	R	P
5	¬F→¬M	P/¬F∧M
6	P	3,4→E
7	Q∧¬S	1,6→E
8	¬S	7∧E
9	¬F	2,8MT
10	¬M	5,9→E
11	¬F∧¬M	9,10∧I

29. * P→Q, Q→W, P ⊢ W

1	P→Q	P
2	Q→W	P
3	P	P/W
4	Q	1,3→E
5	W	2,4→E

31. * ¬(¬P∧¬Q), ¬Q ⊢ P

1	¬(¬P∧¬Q)	P
2	¬Q	P/P
3	¬¬P∨¬¬Q	1DeM
4	P∨Q	3DNx2
5	P	2,4DS

Medium Proofs

34.   (S↔D)→T, P↔(S∧D), P ⊢ T
35. * ¬(P∨Q), ¬(¬A∨¬B) ⊢ ¬P∧B
36.   B→¬(S∨T), ¬(A∨¬B), ¬S→W ⊢ W
37. * P ⊢ ¬P→¬S
38.   ¬(A∧B), B, (¬A∨S)→¬(D∧T) ⊢ ¬D∨¬T
39. * P, (P∨Q)→W, ¬W ⊢ ¬(P∨Q)

40.   G→M ⊢ ¬M → ¬G
41. *   A→¬(B→C), ¬B ⊢ ¬A
42. *   (B→C)→¬(D→E), C ⊢ ¬E
43. *   (S∨W)→M, (S∧T)↔(R∨P), R ⊢ M
44.   L∧(Y∧¬B), P, (P∨¬R)→Z, [Z∨(S∧T)]→W ⊢ W∨¬M
45. *   R∧(S∧T), (T∨M)→W, (W∨¬P)→(A∧B) ⊢ B
46.   B→D, ¬D ⊢ ¬B∨ D
47. *   ¬P∨(¬Q∨R), ¬P→(W∧S), (¬Q∨R)→(W∧S) ⊢ S
48.   ¬W→(R∨ S), M→¬W, ¬S∧M ⊢ R
49. *   (A↔B) ⊢ A→B
50.   [P↔(L∨M)]→W, P, L∨M ⊢ W
51.   A, B ⊢ A↔B
52.   A↔B ⊢ A→A
53. *   C∧(D∨B) ⊢ (C∧D)∨(C∧B)
54.   (C∧D)∨(¬C&¬D) ⊢ (C∧D)∨¬C
55. *   F∨[(G∧D)∧M] ⊢ (F∨M)∨R
56.   A→B, ¬A→C ⊢ ¬B→C
57.   M→¬S, ¬M→W ⊢ S→W
58.   (A∨B)→¬D, ¬(A∨B)→R ⊢ D→R
59.   P ⊢ ¬¬P∨P
60. *   (Q∨B) ⊢ (B∨Q)
61.   (A∧B)∧C ⊢ A∧(B∧C)
62.   ¬(A∨B)→D, ¬D ⊢ A∨B
63.   ¬[(A∧B)∧C]→R, ¬R ⊢ (B∧A)∧C
64.   R↔[(M→T)→Z], S∧[¬P∧(¬Q∧¬S)] ⊢ Z
65.   P∧[S∧(R∧¬P)], R→[(M→T)→W] ⊢ W
66.   ¬¬(¬R→¬R)∨B ⊢ (R→R)∨¬¬B

Solutions to Starred Medium Proofs

35. *   ¬(P∨Q), ¬(¬A∨¬B) ⊢ ¬P∧B

1	¬(P∨Q)	P
2	¬(¬A∨¬B)	P/¬P∧B
3	¬P∧¬Q	1DeM
4	¬¬A∧¬¬B	2DeM
5	¬P	3∧E
6	¬¬B	4∧E
7	B	6DN
8	¬P∧B	5,7∧I

37. * P ⊢ ¬P→¬S

    1   P   P/¬P→¬S
    2   P∨¬S   1∨I
    3   ¬¬P∨¬S   2DN
    4   ¬P→¬S   3IMP

alternatively

    1   P   P/¬P→¬S
    2   ¬P   A/¬S
    3   S   A/contra
    4   P   1R
    5   ¬P   2R
    6   ¬S   3–5¬I
    7   ¬P→¬S   2–6→I

39. * P, (P∨Q)→W, ¬W ⊢ ¬(P∨Q)

    1   P   P
    2   (P∨Q)→W   P
    3   ¬W   P/¬(P∨Q)
    4   P∨Q   1∨I
    5   ¬(P∨Q)   2,3MT

41. * A→¬(B→C), ¬B ⊢ ¬A

    1   A→¬(B→C)   P
    2   ¬B   P/¬A
    3   A   A/contra
    4   ¬(B→C)   1→E
    5   ¬(¬B∨C)   4IMP
    6   ¬¬B∧¬C   5DeM
    7   ¬¬B   6∧E
    8   B   7DN
    9   ¬B   2R
   10   ¬A   3–9¬I

42. * (B→C)→¬(D→E), C ⊢ ¬E

    1   (B→C)→¬(D→E)   P
    2   C   P/¬E
    3   ¬B∨C   2∨I
    4   ¬(D→E)   1,3→E
    5   ¬(¬D∨E)   4IMP
    6   ¬¬D∧¬E   5DeM
    7   ¬E   6∧E

43.   *   (S∨W)→M, (S∧T)↔(R∨P), R ⊢ M

1	(S∨W)→M		P
2	(S∧T)↔(R∨P)		P
3	R		P/M
4		¬M	A/contra
5		¬(S∨W)	1,4MT
6		¬S∧¬W	5DeM
7		¬S	6∧E
8		R	3R
9		R∨P	8∨I
10		S∧T	2,9↔E
11		S	10∧E
12	M		4–11¬E

45.   *   R∧(S∧T), (T∨M)→W, (W∨¬P)→(A∧B) ⊢ B

1	R∧(S∧T)	P
2	(T∨M)→W	P
3	(W∨¬P)→(A∧B)	P/B
4	S∧T	1∧E
5	T	4∧E
6	T∨M	5∨I
7	W	2,6→E
8	W∨¬P	7∨I
9	A∧B	3,8→E
10	B	9∧E

47.   *   ¬P∨(¬Q∨R), ¬P→(W∧S), (¬Q∨R)→(W∧S) ⊢ S

1	¬P∨(¬Q∨R)		P
2	¬P→(W∧S)		P
3	(¬Q∨R)→(W∧S)		P
4		¬P	A
5		W∧S	1,4→E
6		S	5∧E
7		(¬Q∨R)	A
8		W∧S	3,7→E
9		S	8∧E
10	S		1,4–6,7–9∨E

49. *   (A↔B) ⊢ A→B

    1   A↔B   P/A→B
    2   │ A   A/B
    3   │ B   1,2↔E
    4   A→B   2–3→I

53. *   C∧(D∨B) ⊢ (C∧D)∨(C∧B)

    1   C∧(D∨B)   P/(C∧D)∨(C∧B)
    2   C   1∧E
    3   D∨B   1∧E
    4   │ ¬[(C∧D)∨(C∧B)]   A
    5   │ ¬(C∧D)∧¬(C∧B)   4DeM
    6   │ ¬(C∧D)   5∧E
    7   │ ¬(C∧B)   5∧E
    8   │ ¬C∨¬D   6DeM
    9   │ ¬C∨¬B   7DeM
    10   │ ¬D   2,8DS
    11   │ ¬¬C   2DN
    12   │ ¬B   2,9
    13   │ D   3,12DS
    14   (C∧D)∨(C∧B)   4–13¬E

55. *   F∨[(G∧D)∧M] ⊢ (F∨M)∨R

    1   F∨[(G∧D)∧M]   P/(F∨M)∨R
    2   │ ¬[(F∨M)∨R]   A
    3   │ ¬(F∨M)∧¬R   2DeM
    4   │ ¬(F∨M)   3∧I
    5   │ ¬R   3∧I
    6   │ ¬F   4DeM
    7   │ ¬M   4DeM
    8   │ (G∧D)∧M   1,6DS
    9   │ M   8∧E
    10   (F∨M)∨R   2–9¬E

60. * $(Q\lor B) \vdash (B\lor Q)$

1	Q∨B		P/B∨Q
2		¬(B∨Q)	A/contra
3		¬B∧¬Q	2DeM
4		¬B	3∧E
5		¬Q	3∧E
6		Q	1,4DS
7	B∨Q		2–6¬E

Hard Proofs

67. * ¬{A∨[¬(B→R)∨¬(C→R)]}, ¬A↔(B∨C) ⊢ R
68. (A→A)→B ⊢ B
69. * A∨B, R∨¬(S∨M), A→S, B→M ⊢ R
70. ¬B∧C ⊢ ¬(B∨¬C)∨(F→M)
71. * A ⊢ ¬(¬A∧¬B)
72. ¬¬(C∨¬¬D), ¬D ⊢ ¬(¬C∧F)∨M
73. * A→B, D→E, (¬B∨¬E)∧(¬A∨¬B) ⊢ ¬A∨¬D
74. A→¬B, D→¬E, F→¬G, H→¬J, D→G, E→B, F∨A ⊢ ¬D∨¬E
75. * (A∨B)→(D∨E), [(D∨E)∨F]→(G∨H), (G∨H)→¬D, E→¬G, B ⊢ H
76. A→B, A∨(B∨¬D), ¬B ⊢ ¬D∧¬B
77. * A→(B→D), ¬(D→Y)→¬K, (Z∨¬K)∨¬(B→Y) ⊢ ¬Z→¬(A∧K)
78. ¬(A→B), A→D, E→B ⊢ ¬(D→E)
79. * ¬(P→Q)⊣⊢ (P∧¬Q)
80. (P→Q)⊣⊢ ¬(P∧¬Q)
81. P∨Q⊣⊢ Q∨P,commutation
82. P∧Q⊣⊢ Q∧P, commutation
83. P∨(Q∨R)⊣⊢ (P∨Q)∨R, association
84. P∧(Q∧R)⊣⊢ (P∧Q)∧R, association
85. P∨(Q∧R)⊣⊢ (P∨Q)∧(P∨R), distribution
86. * P∧(Q∨R)⊣⊢ (P∧Q)∨(P∧R), distribution
87. (P∧Q)→R⊣⊢ P→(Q→R), exportation
88. * P→Q⊣⊢ ¬Q→¬P, contraposition
89. S→R, P∨S, P→R ⊢ R∨M
90. P→W, W→S, S→T, ¬T ⊢ ¬P
91. A→B ⊢ (¬A∨B)∨R
92. ¬(A→B) ⊢ (A∧¬B)∨R
93. F ⊢ ¬F→W

94. (¬A∧¬B)→¬C, ¬A ⊢ ¬B→¬C
95. (¬A∧¬B)→¬D, ¬A ⊢ D→B
96. A↔(B→C), A, (¬B∨C)→D, S→¬(D∨E), (¬S↔A)→L ⊢ L
97. P ⊢ [P↔(P∨Q)]↔P
98. ¬(A→R)∨¬(R→A) ⊢ A∨R
99. [(P→L)∨S]→(R↔S, L, R ⊢ S↔L
100. Z ⊢ [P→(Z∨¬Z)]↔[Q→(Z∨¬Z)]
101. Z↔F ⊣ ⊢ (Z→F)∧(F→Z)

Solutions to Starred Hard Proofs

67. * ¬{A∨[¬(B→R)∨¬(C→R)]}, ¬A↔(B∨C) ⊢ R

1	¬{A∨[¬(B→R)∨¬(C→R)]}		P
2	¬A↔(B∨C)		P/R
3	¬A∧¬[¬(B→R)∨¬(C→R)]		1DEM
4	¬A		3∧E
5	¬[¬(B→R)∨¬(C→R)]		3∧E
6	¬¬(B→R)∧¬¬(C→R)		5DEM
7	(B→R)∧(C→R)		6DNx2
8	B→R		7∧E
9	C→R		7∧E
10	B∨C		2,4↔E
11		B	A/R
12		R	8,11→E
13		C	A/R
14		R	9,13→E
15	R		10–14∨E

69. * A∨B, R∨¬(S∨M), A→S, B→M ⊢ R

1	A∨B		P
2	R∨¬(S∨M)		P
3	A→S		P
4	B→M		P/R
5		A	A/S∨M
6		A→S	3R
7		S	5,6→E
8		S∨M	7∨I

9	B	A/S∨M
10	B→M	4R
11	M	9,10→E
12	S∨M	11∨I
13	S∨M	1,5–8,9–12∨E
14	R	2,13DS

71. * A ⊢ ¬(¬A∧¬B)

1	A	P/¬(¬A∧¬B)
2	¬¬A	1DN
3	¬¬A∨¬¬B	2∨I
4	¬(¬A∧¬B)	3DeM

73. * A→B, D→E, (¬B∨¬E)∧(¬A∨¬B) ⊢ ¬A∨¬D

1	A→B	P
2	D→E	P
3	(¬B∨¬E)∧(¬A∨¬B)	P/¬A∨¬D
4	¬B∨¬E	3∧E
5	¬A∨¬B	3∧E
6	¬(¬A∨¬D)	A/contra
7	A∧D	6DeM
8	A	7∧E
9	D	7∨E
10	B	1,8→E
11	E	2,9→E
12	¬(B∧E)	4DeM
13	B∧E	10,11∧I
14	¬A∨¬D	6–13¬E

75. * (A∨B)→(D∨E), [(D∨E)∨F]→(G∨H), (G∨H)→¬D, E→¬G, B ⊢ H

1	(A∨B)→(D∨E)	P
2	[(D∨E)∨F]→(G∨H)	P
3	(G∨H)→¬D	P
4	E→¬G	P
5	B	P/H
6	A∨B	5∨I
7	D∨E	1,6→E
8	(D∨E)∨F	7∨I
9	G∨H	2,8→E
10	¬D	3,9→E

11	E	7,10DS
12	¬G	4,11→E
13	H	9,12DS

77. *   A→(B→D), ¬(D→Y)→¬K, (Z∨¬K)∨¬(B→Y) ⊢ ¬Z→¬(A∧K)

1	A→(B→D)	P
2	¬(D→Y)→¬K	P
3	(Z∨¬K)∨¬(B→Y)	P/¬Z→¬(A∧K)
4	¬Z	A/¬(A∧K)
5	A∧K	A/Z∧¬Z
6	A	5∧E
7	B→D	1,6→E
8	K	5∧E
9	¬¬K	8DN
10	¬¬(D→Y)	2,9MT
11	D→Y	10DN
12	B→Y	7,11HS
13	¬¬(B→Y)	12DN
14	Z∨¬K	3,13DS
15	¬K	4,14DS
16	K	8R
17	¬(A∧K)	5–16¬I
18	¬Z→¬(A∧K)	4–17→I

79. *   ¬(P→Q)⊣ ⊢ P∧¬Q, two proofs

1	¬(P→Q)	A/P∧¬Q
2	¬P	A
3	¬P∨Q	2∨I
4	P→Q	3IMP
5	¬(P→Q)	1R
6	P	
7	Q	A
8	¬P∨Q	7∨I
9	P→Q	8IMP
10	¬(P→Q)	1R
11	¬Q	7–10→I
12	P∧¬Q	6,11∧I

```
1       P∧¬Q            P/P∧¬Q
2       ¬¬P∧¬Q          1∧E
3       ¬(¬P∨Q)         2DeM
4       ¬(P→Q)          3IMP
```

86. * P∧(Q∨R)⊣ ⊢ (P∧Q)∨(P∧R), distribution

```
1       P∧(Q∨R)                         P/(P∧Q)∨(P∧R)
2           ¬[(P∧Q)∨(P∧R)]              A/¬P∧¬P
3           ¬(P∧Q)∧¬(P∧R)               2DEM
4           ¬(P∧Q)                      3∧E
5           ¬(P∧R)                      3∧E
6           ¬P∨¬Q                       4DEM
7           P                           1∧E
8           ¬¬P                         7DN
9           ¬Q                          6,8DS
10          Q∨R                         1∧E
11          R                           9,10DS
12          ¬P∨¬R                       5DEM
13          ¬¬P                         7DN
14          ¬R                          12,13DS
15          R                           11R
16      (P∧Q)∨(P∧R)                     2–15¬E
```

```
1       (P∧Q)∨(P∧R)         P/P∧(Q∨R)
2           ¬P              A/P∧¬P
3           ¬P∨¬Q           2∨I
4           ¬(P∧Q)          3DEM
5           P∧R             1¬∨E
6           P               5∧E
7           ¬P              2R
8       P                   2–7¬E
9           ¬(Q∨R)          A/P∧¬P
10          ¬Q∧¬R           9DEM
11          ¬Q              10∧E
12          ¬R              10∧E
13          ¬P∨¬Q           11∨I
14          ¬(P∧Q)          13DEM
```

15	P∧R	1,14DS
16	R	15∧E
17	¬R	12R
18	Q∨R	9–17¬E
19	P∧(Q∨R)	8,18∧I

88. * P→Q ⊣ ⊢ ¬Q→¬P, contraposition

1	P→Q		P/¬Q→¬P
2	¬Q		A/¬P
3		P	A/P∧¬P
4		Q	1→E
5		¬Q	2R
6	¬P		3–5¬I
7	¬Q→¬P		2–6→I

1	¬Q→¬P		P/P→Q
2	P		A/Q
3		¬Q	A/P∧¬P
4		¬P	1→E
5		P	2R
6	Q		3–5¬I
7	¬Q→¬P		2–6→I

Zero-Premise Deductions

102.   ⊢ P→P
103.   * ⊢ P∨¬P, law of excluded middle
104.   ⊢ ¬(P∧¬P), principle of noncontradiction
105.   * ⊢ (A∧B)→A
106.   ⊢ A→¬(B∧¬B)
107.   * ⊢ ¬A→(A→B)
108.   ⊢ A→(A∨¬A)
109.   * ⊢ P→(¬Q→P)
110.   ⊢ [(P→Q)→P]→P
111.   * ⊢ ¬[(A→¬A)∧(¬A→A)]
112.   ⊢ A→(A∧A)
113.   * ⊢ [(A→B)∧(A→D)]→[A→(B∧D)]

114. ⊢ ¬P→¬[(P→Q)→P]
115. * ⊢ [P→(Q→R)]→[(P→Q)→(P→R)], axiom scheme 2
116. ⊢ (A→B)∨(B→D)
117. * ⊢ A→[A→(A∨A)]
118. ⊢ ¬(P∧¬P)
119. * ⊢ [(¬A∨B)∧(¬A∨D)]→[¬A∨(B∧D)]
120. ⊢ [(P→Q)→R]→(¬R→P)
121. * ⊢ (A→B)→[¬(B∧D)→¬(D∧A)]
122. ⊢ (A∧B)→A
123. * ⊢ ¬¬A→A
124. ⊢ A→(B→A)
125. * ⊢ (A∨B)→[(¬A∨B)→ B]

Solutions to Starred Zero-Premise Deductions

103. * ⊢ P∨¬P, law of excluded middle

1		¬(P∨¬P)	A/P∨¬P
2		¬P∧¬¬P	1DeM
3		¬P∧P	2DN
4		¬P	3∧E
5		P	3∧E
6	P∨¬P		1–5¬E

105. * ⊢ (A∧B)→A

1		A∧B	A/A
2		A	1∧E
3	(A∧B)→A		1–2→I

107. * ⊢ ¬A→(A→B)

1		¬A		A/A→B
2			A	A/B
3				¬B	A/B∧¬B
4				A	2R
5				¬A	1R
6			B	3–5¬E
7		A→B		2–6→I
8	¬A→(A→B)			1–7→I

109. * ⊢ P→(¬Q→P)

    1 | P   A/¬Q→P

    2 | | ¬Q   A/P

    3 | | P   1R

    4 | ¬Q→P   2–3→I

    5   P→(¬Q→P)   1–4→I

111. * ⊢ ¬[(A→¬A)∧(¬A→A)]

    1 | (A→¬A)∧(¬A→A)   A/P∧¬P

    2 | A→¬A   1∧E

    3 | ¬A→A   A/B∧¬B

    4 | | A   A/¬A∧¬A

    5 | | ¬A   2,4→E

    6 | ¬A   4,5¬I

    7 | A   3,6→E

    8   ¬[(A→¬A)∧(¬A→A)]   1–7¬I

113. * ⊢ [(A→B)∧(A→D)]→[A→(B∧D)]

    1 | (A→B)∧(A→D)   A

    2 | | A   A/B∧D

    3 | | A→B   1∧E

    4 | | A→D   1∧E

    5 | | B   2,3→E

    6 | | D   2,4→E

    7 | | B∧D   5,6∧E

    8 | A→(B∧D)   2–7→I

    9   [(A→B)∧(A→D)]→[A→(B∧D)]   1–8→I

115. * ⊢ [P→(Q→R)]→[(P→Q)→(P→R)]

    1 | P→(Q→R)   A/[(P→Q)→(P→R)]

    2 | | P→Q   A/P→R

    3 | | | P   A/R

    4 | | | Q→R   1,3→E

    5 | | | Q   2,3→E

    6 | | | R   4,5→E

    7 | | P→R   3–6→I

    8 | (P→Q)→(P→R)   2–7→I

    9   [P→(Q→R)]→[(P→Q)→(P→R)]   1–8→I

117. * ⊢ A→[A→(A∨A)]

1	A	A/A→(A∨A)
2	A	A/A∨A
3	A∨A	2∨I
4	A→(A∨A)	2–3→I
5	A→[A→(A∨A)]	1–4→I

119. * ⊢ [(¬A∨B)∧(¬A∨D)]→[¬A∨(B∧D)]

1	(¬A∨B)∧(¬A∨D)	A
2	¬A∨B	1∧E
3	¬A∨D	1∧E
4	¬[¬A∨(B∧D)]	A
5	¬¬A∧¬(B∧D)	4DeM
6	¬¬A	5∧E
7	¬(B∧D)	5∧E
8	B	2,6DS
9	D	3,6DS
10	B∧D	8,9∧I
11	¬A∨(B∧D)	4–10¬E
12	[(¬A∨B)∧(¬A∨D)]→[¬A∨(B∧D)]	1–11→I

121. * ⊢ (A → B) → [¬(B∧D) → ¬(D∧A)]

1	A→B	A / ¬(B∧D)→¬(D∧A)
2	¬(B∧D)	A / ¬(D∧A)
3	D∧A	A / P, ¬P
4	D	3∧E
5	A	3∧E
6	B	1,5→E
7	B∧D	4,6∧I
8	¬(B∧D)	2R
9	¬(D∧A)	3-8¬I
10	¬(B∧D)→¬(D∧A)	2-9→I
11	(A→B)→[¬(B∧D)→¬(D∧A)]	1-10→I

123. * ⊢ ¬¬A → A

1	¬¬A	A / A
2	A	1DN
3	¬¬A→A	1-2→I

125. *   ⊢ (A∨B)→[(¬A∨B)→B]

1	A∨B		A/(¬A∨B)→B
2		¬A∨B	A/B
3			¬B	A/B∧¬B
4			A	1,3DS
5			¬A	2,3DS
6		B	3–5¬E
7	(¬A∨B)→B		2–6→I
8	(A∨B)→(¬A∨B)→B		1–7→I

Conceptual and Application Exercises

1. De Morgan's Laws (DeM) and implication (IMP) are equivalence rules introduced into the existing set of derivation rules so as to simplify proofs involving negated conjunctions, negated disjunctions, and negated conditionals. An additional equivalence rule could be introduced to simplify proofs involving negated biconditionals ('P↔Q'). Using the tree decomposition rule for negated biconditionals as your guide, formulate an equivalence rule for biconditionals or negated biconditionals. Once you have formulated a valid equivalence form, show how this rule can be derived from the preexisting set of derivation rules.

2. In dealing with negated conjunctions and negated disjunctions, the introduction of De Morgan's Laws served to simplify proofs. However, the proof strategy for handling negated conditionals, such as '¬(P→Q),' is slightly cumbersome. It is suggested that IMP be applied, which yields '¬(¬P∨Q),' and then DeM be applied, yielding '¬¬P∧¬Q.' How might this procedure be simplified?

3. Consider the following argument, translate it into the language of PL, test the argument with a truth tree to see whether or not it is valid, and then construct a proof for the argument using the derivation rules.

   If John is guilty of the crime, then he should be put in jail. If John is not guilty, then he shouldn't. If John is not guilty of the crime, then his DNA will not have been at the scene of the crime, the police will not have found the murder weapon in his apartment, and there will not be witnesses testifying that they saw him flee from the scene of the crime. But John's DNA was at the scene of the crime, the police did find the murder weapon in his apartment, and there were witnesses who testified that they saw him flee the scene of the crime. Thus, John is guilty of the crime, and John should be put in jail.

4. Write your own valid argument in English, use a truth tree to show that it is valid, and then prove that argument using the derivation rules. It may be helpful to start first with your conclusion and then work backward to a set of premises.

5. Classical propositional logic is *monotonic*. That is, if a set of wffs 'G' entails 'P,' and 'G' is a subset of 'D,' then 'D' entails 'P.' Another way of thinking about this is that if you can derive a proposition 'P' from a set of propositions, then you can

also derive 'P' from a larger set of propositions 'D.' Can you think of some reasons why one would reject this?

6. Notice that every disjunction can be rewritten as a corresponding conjunction (and vice versa). What does this suggest about the use of the truth-functional operators expressed by '∨'and'∧'? How essential is it to the expressive range of PL that both of these operators are present in the language?

7. Consider the following argument:

All 'Ss' are 'Ps.'
All 'Ps' are 'Qs.'
All 'Ss' are 'Qs.'

This argument appears to be straightforwardly valid. However, now consider the following translation of this argument into PL.

S
P
Q

Notice that this argument is not valid. What does this mean about the expressive power of PL? What would we like a formal language to do concerning the set of valid arguments expressible in English?

LIST OF STRATEGIC RULES

SP#1(E+)   First, eliminate any conjunctions with '∧E,' disjunctions with DS or '∨E,' conditionals with '→E' or MT, and biconditionals with '↔E.' Then, if necessary, use any necessary introduction rules to reach the desired conclusion.

SP#2(B)   First, work backward from the conclusion using introduction rules (e.g.,'∧I,''∨I,''→I,''↔I'). Then, use SP#1(E).

SP#3(EQ+)   Use DeM on any negated disjunctions or negated conjunctions, and then use SP#1(E). Use IMP on negated conditionals, then use DeM, and then use SP#1(E).

SA#1(P,¬Q)   If the conclusion is an atomic proposition (or a negated proposition), assume the negation of the proposition (or the non-negated form of the negated proposition), derive a contradiction, and then use '¬I' or '¬E.'

SA#2(→)   If the conclusion is a conditional, assume the antecedent, derive the consequent, and use '→I.'

SA#3(∧)   If the conclusion is a conjunction, you will need two steps. First, assume the negation of one of the conjuncts, derive a contradiction, and then use '¬I' or '¬E.' Second, in a separate subproof, assume the negation of the other conjunct, derive a contradiction, and then use '¬I' or '¬E.' From this point, a use of '∧I' will solve the proof.

SA#4(∨)   If the conclusion is a disjunction, assume the negation of the whole disjunction, derive a contradiction, and then use '¬I' or '¬E.'

STEP-BY-STEP USE OF STRATEGIC RULES FOR ASSUMPTIONS

First   Start by identifying the main operator of the conclusion.

→   If it is a conditional, assume the antecedent, derive the consequent, and use '→I.'

∧   If it is a conjunction, you will need two steps. First, assume the negation of one of the conjuncts, derive a contradiction, and then use '¬I' or '¬E.' Second, in a separate subproof, assume the negation of the other conjunct, derive a contradiction, and then use '¬I' or '¬E.' From this point, a use of '∧I' will solve the proof.

∨   If it is a disjunction, assume the negation of the whole disjunction, derive a contradiction, and then use '¬I' or '¬E.'

P, ¬Q   If it is an atomic proposition or a negation, assume the negation of the proposition (or negation of the negated proposition), derive a contradiction, and then use '¬I' or '¬E.'

LIST OF DERIVATION RULES FOR PD+

	Derivation Rules for PD+			
1	**Conjunction Introduction (∧I)** From 'P' and 'Q,' we can derive 'P∧Q.' Also, from 'P' and 'Q,' we can derive 'Q∧P.'		P Q P∧Q Q∧P	 ∧I ∧I
2	**Conjunction Elimination (∧E)** From 'P∧Q,' we can derive 'P.' Also, from 'P∧Q,' we can derive 'Q.'		P∧Q P Q	 ∧E ∧E
3	**Conditional Introduction (→I)** From a derivation of 'Q' within a subproof involving an assumption 'P,' we can derive 'P→Q' out of the subproof. P→Q	P . . . Q		A →I
4	**Conditional Elimination (→E)** From 'P→Q' and 'P,' we can derive 'Q.'		P→Q P Q	 →E

5	**Reiteration (R)** Any proposition 'P' that occurs in a proof or subproof may be rewritten at a level of the proof that is equal to 'P' or more deeply nested than 'P.'		P · · · P	R

6	**Negation Introduction (¬I)**
	From a derivation of a proposition 'Q' and its literal negation '¬Q' within a subproof involving an assumption 'P,' we can derive '¬P' out of the subproof.
			| P
			    | ·
			    | ·
			    | ·
			    | ¬Q
			    | Q
			¬P
				A
				¬I

7	**Negation Elimination (¬E)**
	From a derivation of a proposition 'Q' and its literal negation '¬Q' within a subproof involving an assumption '¬P,' we can derive 'P' out of the subproof.
			| ¬P
			    | ·
			    | ·
			    | ·
			    | ¬Q
			    | Q
			P
				A
				¬E

8	**Disjunction Introduction (∨I)** From 'P,' we can validly infer 'P∨Q' or 'Q∨P.'		P P∨Q Q∨P	∨I ∨I

9	**Disjunction Elimination (∨E)**
	From 'P∨Q' and two derivations of 'R'—one involving 'P' as an assumption in a subproof, the other involving 'Q' as an assumption in a subproof—we can derive 'R' out of the subproof.
			P∨Q
			    | P
			    | ·
			    | ·
			    | ·
			    | R
			    | Q
			    | ·
			    | ·
			    | ·
			    | R
			R
				A
				
				A
				
				∨E

10	**Biconditional Introduction (↔I)** From a derivation of 'Q' within a subproof involving an assumption 'P' and from a derivation of 'P' within a separate subproof involving an assumption 'Q,' we can derive 'P↔Q' out of the subproof.		P . . . Q Q . . . P P↔Q	A A ↔I
11	**Biconditional Elimination (↔E)** From 'P↔Q' and 'P,' we can derive 'Q.' And from 'P↔Q' and 'Q,' we can derive 'P.'		P↔Q P Q P↔Q Q P	 ↔E ↔E
12	**Modus Tollens (MT)** From 'P→Q' and '¬Q,' we can derive '¬P.'		P→Q ¬Q ¬P	 MT
13	**Disjunctive Syllogism (DS)** From 'P∨Q' and '¬Q,' we can derive 'P.'		P∨Q ¬Q P P∨Q ¬P Q	 DS DS
14	**Hypothetical Syllogism (HS)** From 'P→Q' and 'Q→R,' we can derive 'P→R.'		P→Q Q→R P→R	 HS
15	**Double Negation (DN)** From 'P,' we can derive '¬¬P.' From '¬¬P,' we can derive 'P.'		P⊣⊢ ¬¬P	DN
16	**De Morgan's Laws (DeM)** From '¬(P∨Q),' we can derive '¬P∧¬Q.' From '¬P∧¬Q,' we can derive '¬(P∨Q).' From '¬(P∧Q),' we can derive '¬P∨¬Q.' From '¬P∨¬Q,' we can derive '¬(P∧Q).'	¬(P∨Q)⊣⊢¬P∧¬Q ¬(P∧Q)⊣⊢¬P∨¬Q	DeM DeM
17	**Implication (IMP)** From 'P→Q,' we can derive '¬P∨Q.' From '¬P∨Q,' we can derive 'P→Q.'	P→Q⊣⊢¬P∨Q	IMP

DEFINITION

Proof   A proof is a finite sequence of well-formed formulas (or propositions), each
of which is either a premise, an assumption, or the result of preceding formulas and a derivation rule.

NOTE

1. These rules are named after British mathematician and logician Augustus De Morgan
(1806–1871).

Chapter Six

Predicate Language, Syntax, and Semantics

6.1 THE EXPRESSIVE POWER OF PREDICATE LOGIC

The language, syntax, and semantics of PL have two strengths. First, logical properties applicable to arguments and sets of propositions have a corresponding applicability in English. So, if one is dealing with a valid argument in PL, then that argument is also valid for English. Second, the semantic properties of arguments and sets of propositions have decision procedures. That is, there are mechanical procedures for testing whether any argument is valid, whether propositions in a set have some logical property (e.g., they are consistent, equivalent, etc.), and whether any proposition is always true (a tautology), always false (a contradiction), or neither always true nor always false (a contingency).

The weakness of PL is that it is not expressive enough. That is, some valid arguments and semantic relationships in English cannot be expressed in propositional logic. Consider the following example:

All humans are mortal.
Socrates is a human.
Therefore Socrates is a mortal.

This argument is clearly valid in English but cannot be expressed as a valid argument in PL. Symbolically, the argument is represented as follows:

M
S
R

The above argument is clearly invalid. In order to bring English arguments like the one above into the domain of symbolic logic, it is necessary to develop a formal language that does not symbolize sentences as wholes (e.g., *John is tall* as 'J'), but symbolizes parts of sentences. That is, a formal language whose basic unit is not a complete sentence but the subject(s) and predicate(s) of the sentence such a language will be more expressive and able to represent the above argument as valid. This is the language of *predicate logic* (sometimes called the logic of relations). We'll symbolize it as RL.

6.2 THE LANGUAGE OF RL

The elementary vocabulary for RL is as follows:

	Elements of the Language	**Abbreviation**
1	Individual constants (names)	Lowercase letters 'a' through 'v,' with or without numerical subscripts.
2	*n*-place predicates	Uppercase letters 'A' through 'Z,' with or without numerical subscripts
3	Individual variables	Lowercase letters *w* through *z*, with or without numerical subscripts
4	Truth-functional operators and scope indicators from PL	$\neg, \wedge, \vee, \rightarrow, \leftrightarrow, (,), [,], \{, \}$
5	Quantifiers	$\forall, \exists$

The following chapter first articulates each of the above items, details the syntax or formation rules of RL, and then explains the semantics of RL (i.e., what it means for a proposition to be true or false in RL).

6.2.1 Individual Constants (Names) and *n*-Place Predicates

In the introductory section of this chapter, we saw that propositional logic is limited because it takes complete sentences as basic. There is thus a need to enhance our logical language by taking internal (or subsentential) features as basic. Let's start with an example:

(1) John is tall.
(2) Liz is smarter than Jane.

There are two internal features of (1): the noun phrase (name) *John* and the verb phrase (predicate) *is tall*. The typical role of proper names like *John* in (1) and *Liz* and *Jane* in (2) is to designate a singular object of some kind (e.g., a person, place, monument, country, etc.). In RL we represent names (individual constants) with lowercase letters 'a' to 'v,' with or without numerical subscripts. For example,

$$
\begin{aligned}
j &= \text{John} \\
l &= \text{Liz} \\
d_{32} &= \text{Jane}
\end{aligned}
$$

One assumption we will make with respect to names is that a name always refers to one and only one object. In other words, in RL, there is no such thing as an "empty name" (i.e., a name that fails to refer to an object). However, while it is the case that every name refers to one object, we will allow for the possibility of multiple names referring to the same object (e.g., 'John Santellano' and 'Mr. Santellano' can designate the same person).

In addition to names, English sentences also involve predicates of different *adicity*. The adicity of a predicate term is the number of individuals the predicate must have in order to express a proposition. For example, (1) contains not only the proper name *John* but also the predicate (or verb phrase) *is tall*. The predicate term *is tall* has an adicity of 1 since it only needs one individual to yield a sentence that expresses a proposition (i.e., a sentence capable of being true or false). Likewise, (2) contains not only the proper names *Liz* and *Jane* but also the predicate term *is smarter than*. The predicate term *is smarter than* has an adicity of 2 since it requires at least two individuals to yield a sentence that expresses a proposition.

In RL we represent predicates and relations with uppercase letters 'A' to 'Z,' with or without numerical subscripts. Again, the numerical subscripts ensure that the vocabulary is infinite. For example,

$$\begin{aligned} T &= \text{is tall} \\ G &= \text{is green} \\ R &= \text{is bigger than} \\ F_{43} &= \text{is faster than} \end{aligned}$$

Thus far, we have articulated two key items in RL: names and predicates of varying adicity. In addition to retaining the use of truth-functional operators that form part of PL, we can now broach how to translate various sentences involving names and predicates from English into RL. First, consider (1) again:

(1) John is tall.

To translate (1) into RL, begin by replacing any names with a blank and assign the name a letter with lowercase letters 'a' to 'v.' Thus,

$$\frac{\underline{\quad\quad}\text{is tall}}{j = \text{John}}$$

Since *John* is the only name in the sentence, we have isolated a predicate term with a single blank. This shows that *is tall* has an adicity of 1. After all names have been removed what remains is called an *unsaturated predicate* or a *rheme*. Finally, we can represent this unsaturated English predicate with the language of predicate logic by replacing it with an uppercase letter 'A' to 'Z.' Thus,

$$T = \underline{\quad\quad}\text{ is tall}$$

To complete the translation into the language of predicate logic, the name is placed to the right of the predicate. Thus, a complete translation of *John is tall* into RL is the following:

Tj

Not all predicates are one-place predicates (i.e., not all predicate terms have an adicity of 1). Consider the following sentence:

John is taller than Frank.

Again, we start by removing all of the singular terms, assigning them lowercase letters 'j' and 'f' and replacing these singular terms with blanks.

____ is taller than ____.

What remains is a two-place predicate, for there are two places where singular terms might be inserted. We can represent this two-place predicate as follows:

R = ____ is taller than ____.

Finally, to complete the translation of *John is taller than Frank*, we reinsert the singular terms. Thus, a complete translation of *John is taller than Frank* is the following:

Rjf

While there are higher-place predicates in English, expressing some of these is awkward in English. Since there is no need to put a restriction on how many objects we can relate together, what remains after all object terms have been removed is an *n-place predicate*, where *n* is the number of blanks or places where an individual constant could be inserted.

Here is a final example:

John is standing between Frank and Mary.

In order to represent the above sentence, begin by deleting all of the singular-referring expressions from the sentence.

____ is standing between ____ and ____

What remains after all of the names have been deleted is the three-place predicate. In the above example, note that 'and' is not a sentential operator but part of the predicate. The next step is to symbolize the three-place predicate as 'S' and complete the translation by inserting abbreviations for individual constants for *John*, *Frank*, and *Mary*.

Sjfm

6.2.2 Domain of Discourse, Individual Variables, and Quantifiers

Consider a number of examples that PL cannot fully express:

(3) All men are mortal.
(4) Some zombies are hungry.
(5) Every man is happy.

Notice that (3) to (5) do not express propositions about singular objects but instead predicate a property to a quantity of objects. To symbolize sentences of this sort, we cannot use the procedure noted in the previous section. For example, the proposition expressed by (4) does not name a zombie that is hungry. Instead, (4) expresses the more indefinite proposition that some object in the universe is both a zombie and is hungry.

In order to adequately represent (3) to (5), the introduction of two new symbols and the notion of a domain of discourse is required. For convenience, let's abbreviate the domain of discourse as D and use lowercase letters w through z, with or without subscripts, to represent individual variables. The *domain of discourse* D is all of the objects we want to talk about or to which we can refer. So, if our discussion is on the topic of positive integers, then we would say that the domain of discourse is just the positive integers. Or, more compactly,

D: positive integers

If our discussion is about human beings, then we would say that the domain of discourse is just those human beings who exist or have existed. That is,

D: living or dead humans

Individual variables are placeholders whose possible values are the individuals in the domain of discourse. Individual variables are said to range over individual particular objects in the domain in that they take these objects as their values. Thus, if our discussion is on the topic of numbers, an individual variable z is a placeholder for some number in the universe of discourse.

As placeholders, we can also use variables to indicate the adicity of a predicate. Previously we indicated this by using blanks (e.g., _____ *is tall*). Rather than representing a predicate as a sentence with a blank attached to it, we will fill in the blanks with the appropriate number of individual variables:

$$
\begin{aligned}
Tx &= x \text{ is tall} \\
Gx &= x \text{ is green} \\
Rxy &= x \text{ is bigger than } y \\
Bxyz &= x \text{ is between } y \text{ and } z.
\end{aligned}
$$

The domain of discourse is determined in two ways, *stipulatively* and *contextually*. When the D is determined stipulatively, the objects in the domain are named. For example, suppose the objects to which variables and names can refer are limited to human beings. If this is the case, then we would write,

D: human beings

Indicating the D in this way means that individual variables do not take nonhuman beings as substitution instances. So, if someone were to say *Everyone is crazy*, this would be a shorthand way of saying, *Every human being is crazy*.

In normal conversation, the domain of discourse is not always stipulated. The second way the D is determined is contextually. That is, whenever you have a conversation, you don't say, 'Let's only talk about colors' or 'Let's only talk about human beings living in the 1900s.' Often, certain features about the subject matter being talked about determine what is and is not part of the domain of discourse. But the domain of discourse can change rapidly and is not always easily determined. If you and your friends are talking about movies, then the D is movies, but the D can quickly switch to books, to mutual friends of yours who behave similarly to characters in books, and so on. For our purposes, whenever we need to translate from predicate logic to English, or vice versa, we will always stipulate the D.

Finally, the domain is *restricted* or *unrestricted*. In the case of arithmetic, the domain of discourse for variables is restricted to numbers. In the case of a conversation about individuals who pay taxes, the domain is restricted to humans. However, in some cases the domain of discourse is unrestricted. This means that the variable can be a placeholder for anything. If I wrote, *Everything is crazy*, this proposition means, for all x, where x can be anything at all, x is crazy. This includes humans, animals, rocks, and numbers. Another example: suppose you were to say, *Everyone is crazy*, in an unrestricted domain. Here, it is implied that you are only referring to human beings. But since you are working in an unrestricted domain, it is necessary to specify this. Thus, *Everyone is crazy* is translated as for any x in the domain of discourse, if x is a human being, then x is crazy.

The domain places a constraint on the possible individuals we can substitute for individual variables. We will call these *substitution instances for variables*, or substitution instances for short. For example, discussing the mathematical equation '$x + 5 = 7$,' the domain consists of integers and not shoes, ships, or people. If someone were to say that a possible substitution instance for x is 'that patch of green over there,' we would find this extremely strange because the only objects being considered as substitution instances are integers and not patches of green. Likewise, if someone were to say, 'Everyone has to pay taxes,' and someone else responded, 'My cat does not pay taxes,' we would take this to be strange because the only objects being considered as substitution instances are people and not animals or numbers or patches of green. Thus, it is important to note that the domain places a limitation on what can be a instance of a variable.

With names, variables, and predicates having been introduced, we now turn to *quantifiers*. RL contains two quantifiers: the *universal quantifier* '∀,' which captures the English use of 'all,' 'every,' and 'any,' and the *existential quantifier* '∃,' which captures the English use of 'some,' 'at least one,' and the indefinite determiner 'a.' Consider the following sentence:

(6) Everyone is mortal.
(7) Someone is happy.

If we assume that the domain of discourse is people and let 'Mx' stand for *x is mortal* and 'Hx' stand for *x is happy*, then (6) can be read as expressing a number of possible expressions:

(6a) For every *x*, *x* is mortal.
(6b) All *x*'s are mortal.
(6c) For any *x*, *x* is mortal.
(6d) Every *x* is mortal.

   Or, equivalently,

(6*) (For every *x*)(*x* is mortal)

By replacing *for every x* with the universal quantifier and *x is mortal* with 'Mx,' we get the following predicate formula:

(6_{RL}) $(\forall x)Mx$

In the case of (7), (7) can be read as expressing a number of possible expressions:

(7a) For some *x*, *x* is happy.
(7b) Some *x*'s are happy.
(7c) For at least one *x*, *x* is happy.
(7d) There is an *x* that is happy.

   Or, equivalently,

(7*) (For at least one *x*)(*x* is happy)

By replacing *for at least one x* with the existential quantifier and *x is happy* with 'Hx,' we get the following predicate formula:

(7_{RL}) $(\exists x)Hx$

6.2.3 Parentheses and Scope of Quantifiers

In PL, parentheses serve the function of removing any ambiguity from complex expressions involving truth-functional operators. For example, without parentheses (or some other conventional way of determining the scope of various operators),

'A∨B∧C' is syntactically ambiguous and not a well-formed formula. In RL, parentheses have two functions: (1) to remove ambiguity from compound expressions involving operators (just like in propositional logic), and (2) to indicate the range or scope of the quantifier.

The '∀' and '∃' quantifiers operate over the propositional contents to the immediate right of the quantifier or over the complex propositional contents to the right of the parentheses. Consider the following four examples:

(1) (∃x)Fx
(2) ¬(∃x)(Fx∧Mx)
(3) ¬(∀x)Fx∧(∃y)Ry
(4) (∃x)(∀y)(Rx↔My)

In the case of (1), the scope of the quantifier ranges over 'Fx,' which is to the immediate right. In the case of (2), the scope of the quantifier is '(Fx∧Mx),' which is in the parentheses to the immediate right. In the case of (3), the scope of (∀x) operates upon 'Fx' while (∃y) operates on 'Ry.' Finally, (∃x) applies to '(∀y)(Rx ↔ My),' while (∀y) applies to '(Rx ↔ My).'

Scope plays an important role in how we translate from English into RL, and vice versa. Consider the following example:

Ix = *x* is intelligent, and Ax = *x* is an alien.

Now consider the following two propositions:

(5) (∃x)(Ix)∧(∃x)(Ax)
(6) (∃x)(Ix∧Ax)

It may appear that (5) and (6) say the same thing, but they express different propositions and are true and false under different conditions. Consider the following paraphrase of (5):

(5*) There exists an *x* that is intelligent, and there exists an *x* that is an alien.

In order for (5) to be true, it is not necessary that there exist something that is both intelligent and an alien. Proposition (5) will be true, for example, if a stupid alien exists and some intelligent dolphin exists. This is distinct from (6). Consider the following paraphrase of (6)

(6*) There exists an *x* that is intelligent and an alien.

In the case of proposition (6), (6*) shows that (6) is true if and only if (iff) there exists something that is both intelligent and an alien.

The difference between (5) and (6) is the result of a difference in the scope of the quantifier. In the case of (5), the main operator is '∧,' and two different propositional forms are existentially quantified. In the case of (5), there are two quantifiers. The first existential quantifier selects an *x* from the universe of discourse that is said to be intelligent. The second existential quantifier selects an *x* from the universe that is said to be an alien. In the case of (5), neither existential quantifier quantifies over the conjunction in 'Ix∧Ax.' This is distinct from the existential quantifier in (6). This existential quantifier operates not only on 'Ix' and 'Ax' but on the conjunction of 'Ix∧Ax.' That is, it selects an *x* from the universe of discourse that is both intelligent and an alien.

In each of the above cases, parentheses are used to determine what is within the scope of a quantifier. There is thus a parallel between how the scope of quantifiers is determined and how the scope of negation is determined. For example, the negation in '¬(P∧Q)' applies to the conjunction 'P∧Q,' while the negation in '¬P∧Q' applies only to the atomic 'P.' This is the same for quantifiers since (∀x) in '(∀x)(Px→Qx)' applies to the conditional 'Px→Qx,' while (∀x) in '(∀x)(Px)∧(∀y)(Qy)' applies only to 'Px.'

To illustrate the notion of scope even further, consider the following three examples:

(7) (∃x)(Px∧Gy)∧(∀y)(Py→Gy)
(8) (∃x)[(Mx→Gx)→(∃y)(Py)]
(9) (∀z)¬(Pz∧Qz)∧(∃x)¬(Px)

In the case of (7), 'Px∧Gy' fall within the scope of (∃x) while 'Py→Gy' falls within the scope of (∀y). This is indicated by the parentheses. In the case of (8), '[(Mx→Gx)→(∃y)(Py)]' falls within the scope of the leftmost quantifier (∃x), while (∃y) has 'Py' within its scope. Again, this is indicated by the parentheses. Finally, in the case of (9), '¬(Pz∧Oz)' is within the scope of (∀z) and '¬(Px)' is within the scope of (∃x).

Exercise Set #1

A. Identify the names and predicates in the following English sentences. Also, identify the adicity of any predicates you find.
   1. * John is tall.
   2.   John is shorter than Liz.
   3. * John is shorter than Liz but taller than Vic.
   4.   If Vic is standing between Liz and Vic, then Vic is not tall.
   5. * If Vic is standing between Sam and Mary, then Vic is tall.
   6.   Liz is taller than John and taller than Vic, except when she is standing next to Sam.
   7. * If Liz is standing next to Vic, Sam, Mary, and John, then Liz is tall.
   8.   All men are mortal.
   9. * Some men are not mortal.
   10. Everyone loves someone.

Solutions to Starred Exercises in Exercise Set #1

A.

   1. * *John* is a name, and *is tall* is a one-place predicate.
   3. * *John*, *Liz*, and *Vic* are names, ____ *is shorter than* ____ is a two-place
        predicate, and ____ *is taller than* ____ is a two-place predicate.
   5. * *Vic*, *Sam*, and *Mary* are names, ____ *standing between* ____ *and* ____ is a
        three-place predicate, and ____ *is tall* is a one-place predicate.
   7. * *Liz*, *Vic*, *Sam*, *Mary*, and *John* are names, ____ *is standing next to* ____ is
        a two-place predicate, and ____ *is tall* is a one-place predicate.
   9. * There are no names; ____ *is a man* and ____ *is a mortal* are one-place
        predicates.

6.3 THE SYNTAX OF RL

In this section, we distinguish between free and bound variables in RL, specify how
to go about finding the main operator of a well-formed formula (wff, pronounced
'woof') in RL, and finally provide the formal syntax of RL.

6.3.1 Free and Bound Variables

In the last section, we saw that in order for a formula (e.g., 'Fx') to be in the *scope* of
a quantifier, it must be circumscribed by parentheses that follow that quantifier. However, we allow for the exception that no parentheses are needed when their presence
would not remove any ambiguity concerning the scope of the quantifier (e.g., $(\forall x)$
Fx) where the universal quantifier has 'Fx' in its scope. Consider the following three
quantified expressions:

(1) $(\forall x)(Fx \rightarrow Bx) \lor Wx$
(2) $(\exists x)Mx \land (\exists y)Ry$
(3) $(\forall y)(Mx \land By)$

   In the case of (1), 'Fx' and 'Bx' are in the scope of the universal quantifier $(\forall x)$,
while 'Wx' is not in the scope. In the case of (2), 'Mx' is in the scope of $(\exists x)$ while
'Ry' is in the scope of $(\exists y)$. Finally, in the case of (3), 'Mx' and 'By' are both in the
scope of $(\forall y)$. In this section, we clarify the distinction between a *bound variable* and
a *free variable*.

Bound variable	When a variable is within the scope of a quantifier that quantifies that specific variable, then the variable is a bound variable.
Free variable	A free variable is a variable that is not a bound variable.

   Consider (3) again:

(3) $(\forall y)(Mx \land By)$

In asking ourselves whether *x* in 'Mx' is a free or bound variable, we need to ask whether 'Mx' is in the scope of a quantifier and whether that quantifier quantifies for *x*. So,

Is 'Mx' in the scope of a quantifier? Yes, it is in the scope of ($\forall$y).
Does ($\forall$y) quantify for *x*? No, it only quantifies for *y*.

Thus, since 'Mx' is not in the scope of the quantifier that quantifies for that specific variable, then the *x* in 'Mx' is not a bound variable, and if it is not a bound variable, then it is a free variable.

To consider this same distinction in a different way, a variable can be free (not bound) in two cases: (1) when it is not contained in the scope of any quantifier, or (2) when it is in the scope of a quantifier, but the quantifier does not specifically quantify for that specific variable. Consider the following example:

(6) [($\forall$x)(Px$\rightarrow$Gy)$\land$($\exists$z)(Pxy$\land$Wz)]$\lor$Rz

In the case of (6), the universal quantifier has 'Px$\rightarrow$Gy' in its scope, but only *x* can be considered a bound variable since ($\forall$x) only specifies the quantity for *x*. Likewise, while 'Pxy$\land$Wz'is within the scope of the existential quantifier, only the *z* is a bound variable since ($\exists$z) only specifies the quantity for *z*. However, the *z* in 'Rz' is a free variable since it does not fall within the scope of a quantifier.

6.3.2 Main Operator in Predicate Wffs

In a previous section, we learned that the scope of quantifiers behaves similarly to truth-functional negation in propositional logic. We now can define the main operator for well-formed predicate formulas: the main operator of a wff in RL is the operator with the greatest scope.

In a previous section, we learned how to identify which variables are within the scope of a given quantifier. In this section, we learn to how to identify the main operator for predicate wffs. Let us consider an example before articulating a list of guidelines for identifying the main operator of a predicate wff. Consider:

(1) ($\exists$x)(Px$\land$Qx)
(2) ($\exists$x)(Px)$\land$($\exists$x)(Qx)
(3) $\neg$($\exists$x)(Px$\land$Qx)
(4) ($\forall$y)($\exists$x)(Rx$\rightarrow$Py)

Each of the above three predicate wffs has a different main operator. In the case of (1), it is the existential quantifier ($\exists$x). The reason for this is that the '$\land$' is within the scope of the quantifier. Thus, ($\exists$x) is the operator with the greatest scope. In the case of (2), the main operator is the '$\land$.' The reason for this is that it does not fall within the scope of a quantifier, and it operates on two separate quantified expressions, that is,'($\exists$x)(Px)' and '($\exists$x)(Qx).' In the case of (3), the main operator is the negation ($\neg$).

The reason it is not the '∧' is that the '∧' falls within the scope of the existential quantifier, and the reason it is not the existential quantifier is that the operator for negation operates upon the existential quantifier. Finally, in the case of (4), the main operator is the universal quantifier. In that expression, while (∃x) quantifies over the propositional form '(Rx→Py),'(∀y) quantifies over '(∃x)(Rx→Py),' making it the quantifier with the greatest scope.

Consider a variety of examples:

Well-Formed Predicate Formula	Main Operator
(∀x)Px	∀x
(∀x)(Px)	∀x
¬(∀x)(Px)	¬
¬(∃y)(Py)	¬
(∀x)(Px→Qx)	∀x
¬(∀x)(Px→Qx)	¬
(∀x)(Px→¬Qx)	∀x
(∀x)(Px)→(∃x)(Qx)	→
¬(∀x)(Px)→(∃x)(Qx)	→
¬[(∀x)(Px)→(∃x)(Qx)]	¬
(∃x)(Px∧Qx)∧¬(∀y)(Py→Qy)	∧
(∃x)¬(Px∧Qx)	∃x
(∃x)(∀y)[(Px∧Qx)→Dy]	∃x
(∀x)¬(∀y)(Px→Qy)	∀x
¬(∀x)(∀y)(Px→Qy)	¬

6.3.3 The Formal Syntax of RL: Formation Rules

A syntactically correct proposition in RL is known as a well-formed formula. The rules that determine the grammatical and ungrammatical ways in which the elements of RL can be combined are known as *formation rules*. Using these rules, it is possible to construct any and every wff in RL. A wff in RL is defined as follows:

(i)   An *n*-place predicate 'P' followed by *n* terms (names or variables) is a wff.

(ii)   If 'P' is a wff in RL, then '¬P' is a wff.

(iii) If 'P' and 'Q' are wffs in RL, then 'P∧Q,' 'P∨Q,' 'P→Q,' and 'P↔Q' are wffs.

(iv)   If 'P' is a wff in RL containing a name 'a,' and if 'P(x/a)' is what results from substituting the variable *x* for every occurrence of 'a' in 'P,' then '(∀x)P(x/a)' and '(∃x)P(x/a)' are wffs, provided 'P(x/a)' is not a wff.

(v)   Nothing else is a wff in RL except that which can be formed by repeated applications of (i) to (iv).

Notice that (i) says 'followed by *n* terms' and that these terms are either names or variables. Thus, if 'P' is a three-place predicate, then 'Pabc' and 'Paaa' are wffs since they are formulas consisting of 'P' followed by three names. Similarly, 'Pxyz,' 'Pxxx,' and 'Pxyx' are wffs since they are formulas consisting of 'P' followed by three variables. In contrast, 'Pab' and 'Pa' are not wffs since they are formulas with 'P' but are not followed by three terms. It is helpful to distinguish between 'Pxyz' (a wff with at least one free variable) and 'Pabc' (a wff consisting only of names). Let's call a wff consisting of an *n*-place predicate 'P' followed by *n* terms where one of those terms is a free variable an *open sentence* or *open formula*. In contrast, let's call a wff consisting of an *n*-place predicate 'P' followed by *n* terms where no term is a free variable a *closed sentence* or *closed formula*.

Open formula	An open formula is a wff consisting of an *n*-place predicate 'P' followed by *n* terms, where one of those terms is a free variable.
Closed formula	A closed formula is a wff consisting of an *n*-place predicate 'P' followed by *n* terms, where every term is either a name or a bound variable.

Notice that (ii) says that if you have a wff '**P**,' then the negation of that wff is also a wff. Thus, if 'Pa,''¬Qab,' and '($\forall$x)Px' are all wffs, then '¬Pa,''¬¬Qab,' and '¬($\forall$x)Px' are also wffs. Notice that (iii) says that if you have two wffs '**P**' and '**Q**,' then placing a truth-functional operator between the two wffs forms a wff. Thus, if 'Pa,''¬Qab,'and '($\forall$x)Px' are all wffs, then 'Pa∧¬Qab,''¬Qab∨($\forall$x)Px,' 'Pa→($\forall$x) Px,' and '¬Qab↔Pa' are wffs (as are many others).

Perhaps the most complicated of the four formation rules is (iv), and so we will consider this rule in two parts. First, take a look at the antecedent of (iv):

If '**P**' is a wff in RL containing a name 'a,' and if '**P**(*x*/a)' is what results from substituting the variable *x* for every occurrence of 'a' in '**P**'

First, '**P**(*x*/a)' symbolizes the substitution (or replacement) of names for variables. For example, take the following wff:

(1) Pb

A replacement 'P(*x*/b)' for (1) is just the substitution of the variable *x* for every 'b' in (1). Thus,

Pb	⟶	Px
	P(*x*/b)	

Here is another example:

(2) Pbb

In the case of (2),'P(*z*/b)' results from substituting every 'b' with *z*. That is,

Pbb	──────────▶	Pzz
	P(*z*/b)	

Notice that this process of substitution alone does not generate a wff. For that we need to turn to the consequent clause of (iv):

then '($\forall$x)**P**(*x*/a)'and '($\exists$x)**P**(*x*/a)'are wffs, provided '**P**(*x*/a)'is not a wff.

The consequent of (iv) says that by putting a universal quantifier or existential quantifier in front of the formula that is the result of substituting a variable *x* for every name 'a,' the resulting formula is a wff. For example, using (1) from above, note that 'Pb' is a wff and contains a name 'b.' Second, take 'P(*x*/b),' which is the formula that results from substituting the variable *x* for every occurrence of 'b' in 'Pb.' This gives us 'Px.' The consequent clause of (iv) says that both of the following will be wffs:

$$(\forall x)Px$$
$$(\exists x)Px$$

Finally, (v) ensures that the only wffs allowed into RL are those that are the result of using (i) to (iv).

To illustrate the use of these rules, we will consider three examples. Consider the following formulas in RL:

(3) Pab$\wedge$Ra
(4) $\neg$Qa$\rightarrow$($\forall$x)Rx
(5) ($\forall$x)Pxx$\rightarrow$$\neg$($\exists$y)Gy

Assume that 'Pxy' is a two-place predicate, and all other predicates are one-place. First, we start by showing that (3) is a wff.

1	'Pab' and 'Ra' are wffs.	Rule i
2	If 'Pab' and 'Ra' are wffs, then 'Pab$\wedge$Ra' is a wff.	Line 1 + rule iii

Notice that 'Pab' and 'Ra' are both wffs by rule (i). Both have the requisite number of names after them. Similar to propositional logic, rule (iii) justifies 'Pab$\wedge$Ra' as a wff in RL.

Moving to a more complex example, consider (4):

1	'$\neg$Qa' is a wff.	Rule i
2	If 'Ra' is a wff, and 'R(*x*/a)' is what results from substituting *x* for every occurrence of 'a,' then '($\forall$x)Rx' is a wff.	Rule iv
3	'Ra' is a wff.	Rule i

4	'(∀x)Rx' is a wff.	Lines 2, 3
5	If '¬Qa' and '(∀x)Rx' are wffs, then '¬Qa→(∀x)Rx' is a wff	Lines 1, 4 + rule iii
6	'¬Qa→(∀x)Rx' is a wff.	Lines 1, 4, 5

Using the formation rules, the above shows that '¬Qa→(∀x)Rx' is a wff in RL. Perhaps key to showing this is the use of rule (iv) at line 2. Line 2 states that if 'Ra' is a wff and we can substitute the variable x for every 'a' in 'Ra,' then the resulting universally quantified proposition is a wff. Since 'Ra' is clearly a wff, it follows that '(∀x)Rx' is a wff.

Finally, consider the initial use of formation rules to show that (5) is a wff:

1	'Paa' and 'Ga' are wffs.	Rule i
2	If 'Paa' is a wff, and 'P(x/a)' is what results from substituting x for every occurrence of 'a,' then '(∀x)Pxx' is a wff.	Rule iv
3	'(∀x)Pxx' is a wff.	Lines 1, 2
4	If 'Ga' is a wff, and 'G(y/a)' is what results from substituting y for every occurrence of 'a,' then '(∃y)Gy' is a wff.	Rule iv
5	'(∃y)Gy' is a wff.	Lines 1,5
6	If '(∀x)Pxx' and '(∃y)Gy' are wffs, then '(∀x)Pxx→¬(∃y)Gy' is a wff.	Rule iii
7	'(∀x)Pxx→¬(∃y)Gy' is a wff.	Lines 3, 6, 7

Exercise Set #2

A   Identify the main operator in the following predicate wffs:
   1. *   (∀x)(Px∨Qx)
   2.   (∃y)(Py)∧(∃z)(Pz)
   3. *   ¬(∃y)(Py)∧¬(∃z)(Pz)
   4.   ¬(∃y)(Py)∨¬(∃z)(Pz)
   5. *   ¬(∃y)[Py∧(∀z)(Pz)]
   6.   (∀x)[(Px→(Qx∧Mx)]
   7. *   [(∃x)(Px)∧(∃y)(Py)]∨∀z(Qz)
   8.   ¬(∀x)[(Pb→Qb)↔Px]
   9. *   (∀x)(∃y)¬(∀z)(Px→Qyz)
   10.   (∀x)(∀y)(∀z)(Pxyz∧Rxyz)→(∃x)Px

B. Identify bound variables, quantifiers, scoped variables, free variables, and names. Also determine whether a given proposition has constant truth value.
   1. *   (∃x)(Rx→Ga)
   2.   (∀y)(Mx→Py)∨Ga
   3. *   (∃x)(Mx)∨(∀x)(Rx)
   4.   Rz∧(∀z)[(Rz∧My)→ Qz]
   5. *   Pa∧(∃w)(Vw∧Lx)
   6.   (∀x)[Fx→(∃z)(Mz)]

7. * ¬(∃x)(Fx)∧(∃x)(Fx)
8.   (∃x)(¬Fx)∧(∃x)(Fx)
9.   (Tb∧Qa)→(∀x)(Fx→Gy)
10.   Tb∧¬Tb
11.   (Ta∧Ra)∨¬(Ta∧Ra)
12.   Rx→(∀x)(Px)

C. Using the formation rules, show that the following propositions are wffs in RL, where 'Pxy' is a two-place predicate and 'Rx' and 'Zx' are one-place predicates:

1. * Ra∧Paa
2.   Ra→Paa
3. * (∀x)Pxx
4.   (∃x)Pxx
5. * ¬(∃y)Pyy
6.   ¬(∀x)Pxx∧(∃x)Zx
7. * (∃x)(∀y)Pxy
8.   (¬Pab∧Rb)→Ra
9.   (∃x)Pxx↔¬(∀z)Pzz
10.   (∃x)(∀y)(Pxx∧Zy)
11.   (∃x)Pxy→(∀y)Ry
12.   ¬¬(∀x)Pxx

Solutions to Starred Exercises in Exercise Set #2

A.

1. * ∀
3. * ∧
5. * ¬
7. * ∨
9. * (∀x)

B.

1. * (∃x)(Rx → Ga)

   There are no free variables. The 'a' in 'Ga' is a name. The *x* in 'Rx' is a bound variable and falls in the scope of (∃x). The proposition has a constant truth value.

3. * (∃x)(Mx)∨(∀x)(Rx)

   There are no free variables and no names. The *x*'s in 'Rx' and 'Mx' are both bound variables. The *x* in 'Mx' falls within the scope of (∃x), and the *x* in 'Rx' falls within the scope of (∀x). The proposition has a constant truth value.

5. * Pa∧(∃w)(Vw∧Lx)

   The *x* in 'Lx' is free, and the 'a' in 'Pa' is a name. The *w* in 'Vw' is within the scope of (∃w) and is bound by 'Vw.' The proposition does not have a constant truth value.

7. * ¬(∃x)(Fx)∧(∃x)(Fx)

There are no free variables and no names. The *x*'s in both the left and right 'Fx' are bound and scoped. The proposition has a constant truth value.

C.

1. * 'Ra∧Paa' is a wff. *Proof:* 'Ra' is a one-place predicate followed by one name (rule i). 'Paa' is a two-place predicate followed by two names (rule i). If 'Ra' and 'Paa' are wffs, then 'Ra∧Paa' is a wff (rule iii).

3. * (∀x)Pxx. *Proof:* If 'Paa' is a wff in RL containing a name 'a,' and 'Pxx' is what results from substituting *x* for every occurrence of 'a' in 'Paa,' then '(∀x)Pxx' is a wff (rule iv).

5. * ¬(∃y)Pyy. *Proof:* If 'Paa' is a wff in RL containing a name 'a,' and 'Pyy' is what results from substituting *y* for every occurrence of 'a' in 'Paa,' then '(∃y)Pyy' is a wff (rule iv). If '(∃y)Pyy' is a wff, then '¬(∃y)Pyy' is a wff (rule ii).

7. * (∃x)(∀y)Pxy. *Proof:* If 'Pab' is a wff in RL containing a name 'b,' and 'Pay' is what results from substituting *y* for every occurrence of 'b' in 'Pab,' then '(∀y)Pay' is a wff (rule iv). If '(∀y)Pay' is a wff in RL containing a name 'a,' and 'Pxy' is what results from substituting *x* for every occurrence of 'a' in 'Pay,' then '(∃x)(∀y)Pxy' is a wff (rule iv).

6.4 PREDICATE SEMANTICS

The following provides the semantics of RL. This is done by (1) explaining the nature of a set and set membership, then (2) articulating what it means to say that a formula is true in RL.

6.4.1 A Little Set Theory

In this section, the very basics of set theory are introduced. Naively speaking, a *set* is a collection of objects considered without concern for order or repetition. The objects that constitute a set are known as its *members* or *elements*. Consider a set 'M' consisting of odd integers between 1 and 5. Such a set consists of three members. That is, 1 is in 'M,' 3 is in 'M,' and 5 is in 'M.' It is often convenient to put the names of the members of a set between braces. For example, M = {1, 3, 5}. In addition, we can indicate that certain elements are members or are not members of a set. That is, some of integers belong to the set while others do not belong. To indicate membership in the set, the following symbols are used: '∈' (membership) and '∉' (nonmembership). So, in the case of 'M,' '1∈M,' '3∈M,' '5∈M,' but '2∉M' and '8∉M.'

The members of a set are considered without regard to ordering or repetition. Thus, consider that a single set 'T' can be represented as having three members: Mary, John, and Sally. That is, 'T' can be represented in the following ways:

$$T = \{Mary, John, Sally\}$$
$$T = \{Mary, Sally, John\}$$

$$T = \{Sally, Mary, John\}$$
$$T = \{Sally, John, Mary, Mary, Mary, John\}$$

The above symbolizations of sets are identical.

Some sets are too big to list all of the objects by naming them. Infinite or large sets can be specified by predication. For example, suppose that a set consists of all politicians. Such a set could be represented as follows:

$$P = \{x \mid x \text{ is a politician}\}.$$

Here is another example. Consider the set of positive even integers. Such a set is infinite, and so we cannot simply list all of the integers (since there are an infinite number). To represent this set, we can simply write,

$$E = \{x \mid x \text{ is a positive even integer}\}.$$

6.4.2 Predicate Semantics

In PL, where the basic unit of representation is the proposition, we straightforwardly interpret the proposition relative to the world; as such, the sentences are interpreted as being simply true or false. In predicate logic, our basic units of representation are parts of sentences, and these parts cannot simply be assigned a truth value since names like *John* and predicates like *is red* are not things that can be true or false. While the goal of predicate logic is, much like in propositional logic, to assign sentences truth values, the manner in which this is done is more complicated since it takes into consideration the contribution the subsentential parts make to the truth or falsity of a sentence.

The truth or falsity of a wff in RL is determined relative to a model. A model is a two-part structure. There is the part that stipulates a domain, and there is the part that interprets the RL wff relative to the domain. In other words, a *model* consists of a *domain* (D) and an *interpretation function* (I).

Model   A model in RL is a structure consisting of a domain and an interpretation function.

An interpretation function assigns (1) objects in D to names, (2) a set of n-tuples in D to n-place predicates, and (3) truth values to closed formula.

Interpretation-function   An interpretation-function is an assignment of (1) objects in D to names, (2) a set of n-tuples in D to n-place predicates, and (3) truth values to sentences.

The notions of an interpretation and a domain of discourse both need further elaboration. An interpretation of a language requires the stipulation (or selection) of a domain of discourse. The domain of discourse consists of all of the things that a language can meaningfully refer to or talk about.

Domain   The domain of discourse (D) consists of all of the things that a language can meaningfully refer to or talk about.

We assume that domains are never empty. There is always at least one thing that we can meaningfully talk about. A specification of D is achieved in one of two ways. First, D can be specified simply by listing the individual objects in the domain. For example, imagine a domain consisting of three people: John, Sally, and Mary. The domain consisting of these three people can be represented as follows:

$$D = \{John, Sally, Mary\}$$

Second, another way of specifying the domain is by stating a class of objects. For example, suppose that the domain consists of living human beings. It would be extremely cumbersome to list all of them by name. Instead, the domain consisting of living human beings can be represented as follows:

$$D = \{x \mid x \text{ is a living human being}\}.$$

Next, we turn to the interpretation of the names (object constants) in RL. In the case of names, *I* assigns each name an object in D. This assignment of an element in D to a name determines the meaning of the name by giving its *extension*. That is, it determines the meaning of a name by assigning it an element in D. In RL, if a name lacks an interpretation, then it is known as a *nonreferring* (or *uninterpreted*) *term*.

Let's consider this more concretely with an example. Imagine the following D:

$$D = \{Alfred, Bill, Corinne\}$$

Now consider the following set names in RL: 'a,' 'b,' 'c.' In order for these names to mean anything, they must be interpreted. Interpreting names consists of assigning them an object in D. Thus, we might write,

> *I* (a) Alfred
> *I* (b) Bill
> *I* (c) Corinne

That is, Alfred in D is assigned to 'a,' Bill in D is assigned to 'b,' and Corinne in D is assigned to 'c.'

Not only does an interpretation function assign objects in D to names (individual constants), but it also assigns *n*-tuples to *n*-place predicates. To get a clearer understanding of what an *n*-tuple is, consider the following *n*-place predicates:

> Sx: *x* is short
> Bx: *x* is bad

Notice that both of the above predicates are one-place predicates. One way we might go about interpreting *n*-place predicates is by assigning each *n*-place predicate to a set (or collection) of objects in D. Thus,

I (Sx):   objects in D that are short (i.e.,{Alfred, Bill})
I (Bx):   objects in D that are bad (i.e.,{Corinne, Alfred})

However, consider the following two-place predicate:

Lxy: *x* loves *y*

Here the interpretation function does not simply tell us about a collection of single objects (i.e., about which objects are loved or which objects do the loving). Instead, an interpretation of this two-place 'Lxy' tells us something about pairs of objects (i.e., about which objects love which objects). It tells us who loves whom. Thus, rather than saying an interpretation of an *n*-place predicate tells us something about a collection of objects, we say that it tells us something about a set (or collection) of *n-tuples*, where an *n*-tuple is a sequence of *n*-objects. In the above case, an interpretation of 'Lxy' is an assignment of a set of 2-tuples to 'Lxy.'

For convenience, we will write tuples by listing the elements (objects) within angle brackets ($<>$) and separate elements with commas. For example,

<Alfred, Corinne>

denotes a 2-tuple, while

<Alfred, Corinne, Bill>

denotes a 3-tuple.

Thus, suppose that in D, Bill loves Corinne (and no one else), Corinne loves Alfred (and no one else), and Alfred loves no one. The interpretation of 'Lxy' in D would be represented by the following set of 2-tuples:

I (Lxy): {<Bill, Corinne>, <Corinne, Alfred>}

This says that the interpretation of the predicate *x loves y* relative to D consists of a set with two 2-tuples:one 2-tuple is <Bill, Corinne> and the other is <Corinne, Alfred>.

Finally, now that we have an understanding of how the interpretation function assigns objects to names and *n*-tuples to *n*-place predicates, it is possible to define when a wff is true in RL. We will call an interpretation where a truth value is assigned to a wff a *valuation* (v). For this, let '**R**' be an *n*-place predicate, let 'a_1,'. . .'a_n' be a finite set of names in RL, and let 'a_i' be a randomly selected name.

(1) $v(\mathbf{R}a_i) = $ T if and only if the interpretation of 'a_i' is in '**R**.'

This says that we can assign a value of true to 'Ra_i' if and only if our interpretation of 'a_i' is in the interpretation of 'R.' To consider this concretely, we examine two examples. First, consider the following wff:

Sa

Let's say that 'Sa' is the predicate logic translation of *Alfred is short*. According to (1), 'Sa' is true if and only if 'a' is in 'S,' that is, if and only if an interpretation of the predicate 'short' includes an interpretation of the name *Alfred*. Earlier, we said that

I (Sx): objects in D that are short (i.e., {Alfred, Bill})

And so, 'Sa' is true in the model since Alfred belongs to the collection of objects that are short. Second, consider the following more complex wff:

Lca

'Lca' is a predicate logic translation of 'Corinne loves Alfred.' According to (1), 'Lca' is true if and only if the predicate 'loves' includes the ordered pair <Corinne, Alfred>. Since it does, the interpretation function assigns a value of true to 'Lca.' That is,

$$v(\text{Lca}) = \text{T}$$

The interpretation of wffs that involve truth-functional operators as their main operators is the same as in propositional logic and straightforward given that we know the truth values of their components. Given that '**P**' and '**Q**' in (2) to (6) are well-formed formulas, then relative to a model,

2   $v(\neg \mathbf{P}) = \text{T}$ iff $v(\mathbf{P}) = \text{F}$
    $v(\neg \mathbf{P}) = \text{F}$ iff $v(\mathbf{P}) = \text{T}$
3   $v(\mathbf{P} \wedge \mathbf{Q}) = \text{T}$ iff $v(\mathbf{P}) = \text{T}$ and $v(\mathbf{Q}) = \text{T}$
    $v(\mathbf{P} \wedge \mathbf{Q}) = \text{F}$ iff $v(\mathbf{P}) = \text{F}$ or $v(\mathbf{Q}) = \text{F}$
4   $v(\mathbf{P} \vee \mathbf{Q}) = \text{T}$ iff either $v(\mathbf{P}) = \text{T}$ or $v(\mathbf{Q}) = \text{T}$
    $v(\mathbf{P} \vee \mathbf{Q}) = \text{F}$ iff $v(\mathbf{P}) = \text{F}$ and $v(\mathbf{Q}) = \text{F}$
5   $v(\mathbf{P} \rightarrow \mathbf{Q}) = \text{T}$ iff either $v(\mathbf{P}) = \text{F}$ or $v(\mathbf{Q}) = \text{T}$
    $v(\mathbf{P} \rightarrow \mathbf{Q}) = \text{F}$ iff $v(\mathbf{P}) = \text{T}$ and $v(\mathbf{Q}) = \text{F}$
6   $v(\mathbf{P} \leftrightarrow \mathbf{Q}) = \text{T}$ iff either $v(\mathbf{P}) = \text{T}$ and $v(\mathbf{Q}) = \text{T}$ or $v(\mathbf{P}) = \text{F}$ and $v(\mathbf{Q}) = \text{F}$
    $v(\mathbf{P} \leftrightarrow \mathbf{Q}) = \text{F}$ iff either $v(\mathbf{P}) = \text{T}$ and $v(\mathbf{Q}) = \text{F}$ or $v(\mathbf{P}) = \text{F}$ and $v(\mathbf{Q}) = \text{T}$

Let's take stock. Assigning truth values using a model has involved (1) the selection of a domain, (2) an interpretation function that assigns objects in the domain to names, (3) an interpretation function that assigns a set of n-tuples to n-place predicates, and (4) an interpretation function (valuation) that assigns the value true to an atomic sentence if and only if the object or tuple that the name designate is in the set of n-tuples designated by the n-place predicate. Missing from the above is a valuation of wffs of the form '$(\forall x)\mathbf{P}$' and '$(\exists x)\mathbf{Q}$.' In what follows, we assume that formulas of the form '$(\forall x)\mathbf{P}$' and '$(\exists x)\mathbf{Q}$' are closed; that is, they do not contain any free variables, and x and only x occurs free in '**P**' and '**Q**.' In what follows, we will define truth values of quantified formulas by relying on the truth values of simpler values of nonquantified formula. This method requires a little care for, at least initially, we might say that '$(\exists x)$ Px' is true if and only if 'Px' is true, given some replacement of x with an object or a name (object constant) is true. Likewise, a wff like '$(\forall x)$Px' is true if and only if 'Px' is true, given that every replacement of *x* with an object or name yields a true proposition. However, this will not work without some further elaboration. On the one hand, we cannot replace variables with objects from the domain since variables are linguistic items, and replacing a variable with an object won't yield a wff, or even a proposition. On the other hand, we cannot replace variables with names from our logical vocabulary because this falsely assumes that we have a name for every object in the domain. It might be the case that some objects in the domain are unnamed.

The solution to this problem is not simply to expand our logical vocabulary so that there is a name for every object but to consider the multitude of different ways in which a single name can be interpreted relative to the domain, that is, to consider the many different ways that an object in the domain can be assigned to a name.

To see this more clearly, consider the following domain:

$$D: \{John, Vic, Liz\}$$

Now let's consider the following interpretation *I* of 'a' relative to D:

$$I\,(a): John$$

This is a perfectly legitimate interpretation, but we might think of a variant interpretation of *I*, such as

$$I_1(a): Vic$$

Further, we might even think of another variant interpretation of *I*, such as

$$I_2(a): Liz$$

Let's say that for any name '**a**,' an interpretation 'I_a' is *a-variant* or *a-varies* if and only if 'I_a' interprets '**a**' (i.e., it assigns it an object in D) and either does not differ from *I* or differs only in the interpretation it assigns to '**a**' (i.e., it doesn't differ on the interpretation of any other feature of RL). Thus, *I*, I_1, and I_2 are all **a**-variant interpretations of *I* since they all assign '**a**' to an object in the domain and either do not differ from *I* or differ only in the interpretation they assign to '**a**.'

Using the notion of a variant interpretation, we can define what it means for a quantified formula to be true or false.

7    $v(\forall x)\mathbf{P} = T$ iff for every name '**a**' not in '**P**' and every a-variant interpretation '$\mathbf{P}(a/x) = T$.'
     $v(\forall x)\mathbf{P} = F$ iff for at least one '**a**' not in '**P**' and at least one a-variant interpretation '$\mathbf{P}(a/x) = F$.'

'8 $v(\exists x)\mathbf{P} = T$ iff for at least one name '**a**' not in '**P**' and at least one a-variant interpretation '$\mathbf{P}(a/x) = T$'.
     $v(\exists x)\mathbf{P} = F$ iff for every name '**a**' not in '**P**' and every a-variant interpretation '$\mathbf{P}(a/x) = F$.'

Using the notion of a variable interpretation, we have a solution to the problem of defining quantified formulas by replacing variables with names from our logical vocabulary since our solution does not falsely assume that we have a name for every object in the domain. Instead, it assumes that there are many ways in which a name can be interpreted, and so while there may not always be a name for every object in the domain, there is always at least one variant interpretation that assigns a name to the previously unnamed object. Using this notion, a universally quantified proposition '(∀x)**P**' is true if and only if for every name '**a**' not in '**P**' and every a-variant interpretation, it is the case that '**P**(a/x)' is true. In other words, it is true for every formula that is the result of replacing *x* with '**a**'; for instance,'(∀x)Px' is true if and only if 'Pa,' 'Pb,' 'Pc,' and so on are true. Likewise, an existentially quantified proposition '(∃x)**P**' is true if and only if for at least one name '**a**' not in '**P**' and at least one a-variant interpretation, it is the case that '**P**(a/x)'is true (the formula that is the result of replacing *x* with '**a**').

Exercise Set #3

A. Let D = {a, b, c}, *I* (a) = a, *I* (b) = b, *I* (c) = c, *I* (H) = {<a>, <b>, <c>}, *I* (L) = {<a,b>, <c,b>}. Using this interpretation, determine the truth values of the following wffs:
1. *   Lab
2.   Lba
3. *   Lac
4.   Lab∧Lcb
5. *   Lba∧Lbc
6.   (∃x)(∃y)Lxy
7.   (∀x)Hx
8.   (∀x)Hx→(∀x)Lxx
9.   (∃x)Lxx

B. Let D = {a, b, c, d}, *I* (a) = a, *I* (b) = b, *I* (c) = c, *I* (d) = d,*I* (H) = {<a>, <b>, <c>}, *I* (L) = {<a,a>, <a,b>, <b,b>, <c,c>}. Using this interpretation, determine the truth values of the following wffs:
1. *   (∀x)Lxx
2.   (∀x)(∀y)Lxy
3. *   (∃x)¬Lxx
4.   (∃x)Lxx→Lca
5. *   (∀x)Hx∧(∃x)Lxx
6.   (∃x)(∃y)Lxy
7.   (∀x)Hx
8.   (∀x)Hx→(∀x)Lxx
9.   (∃x)Lxx

Solutions to Starred Exercises in Exercise Set #3

A.
     1. * $v(Lab) = T$
     3. * $v(Lac) = F$
     5. * $v(Lba \land Lbc) = F$

B.
     1. * $v(\forall x)Lxx = F$
     3. * $v(\exists x)\neg Lxx = F$
     5. * $v((\forall x)Hx \land (\exists x)Lxx) = F$

6.5 TRANSLATION FROM ENGLISH TO PREDICATE LOGIC

In this section, we explore various facets of translating from English into predicate logic and vice versa. This section is designed to give you a basic understanding of how to translate from natural language into RL.

6.5.1 Translation Keys

The first step to any translation is to construct a *translation key*. A translation key does three things: (1) it stipulates the domain of discourse, (2) it symbolizes all singular terms, and (3) it symbolizes all *n*-place predicates. Consider the following key:

> D: living human beings
> j: John
> f: Frank
> Txy: x is taller than y

    The purpose of a translation key is to allow us to translate English sentences into the language of predicate logic and predicate logic expressions into English sentences. Consider a second translation key:

> D: human beings (alive or undead)
> j: John
> f: Frank the zombie
> Txy: x is taller than y
> Fxy: x wants to eats the brains of y
> Hx: x is hungry
> Lx: x is living

    Using the above translation key, we can translate the following sentences:

(1) John is taller than Frank the zombie.
(2) Frank the zombie wants to eat the brains of John.
(3) (3) Frank the zombie is hungry, and John is living.

(1) to (3) can be translated using the key and the conventions for translating propositions involving predicates and singular terms:

(1*) Tjf
(2*) Ffj
(3*) Hf∧Lj

It is not immediately obvious how to translate the following expressions:

(4) Someone is taller than someone.
(5) Everyone is hungry, and someone is living.
(6) John is taller than everyone.

Our next step is to develop techniques and conventions for using a translation key to translate expressions with quantifiers.

6.5.2 The Universal Quantifier (∀)

Rather than examining a number of English sentences and trying to determine their predicate logic translation, it is easier to start by examining a number of predicate logic wffs, to develop what is known as a *bridge translation*, and then to use this bridge translation to translate into more colloquial English.

To begin, consider the following translation key:

$$
\begin{aligned}
&\text{D:} && \text{human beings (living or dead)} \\
&\text{H}x: && x \text{ is happy} \\
&\text{Z}x: && x \text{ is a zombie} \\
&\text{M}x: && x \text{ is mortal} \\
&\text{R}x: && x \text{ is a murderer} \\
&\text{W}x: && x \text{ is wrong}
\end{aligned}
$$

Now consider the following predicate wffs:

(1) (∀x)Hx
(2) (∀x)¬Zx
(3) (∀x)(Zx→Hx)
(4) (∀x)(Zx→¬Hx)
(5) ¬(∀x)(Zx→Hx)

Let's consider a translation of (1) by taking one part of the formula at a time.

(∀x) is translated as *for every x*, *for all x's*, or *for each x*. The second part of (1) says that *x is H* or *x is happy*. Putting these two parts together, we get a bridge translation. A bridge translation is not quite English and not quite predicate logic. Here is a bridge translation of (1):

(1$_B$) For every *x*, *x* is happy.

Using this bridge translation, we can more easily translate (1) into colloquial English:

(1$_E$) Everyone is happy.

Consider a bridge translation of (2):

(2$_B$) For every x, x is not a zombie.

Using (2$_B$) we can render (2) into something more natural:

(2$_E$) Everyone is a not a zombie.

Consider a bridge translation of (3):

(3$_B$) For every x, if x is a zombie, then x is happy.

It may not be immediately obvious how to render (3$_B$) into English. If it isn't, then you can try to make (3$_B$) more concrete by expanding the bridge translation as follows:

(3$_B$*) Choose any object you please in the domain of discourse; if that object is a zombie, then it will be also be happy.

Rendered into standard English, (3$_B$) and (3$_{B*}$) read,

(3$_E$) Every zombie is happy.

Consider a bridge translation of (4):

(4$_B$) For every x, if x is a zombie, then x is not happy.

An additional bridge translation is the following:

(4$_B$*) Choose any object you please in the domain of discourse consisting of human beings (living or dead); if that object is a zombie, then it will not be happy.

In colloquial English, this is the following:

(4$_E$) No zombies are happy.

Notice that in the case of (5), which is '$\neg(\forall x)(Zx \to Hx)$,' the main operator is negation. One way to translate this is by translating '$(\forall x)(Zx \to Hx)$' as follows:

Every zombie is happy.

Next, translate the negation into English by putting 'not' in front of this expression. That is, (5) reads,

(5$_E$) Not every zombie is happy.

Finally, consider universally quantified expressions not involving '→' as the main operator

(6) $(\forall x)(Zx \land Hx)$
(7) $(\forall x)(Zx \lor Hx)$
(8) $(\forall x)(Zx \leftrightarrow Hx)$

In translating (6), it is helpful first to use a bridge translation:

(6$_B$) For all x in the universe of discourse, x is a zombie and happy.

In other words,

(6$_E$) Everyone is a happy zombie.

(7) and (8) are, respectively,

(7$_E$) Everyone is either a zombie or happy.
(8$_E$) Everyone is a zombie if and only if he or she is happy.

6.5.3 The Existential Quantifier (∃)

In this section, we consider translations involving the existential quantifier (∃). To begin, consider the following translation key:

> D: human beings (living or dead)
> Hx: x is happy
> Zx: x is a zombie
> Mx: x is mortal
> Rx: x is a murderer
> Wx: *x is wrong*

Now consider the following predicate wffs:

(1) $(\exists x)Hx$
(2) $(\exists x)\neg Zx$
(3) $\neg(\exists x)Zx$
(4) $(\exists x)(Zx \land Hx)$
(5) $(\exists x)Zx \land (\exists x)Hx$

Let's consider a translation of (1) by taking one part of the formula at a time.

(∃x) is translated as *for some x*, *there exists an x*, or *there is at least one x*. The second part of (1) says that *x is H* or *x is happy*. Putting these two parts together, we get a bridge translation. Again, a bridge translation is not quite English and not quite predicate logic. Here is a bridge translation of (1):

(1$_B$) For some *x*, *x* is happy.

(1) says that there is at least one object in D that has the property of being happy. Using the bridge translation (1$_B$), we can more easily translate (1) into colloquial English:

(1$_E$) Someone is happy.

Consider (2). Again, we can use a bridge translation.

(2$_B$) For some *x*, *x* is not a zombie.

(2$_B$) can be translated into colloquial English as follows:

(2$_E$) Someone is not a zombie.

In the case of (3), note that the negation has wide scope. Thus, we can translate '(∃x)Zx' first and then translate '¬(∃x)Zx.' That is, '(∃x)Zx' translates into *Someone is a zombie*, and '¬(∃x)Zx' translates as

(3$_E$) It is not the case that someone is a zombie.

Notice that (2) and (3) say something distinct. (2) says that something exists that is not a zombie, while (3) says that zombies do not exist. Let's consider (4) and (5) together. The bridge translations for (4) and (5) are as follows:

(4$_B$) For some *x*, *x* is a zombie, and *x* is happy.
(5$_B$) For some *x*, *x* is a zombie, and for some *x*, *x* is happy.

Notice that these two propositions do not say the same thing. (4) asserts that there is something that is both a zombie and happy, while (5) asserts that there is a zombie, and there is someone who is happy.

Finally, consider some propositions where '∧' is not the main operator.

(6) (∃x)(Zx→Hx)
(7) (∃x)(Zx∨Hx)
(8) (∃x)(Zx↔Hx)

The bridge translations for these are

(6_B) For some x, if x is a zombie, then x is happy.
(7_B) For some x, x is a zombie, or x is happy.
(8_B) For some x, x is a zombie if and only if x is happy.

These can be translated into the following English expressions:

(6_B) There exists something such that if it is a zombie, then it is happy.
(7_B) There exists something that is either a zombie or happy.
(8_B) There exists something that is a zombie if and only if it is happy.

6.5.4 Translation Walk Through

This section provides a walkthrough of a translation of the following:

(1) Some rich people are not miserly, and some miserly people are not rich.

The first step is to create the translation key:

$$D:\ \text{unrestricted}$$
$$Rx:\ x \text{ is rich}$$
$$Px:\ x \text{ is people}$$
$$Mx:\ x \text{ is miserly}$$

The second step is to determine the main operator of the sentence. In this case, the proposition is a complex conjunction.

(1*) [Proposition]$\land$[Proposition]

The third step is to determine the subject of each of the constituent propositions. The proposition to the left of the conjunct is about people, and the proposition to the right is about people.

(1**) $Px \land Px$

Next determine what is said about the subject. In the left conjunct, the proposition says that people who are rich are not miserly. In the right conjunct, the proposition says that people who are miserly are not rich.

(1***) $[Px \land Rx \land \neg Mx] \land [Px \land Mx \land \neg Rx]$

Finally, for each proposition, determine the scope of the quantifier.

(1****) $(\exists x)[(Px \land Rx) \land \neg Mx] \land (\exists x)[(Px \land Mx) \land \neg Rx]$

6.5.5 Sample Translations

Universal Quantifier	
Not everything is moveable.	$\neg(\forall x)Mx$
Everything is movable.	$(\forall x)Mx$
Nothing is moveable.	$(\forall x)\neg Mx$
Everything is immoveable.	$(\forall x)\neg Mx$
It is not true that everything is immoveable.	$\neg(\forall x)\neg Mx$
Honey tastes sweet.	$(\forall x)(Hx \rightarrow Tx)$
If something is honey, then it tastes sweet.	$(\forall x)(Hx \rightarrow Tx)$
Everything is either sweet or gross.	$(\forall x)(Sx \vee Gx)$
Either everything is sweet, or else everything is gross.	$(\forall x)(Sx) \vee (\forall x)(Gx)$
Existential Quantifier	
Some people are living.	$(\exists x)(Px \wedge Lx)$
Some people are not living.	$(\exists x)(Px \wedge \neg Lx)$
Some living people are mistreated.	$(\exists x)[(Px \wedge Lx) \wedge Mx]$
Some dead people are mistreated.	$(\exists x)[(Px \wedge \neg Lx) \wedge Mx]$
Some people are liars and thieves.	$(\exists x)[Px \wedge (Lx \wedge Tx)]$
It is not true that some people are honest.	$\neg(\exists x)(Px \wedge Hx)$
Some people are neither honest nor truthful.	$(\exists x)[Px \wedge \neg(Hx \vee Tx)]$
Some people are liars, and some are thieves.	$(\exists x)(Px \wedge Lx) \wedge (\exists x)(Px \wedge Tx)$
Some thieving liars are caught, and some are not.	$(\exists x)[(Tx \wedge Lx) \wedge Cx] \wedge (\exists x)$ $[(Tx \wedge Lx) \wedge \neg Cx]$

Exercise Set #4

A. Using the following translation key, translate the predicate logic expressions be-
    low into English: D: living humans, Hxy: *x* hates *y*, s: Sally, b: Bob, Lxy: *x* loves *y*
    1. * $(\forall x)Lxb$
    2. $(\exists x)Hxs$
    3. * $(\forall x)(Lxb \rightarrow \neg Hxs)$
    4. $(\exists x)(Lxb \wedge Hxs)$
    5. * $[(\exists x)(Lbs \wedge Hbx)] \rightarrow Lbs$
    6. $(\forall x)Lxx \wedge (\exists y)Hyb$
    7. * $(\exists x)Lxx \wedge (\forall y)Hyy$
    8. $[(\exists x)Lxb \wedge (\exists x)Lxs] \wedge [(\exists x)\neg Lxb \wedge (\exists x)\neg Lxs]$
    9. * $(\exists x)Lxb \wedge (\exists x)Lbx$
    10. $[(\exists x)(\neg Hxs) \wedge (\exists x)(Lxs)] \wedge (\forall x)(Lxb)$

Solutions to Starred Exercises in Exercise Set #4

1. * (∀x)Lxb; everyone loves Bob.
3. * (∀x)(Lxb→¬Hxs); everyone who loves Bob does not hate Sally.
5. * [(∃x)(Lbs∧Hbx)]→Lbs; if Bob loves Sally and hates someone, then Bob loves Sally.
7. * (∃x)Lxx∧(∀y)Hyy; someone loves himself or herself, and everyone hates him or herself.
9. * (∃x)Lxb∧(∃x)Lbx; someone loves Bob, and Bob loves someone.

6.6 MIXED AND OVERLAPPING QUANTIFIERS

In this section, quantifiers with overlapping scope are considered. Translation from English into RL is an art, and so there is no foolproof method or decision procedure for translating from one language to the other. In what follows, a four-step procedure is outlined, and a number of examples are provided.

It is helpful to start with a simple case. Consider the following English expression in a domain of discourse consisting of persons:

(1) Someone loves someone.

Step 1 of our translation procedure begins by identifying and symbolizing any English expressions that represent quantifiers (and their bound variables) and propositional operators. In the case of (1), there are two instances of *someone*, and these can be represented using the existential quantifier. In addition, note that the existential quantifier will bind two different variables x and y since (1) does not say that someone loves him- or herself (i.e., x loves x); nor does it preclude the possibility that the person that someone loves is him- or herself (i.e., x loves y and $x = y$).

(1)	Someone loves someone.	
Step 1	Identify and symbolize any English expressions that represent quantifiers (and their bound variables) and propositional operators.	(∃x) and (∃y)

Step 2 of our translation procedure suggests translating any ordinary language predicates into predicates of RL. In the case of (1), there is the two-place predicate *loves* that can be symbolized as 'Lxy.'

(1)	Someone loves someone.	
Step 2	Translate any ordinary language predicates into predicates of RL.	Lxy

Step 3 is the most difficult. Here, you are asked to use the quantifiers from step 1 and the predicates from step 2 to represent the proposition that (1) expresses. (1) expresses the proposition that at least one person in the domain loves at least one person in the domain; that is, $(\exists x)(\exists y)Lxy$.

(1)	Someone loves someone.	
Step 3	Use the quantifiers from step 1 and the predicates from step 2 and represent the proposition that (1) expresses.	$(\exists x)(\exists y)Lxy$

Finally, it can be helpful to add a fourth step that is used to check the translation. Step 4 suggests that you read the RL wff in English and check this reading against the literal meaning of the English sentence. In the case of (1), '$(\exists x)(\exists y) Lxy$' is read as

There exists an x and there exists a y such that x loves y.

(1)	Someone loves someone.	
Step 4	Read the predicate logic wff in English and check to see whether it captures the meaning of the sentence undergoing translation	There exists an x and there exists a y such that x loves y.

These four steps will aid you in your efforts to translate various English sentences into predicate logic wffs. To gain further clarity and practice, consider the following sentence:

(2) Every zombie loves every human.

Again, let's take our translation of (2) one step at a time. (2) expresses that every single member of one set of objects (zombies) loves every single member of another set of objects (humans). First, notice that (2) has two English expressions (*every*) that can be captured by the universal quantifier, and so we can replace these with two instances of the universal quantifier. Second, we need to isolate two different sets of objects (all of the zombies and all of the humans), for (1) expresses that all of the zombies love all of the humans and not that all of these zombies love themselves. In order to isolate these sets, we will ultimately predicate the respective properties of being a zombie and being a human to two different bound variables x and y.

(2)	Every zombie loves every human.	
Step 1	Identify and symbolize any English expressions that represent quantifiers (and their bound variables) and propositional operators.	$(\forall x)$ and $(\forall y)$

Next, translate any ordinary language predicates into predicates of RL, making sure to pay attention to their adicity.

(2)	Every zombie loves every human.	
Step 2	Translate any ordinary language predicates into predicates of RL.	Zx, Hy, Lxy

Third, use the quantifiers from step 1 and the predicates from step 2 to try to capture the meaning of (2).

(2)	Every zombie loves every human.	
Step 3	Use the quantifiers from step 1 and the predicates from step 2 and represent the proposition that (1) expresses.	$(\forall x)(\forall y)[(Zx \wedge Hy) \rightarrow Lxy]$

Finally, check the predicate logic formula against the original English translation by reading off the predicate wff in English.

(2)	Every zombie loves every human.	
Step 4	Read the predicate logic wff in English and check to see whether it captures the meaning of the sentence undergoing translation.	For all x and for all y, if x is a zombie and y is a human, then x loves y.

This step-by-step method can likewise capture the meaning of a number of other English sentences.

English Sentence	Translation into RL
Some zombie loves some human.	$(\exists x)(\exists y)((Zx \wedge Hy) \wedge Lxy)$
Some zombie loves every human.	$(\exists x)(\forall y)((Zx \wedge Hy) \wedge Lxy)$
Every zombie loves some human.	$(\forall x)(\exists y)[(Zx \wedge Hy) \rightarrow Lxy]$
Some humans don't love some zombie.	$(\exists x)(\exists y)((Hx \wedge Zy) \wedge \neg Lxy)$
No human loves some zombie.	$(\forall x)(\exists x)[(Hx \wedge Zy) \rightarrow \neg Lxy]$

It is important to point out that while sometimes the order of the quantifiers does not matter, in other cases it is significant. Consider the following predicate logic expression in a domain of discourse consisting of persons where 'Lxy' is the relational predicate that stands for *x loves y*:

$$(\forall x)(\forall y)Lxy$$
$$(\forall y)(\forall x)Lxy$$

These expressions are very similar and, in fact, express the same proposition. That is, both predicate logic wffs express the proposition that *Everyone loves every-*

one. In addition, consider the following predicate logic expressions involving the existential quantifier:

$$(\exists x)(\exists y)Lxy$$
$$(\exists y)(\exists x)Lxy$$

Again, both expressions appear similar, and both express the same proposition that *someone loves someone*. Examples like those above may give you the impression that the order of the quantifiers does not matter when you are either translating a predicate wff into English or interpreting the expression. This, however, is not the case for the following two wffs:

$$(\exists x)(\forall y)Lxy$$
$$(\forall x)(\exists y)Lxy$$

While these expressions appear similar, they express different propositions. In English, '$(\forall x)(\exists y)Lxy$' expresses the proposition that *Someone loves everyone*. This proposition is true just in the case that there is at least one person who loves every person. In contrast, '$(\forall y)(\exists x)Lxy$' expresses the proposition that *Everyone loves someone*. This proposition is true just in the case that every individual in the domain of discourse loves at least one person. This shows that the order in which the quantifiers are arranged has an effect on how the formula is interpreted and how we ought to translate the expression from one language into the other.

Exercise Set #5

A. Using the following translation key, translate the following English sentences into the language of predicate logic: D: persons, b: Bob, Zx: *x* is a zombie, Ex: *x* eats *y*, Kx: *x* kills *y*, Hx: *x* is a human, Lxy: *x* loves *y*
  1. * All zombies are human.
  2.   No humans are zombies.
  3. * Everyone is a zombie, or everyone is a human.
  4.   Some zombies eat some humans, but no human eats a zombie.
  5. * If Bob is not a zombie, then some zombie has not eaten some human.
  6.   All zombies eat humans unless some human kills every zombie.
  7. * If Bob is a zombie, then some zombie ate some human.
  8.   No humans eat zombies, but some zombies eat humans.
  9. * If some zombie kills Bob and Bob eats some human, then some zombie eats some human.
  10.   If every zombie eats every human, then there are no humans.

B. Using the following translation key, translate the predicate logic arguments below into English: D: persons, Px: *x* is all-powerful, Ex: *x* is evil, Kx: *x* is all-knowing, Hx: *x* is a human, Lxy: *x* loves *y*, s: Sally
  1. * Someone loves Sally. Therefore, someone loves someone.
  2.   There exists something that is all-knowing. There exists something that is all-loving. Therefore, there exists something that is all-knowing and all-loving.

3. * Something is all-knowing and all-powerful. Therefore, something is all-knowing, and something is all-powerful.
4. There exists something that is all-knowing or something that is all-powerful. There is not something that is all-powerful. Therefore, there is something that is all-knowing.
5. * If something is all-knowing, then there is someone who loves everyone. But there is not someone who loves everyone. Therefore, there is not something that is all-knowing.
6. If something loves everyone, then there will not be something that is evil. But there is something that is evil. Therefore, there is not something that loves everyone.
7. * Something loves every human. But some human does not love some human. Therefore, everyone does not love everyone.
8. If there exists someone who is all-powerful and all-knowing, then there exists someone who loves everyone. There does exist someone who is all-powerful and all-knowing. Therefore, someone loves everyone.
9. If there exists something that is all-powerful and all-knowing and loves everyone, then evil does not exist. But evil does exist. Therefore, it is not the case that there is something that is all-powerful and all-knowing and loves everyone.

Solutions to Starred Exercises in Exercise Set #5

A.
1. * $(\forall x)(Zx \rightarrow Hx)$
3. * $(\forall x)Zx \lor (\forall x)Hx$
5. * $\neg Zb \rightarrow \{(\exists x)(\exists y)[(Zx \land Hy) \land \neg Exy]\}$
7. * $Zb \rightarrow \{(\exists x)(\exists y)[(Zx \land Hy) \land Exy]\}$
9. * $[(\exists x)(Zx \land Kxb) \land (\exists x)(Hx \land Ebx)] \rightarrow \{(\exists x)(\exists y)[(Zx \land Hy) \land Exy]\}$

B.
1. * $(\exists x)Lxs \vdash (\exists x)(\exists y)Lxy$
3. * $(\exists x)(Kx \land Px) \vdash (\exists x)Kx \land (\exists x)Px$
5. * $(\exists x)Kx \rightarrow (\exists x)(\forall y)Lxy, \neg(\exists x)(\forall y)Lxy \vdash \neg(\exists x)Kx$
7. * $(\exists x)(\forall y)(Hy \land Lxy), (\exists x)(\exists y)[(Hx \land Hy) \land \neg Lxy] \vdash (\forall x)(\forall y)\neg Lxy$

DEFINITIONS

Bound variable	A bound variable is a variable that is within the scope of a quantifier that quantifies for that variable.
Free variable	A free variable is a variable that is not a bound variable.
Open formula	An open formula is a wff consisting of an *n*-place predicate '**P**' followed by *n* terms, where one of those terms is a free variable.
Closed formula	A closed formula is a wff consisting of an *n*-place predicate '**P**' followed by *n* terms, where every term is a name or a bound variable.

Domain	The domain of discourse (D) consists of all of the things that a language can meaningfully refer to or talk about.
Interpretation-function	An interpretation-function is an assignment of (1) objects in D to names, (2) a set of n-tuples in D to n-place predicates, and (3) truth values to sentences.
Model	A model in RL is a structure consisting of a domain and an interpretation function.

Chapter Seven

Predicate Logic Trees

In this chapter, we explore the truth-tree method for predicate trees. Unlike proposi-tional logic, the system of predicate logic is *undecidable*. As such, there is no decision procedure like truth tables or trees that always produces a yes or no answer about whether a given proposition, set of propositions, or argument has a logical property. However, the truth-tree method does offer a partial decision procedure for predicate logic formulas in that it will give an answer for a number of propositions, sets of propositions, and arguments.

7.1 FOUR NEW DECOMPOSITION RULES

In propositional logic, there are nine proposition types that can undergo decomposition.

Nine Decomposable Proposition Types	
Conjunction	$P \land R$
Disjunction	$P \lor R$
Conditional	$P \rightarrow R$
Biconditional	$P \leftrightarrow R$
Negated conjunction	$\neg(P \land R)$
Negated disjunction	$\neg(P \lor R)$
Negated conditional	$\neg(P \rightarrow R)$
Negated biconditional	$\neg(P \leftrightarrow R)$
Double negation	$\neg\neg P$

Propositions of this form are capable of undergoing decomposition using the rules formulated for propositional truth trees.

Stacking	Branching	Stacking & Branching

P∧Q
P ∧D
Q ∧D
	¬(P∧Q)
	   ╱╲
	¬P **¬Q** ¬∧D
		P↔Q
		   ╱╲
		P **¬P** ↔D
		Q **¬Q** ↔D

¬(P∨Q)
¬P ¬∨D
¬Q ¬∨D
	P∨Q
	   ╱╲
	P **Q** ∨D
		¬(P↔Q)
		   ╱╲
		P **¬P** ¬↔D
		¬Q **Q** ¬↔D

¬(P→Q) **P** ¬→D **¬Q** ¬→D	**P→Q** ╱╲ **¬P** **Q** →D	
¬¬P **P** ¬¬D		

In predicate logic, there are four additional proposition types that can undergo decomposition. These are the following:

Four Decomposable Proposition Types	
Existential	(∃x)**P**
Universal	(∀x)**P**
Negated existential	¬(∃x)**P**
Negated universal	¬(∀x)**P**

For each of these proposition types, there is a corresponding decomposition rule. That is, there is one for universally quantified propositions (∀D), one for existentially quantified propositions (∃D), and one for each of their negated forms, (¬∀D) and (¬∃D).

First, we begin with the two rules for negated quantified expressions.

Negated Existential Decomposition (¬∃D)	Negated Universal Decomposition (¬∀D)
¬(∃x)**P**✓	¬(∀x)**P**✓
(∀x)¬**P**	(∃x)¬**P**

In the case of (¬∃D), a negated existentially quantified proposition '¬(∃x)**P**' is decomposed into a universally quantified negated proposition '(∀x)¬**P**.' In the case of (¬∀D), a negated universally quantified proposition '¬(∀x)**P**' is decomposed into an existentially quantified negated proposition '(∃x)¬**P**.' As an illustration, consider the following truth tree:

1	¬(∃x)Px✓	P
2	¬(∀y)Wy✓	P
3	(∀x)¬Px	1¬∃D
4	(∃y)¬Wy	2¬∀D

In the above example, notice that negated quantified expressions fully decompose. In the case of line 1, the negated existential proposition is decomposed into a universally quantified proposition that ranges over a negated formula. In the case of line 2, the negated universal proposition is decomposed into an existentially quantified proposition that ranges over a negated formula. Again, whenever a proposition is fully decomposed, a checkmark (✓) is placed next to that proposition to indicate that the proposition cannot be further decomposed.

Consider another example:

1	¬(∃x)(∀y)Pxy✓	P
2	¬(∀y)Wy∧(∃z)Pz✓	P
3	(∀x)¬(∀y)Pxy	1¬∃D
4	¬(∀y)Wy✓	2∧D
5	(∃z)Pz	2∧D
6	(∃y)¬Wy	4¬∀D

Notice the use of (¬∃D) at line 3. Line 1 is a negated existentially quantified proposition, and a use of (¬∃D) results in a universally quantified proposition that ranges over a negated universally quantified proposition. Also notice that since line 2 is a conjunction, (∧D) is applied to line 2, and then (¬∀D) is applied to '¬(∀y)Wy.'

The two rules for the decomposition of quantified expressions are as follows:

Existential Decomposition (∃D)	Universal Decomposition (∀D)
(∃x)**P**✓	(∀x)**P**
P(a/x)	**P**(a/x)
where 'a' is an individual constant (name) that does not previously occur in the branch.	where 'a' is any individual constant (name).

According to (∃D) and (∀D), an individual constant (name) is substituted for a bound variable in a quantified expression. This procedure is symbolized as 'P(a/x)' (i.e., replace x with 'a'). Thus, if there is a quantified expression of the form '(∀x)P' or '(∃x)P,' a substitution instance of 'P(a/x)' replaces x's bound by the quantifier with 'a.'

In order to consider these decomposition rules more closely, consider (∀D). The decomposition rule (∀D) can be more explicitly stated as follows:

Universal Decomposition (∀D)

(∀x)**P**

P(a . . . v/x)

Consistently replace every bound *x* with any individual constant (name) of your choosing (even if it already occurs in an open branch) under any (not necessarily both) open branch of your choosing.

The motivation behind this formulation of universal decomposition is that if '(∀x) **P**' is true, then every substitution instance for '**P**(a/x)' is true. That is, if *Everything is a monkey*, then every substitution instance '**P**(a/x),' '**P**(b/x),' '**P**(c/x)' in the domain is true. In finite domains, there is a finite number of substitution instances, and so universally quantified propositions can be decomposed by going through every object in the domain and checking to see whether it has the property '**P**.' However, in infinite domains, a universal quantifier ranges over an infinite number of objects, and so universally quantified propositions '(∀x)**P**' cannot be decomposed, in a finite number of steps, by writing out various substitution instances—that is, '**P**(a/x),' '**P**(b/x),' '**P**(c/x),' and so on. It thus never receives a checkmark (✓) to indicate that it has been fully decomposed (see above).

There are two important features of (∀D). The first is that a universally quantified proposition '(∀x)**P**' is partially decomposed by writing any substitution instance '**P**(a/x)' on one or more open branches that '(∀x)**P**' contains. Thus, consider the following tree:

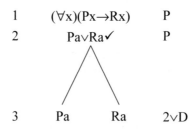

In the above case, we can partially decompose '(∀x)(Px→Rx)' by writing any substitution instance '**P**(a/x)' under any branch that '(∀x)(Px→Rx)' contains. In other words, '(∀x)(Px→Rx)' can be decomposed under the left branch by replacing each x with an 'a':

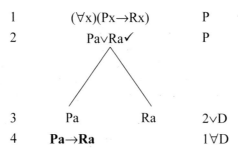

Or it can be decomposed under the right branch by replacing each *x* with an 'a':

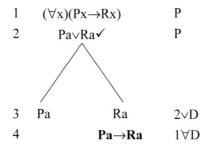

Or it can be decomposed under both branches by replacing each *x* with an 'a':

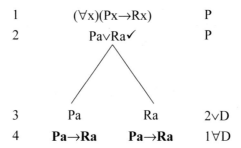

In addition, it is important to keep in mind that when we decompose '($\forall$x)(Px→Rx),' we can write *any* substitution instance '**P**(a/*x*)' under *any* branch that '($\forall$x)(Px→Rx)' contains (provided the substitution is done consistently). In other words, rather than making use of the variable replacement '**P**(a/*x*),' we could have used '**P**(b/*x*),' '**P**(c/*x*),' and so on.

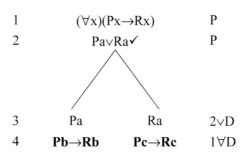

In the left branch on line 4, '$(\forall x)(Px \rightarrow Rx)$' is decomposed using '**P**(b/x),' while on the right branch '$(\forall x)(Px \rightarrow Rx)$' is decomposed using '**P**(c/x).' Notice, however, that line 1 is still not checked off since universally quantified propositions never fully decompose when a domain is infinite. This means that we can decompose '$(\forall x)(Px \rightarrow Rx)$' again and again until all of the objects in the domain have been accounted for. Thus,

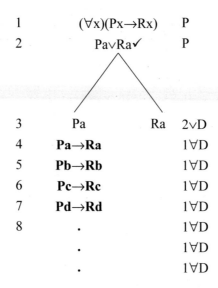

1	$(\forall x)(Px \rightarrow Rx)$	P
2	Pa∨Ra✓	P
3	Pa Ra	2∨D
4	**Pa→Ra**	1∀D
5	**Pb→Rb**	1∀D
6	**Pc→Rc**	1∀D
7	**Pd→Rd**	1∀D
8	.	1∀D
	.	1∀D
	.	1∀D

But this is not to say that the use of $(\forall D)$ will never yield a completed tree since there are many truth trees that will close. For example, consider a tree involving the following set of propositions:

$$\{(\forall x)(Pxab \rightarrow Rmxd), Psab \land \neg Rmsd\}$$

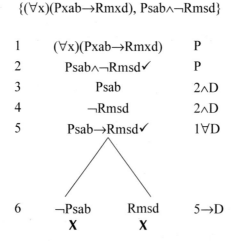

1	$(\forall x)(Pxab \rightarrow Rmxd)$	P
2	Psab∧¬Rmsd✓	P
3	Psab	2∧D
4	¬Rmsd	2∧D
5	Psab→Rmsd✓	1∀D
6	¬Psab Rmsd	5→D
	X **X**	

Notice that in the above example, the replacement of x with s in line 5 ultimately results in the tree closing.

Next consider (∃D).

Existential Decomposition (∃D)
(∃x)P✓
P(a/*x*)
where 'a' is an individual constant (name) that does not previously occur in the branch.

The decomposition of an existential proposition '(∃x)**P**' involves removing the quantifier and then replacing the bound variable with an individual constant that does not previously occur in the branch. The motivation behind this formulation is that if '(∃x)**P**' is true, then there is at least one thing in the domain of discourse that has the property picked out by '**P**.' It is important to note, however, that '(∃x)**P**' does not tell us which thing in the domain has the property '**P**.' Thus, it is important to note that the substitution instance '**P**(a/*x*)' that we choose must be foreign to the branch (more on this later).

As a simple example, consider the following set of propositions:

$$(\exists x)Px, \ Pa$$

1	(∃x)Px✓	P
2	Pa	P
3	**Pb**	1∃D

Notice that a use of (∃D) involves removing the existential quantifier and replacing the bound variable with an individual constant foreign to the branch. Since 'a' already occurs in the branch containing '(∃x)Px,' we choose the variable replacement '**P**(b/*x*),' but we could have chosen '**P**(c/*x*),' '**P**(d/*x*),' '**P**(e/*x*),' and so on.

Consider another, slightly more complicated example involving the following set of propositions :

$$\{(\exists y)(Py{\rightarrow}Ry), \ (\exists x)Px{\vee}(\exists z)(Qz),Pa\}$$

1	(∃y)(Py→Ry)✓	P
2	(∃x)Px∨(∃z)Qz	P
3	Pa	P
4	**Pb→Rb**	1∃D

Notice that in decomposing line 1, the bound *y*'s were not replaced with 'a' since this would violate the restriction on (∃D). Namely, it would violate the restriction that states the individual constant used to replace the quantified variable must not occur previously in the branch. Continuing the tree,

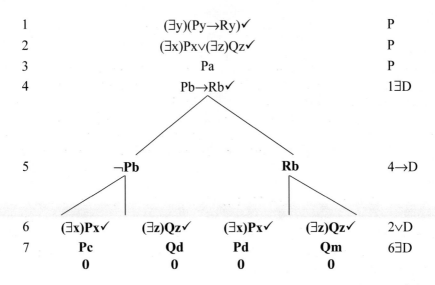

1   (∃y)(Py→Ry)✓   P
2   (∃x)Px∨(∃z)Qz✓   P
3   Pa   P
4   Pb→Rb✓   1∃D

5   ¬Pb   Rb   4→D

6   (∃x)Px✓   (∃z)Qz✓   (∃x)Px✓   (∃z)Qz✓   2∨D
7   Pc   Qd   Pd   Qm   6∃D
        0   0   0   0

Notice that the above tree is completed and that the decomposition of line 6 involves replacing existentially bound variables with a variety of different object constants. Note that since each proposition occurs in a different branch, all of these could
be replaced with the same object constant (e.g., 'c'). It is important to see that the only
restriction on using (∃D) is that you cannot replace a variable with an object constant
that already occurs in that branch.

The reason for the restriction on the use of (∃D) can be explained with an example.
Consider the following tree for the following set of propositions:

$$\{(∃x)(Px), (∃x)(Qx)\}$$

1   (∃x)Px   P
2   (∃x)Qx   P
3   Pa   1∃D
4   Qa   2∃D—**NO!**

In the above case, there are two propositions: '(∃x)Px' says that some x in D has
property 'P,' while '(∃x)Qx' says that some x in D has property 'Q.' These two
propositions do not say that some one object is both 'P' and 'Q.' That is, the condition
under which '(∃x)Px' and '(∃x)Qx' are true is not the same as the condition under
which '(∃x)(Px∧Qx)' is true. The truth conditions of '(∃x)Px' are represented by selecting some unique and arbitrary individual 'a' in the universe of discourse, and the
truth conditions of '(∃x)Qx' are represented by selecting some unique and arbitrary
individual 'b' in the universe of discourse. Following the restriction produces the following tree:

1   (∃x)Px   P
2   (∃x)Qx   P

3	Pa	1∃D
4	Qb	2∃D

Line 4 in the first tree is incorrect, but line 4 in the second tree is correct because when a substitution instance for '(∃x)Qx' is chosen, 'a' cannot be chosen since '(∃x)Px' and '(∃x)Qx' are not true if and only if 'Pa' and 'Qa' are true.

In order to protect against the unwarranted assumption that each proposition is referring to the same object, the use of (∃D) is restricted by only allowing for substitution instances of individual constants (names) that do not previously occur in the branch.

7.2 STRATEGIES FOR DECOMPOSING TREES

In formulating a set of strategic rules for predicate truth trees, all of the previous strategic rules are imported, and additional strategic rules are added specifically for decomposing quantified expressions.

Strategic Rules for Decomposing Predicate Truth Trees
1   Use no more rules than needed.
2   Decompose negated quantified expressions and existentially quantified expressions first.
3   Use rules that close branches.
4   Use stacking rules before branching rules.
5   When decomposing universally quantified propositions, use constants that already occur in the branch.
6   Decompose more complex propositions before simpler propositions.

Of new interest are rules (2) and (5). Rule (2) gives priority to (¬∃D), (¬∀D), and (∃D) over any use of (∀D). Rule (5) is present to avoid overly complex truth trees. Consider the following example:

1	(∀x)(∀y)¬Pxy	P
2	(∃y)Pay✓	P
3	Pab	2∃D
4	(∀y)¬Pay	1∀D
5	¬Pab	4∀D
	X	

The above truth tree closes. This is shown by first using (∃D) at line 3, and then two instances of (∀D). Notice that the substitution instances for '(∀x)(∀y)¬Pxy' are 'P(a/x)' and 'P(b/y).' This follows strategic rule (5) whereby constants are chosen based on whether they already occur in the branch.

Now consider what would happen if we ignored the strategic rules and first used (∀D) and then (∃D):

1	(∀x)(∀y)¬Pxy	P
2	(∃y)Pay✓	P
3	(∀y)¬Pmy	1∀D
4	¬Pmb	3∀D
5	Pac	2∃D
6	(∀y)¬Pay	1∀D
7	¬Pac	6∀D
	X	

In the above example, lines 3 and 4 turn out to be unhelpful. Since our use of (∃D) at line 5 has the restriction that the substitution instance cannot already occur previously in the branch, we cannot substitute 'b' for y. In the above example, when using (∃D) at line 5, the substitution form is '**P**(c/y).' Since a universally quantified proposition never fully decomposes, we must decompose line 1 again, and this time our choice of substitutions is guided by 'Pac,' which was obtained by (∃D) at line 5.

Exercise Set #1

A. Construct a predicate truth tree for the following sets of propositions. We have not formulated all of the necessary definitions to determine whether the tree has a completed open branch, so focus on trying to use the rules correctly.
  1. * (∃x)Px, (∀x)¬Px
  2.   ¬(∀x)(Px), Pb
  3. * (∃x)(Px∧Qx), (∀x)Px→(∀x)Qx
  4.   ¬(∀x)Px, ¬(∀y)(Py∧Gy), (∀z)(Pz∧¬Gz)
  5. * ¬(∀x)(Px∧Qx), (∃y)(Py∧Qy)
  6.   ¬(∀x)¬Fx∧¬(∀x)Fx
  7. * (∃x)(∀y)Pxy, (∀x)¬Pxx
  8.   (∃x)(∃y)Pxy∧(∃z)Pzz, (∀x)(∀y)Pxy
  9.   (∀x)(∀y)Pxyx↔(∀x)(∀y)Pyxy
  10.   ¬[(∃x)Px↔¬(∀x)¬Px]

Solutions to Starred Exercises in Exercise Set #1

  1. * (∃x)Px, (∀x)¬Px

1	(∃x)Px✓	P
2	(∀x)¬Px	P
3	Pa	1∃D
4	¬Pa	2∀D
	X	

It is important to see that we made use of (∃D) before (∀D) here. If we had used (∀D) first, our subsequent use of (∃D) would have had to be an object constant that was foreign to the branch.

3. * $(\exists x)(Px \wedge Qx)$, $(\forall x)Px \rightarrow (\forall x)Qx$

1	$(\exists x)(Px \wedge Qx)\checkmark$	P
2	$(\forall x)Px \rightarrow (\forall x)Qx\checkmark$	P
3	$Pa \wedge Qa\checkmark$	$1\exists D$
4	Pa	$3\wedge D$
5	Qa	$3\wedge D$

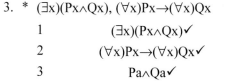

6	$\neg(\forall x)Px\checkmark$	$(\forall x)Qx$	$2\rightarrow D$
7	$(\exists x)\neg Px\checkmark$		$6\neg\forall D$
8	$\neg Pb$		$7\exists D$
9	**O**	Qa	$6\forall D$

At this point, it is unclear whether or not the right-hand branch forms of a completed open branch since '$(\forall x)Qx$' can be decomposed for any number of individual constants in the domain (e.g.,'Qb,''Qc,''Qd_1,''Qd_2,'. . .'Qd_n,').

5. * $\neg(\forall x)(Px \wedge Qx)$, $(\exists y)(Py \wedge Qy)$

1	$\neg(\forall x)(Px \wedge Qx)\checkmark$	P
2	$(\exists y)(Py \wedge Qy)\checkmark$	P
3	$(\exists x)\neg(Px \wedge Qx)\checkmark$	$1\neg\forall D$
4	$Pa \wedge Qa\checkmark$	$2\exists D$
5	$\neg(Pb \wedge Qb)\checkmark$	$3\exists D$
6	Pa	$4\wedge D$
7	Qa	$4\wedge D$

8	$\neg Pb$	$\neg Qb$	$5\neg\wedge D$
	O	**O**	

Notice that at line 5 we use '$P(b/x)$' as a substitution instance. A substitution instance of '$P(a/x)$' would be wrong since 'a' already occurs as an object constant in the branch.

7. * $(\exists x)(\forall y)Pxy$, $(\forall x)\neg Pxx$

1	$(\exists x)(\forall y)Pxy\checkmark$	P
2	$(\forall x)\neg Pxx$	P
3	$(\forall y)Pay$	$1\exists D$
4	Paa	$3\forall D$
5	$\neg Paa$	$2\forall D$
	X	

7.3 LOGICAL PROPERTIES

The two tasks for this section are to use the truth-tree method as a procedure for determining whether a particular proposition, set of propositions, or argument has some logical property (e.g., consistency, validity). But before the truth-tree method is formulated, it is necessary to redefine a *completed open branch* and explain how analyzing trees in RL is different from analyzing trees in PL. In the chapter on propositional trees, a completed open branch was defined as the following: a branch where all the complex propositions in the branch are decomposed into atomic propositions or their literal negations. For trees in predicate logic, a new definition is required.

Completed open branch	A branch is a completed open branch if and only if (1) all complex propositions that can be decomposed into atomic propositions or negated atomic propositions are decomposed; (2) for all universally quantified propositions '(∀x)**P**' occurring in the branch, there is a substitution instance '**P**(a/x)' for each constant that occurs in that branch; and (3) the branch is not a closed branch.
Closed tree	A tree is a closed tree if and only if all branches close.
Closed branch	A branch is a closed branch if and only if there is a proposition and its literal negation (e.g., '**P**' and '¬**P**').

In order to get clearer on the definition of a completed open branch, a few examples are considered below. First, consider the following tree:

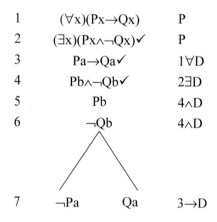

1	(∀x)(Px→Qx)	P
2	(∃x)(Px∧¬Qx)✓	P
3	Pa→Qa✓	1∀D
4	Pb∧¬Qb✓	2∃D
5	Pb	4∧D
6	¬Qb	4∧D
7	¬Pa Qa	3→D

At first glance, it may appear that the tree does contain a completed open branch because there are no closed branches, and every decomposable proposition has been decomposed. However, take a closer look at clause (2) in the definition of a completed open branch:

(2) For all universally quantified propositions '(∀x)**P**' occurring in the branch, there is a substitution instance '**P**(a/x)' for each constant that occurs in that branch.

Notice that 'b' is an object constant occurring in the branch at lines 4 to 6, but there is no substitution instance '**P**(b/x)' for '($\forall$x)(Px→Qx)' occurring in the branch containing 'b.' Otherwise put, we haven't decomposed '($\forall$x)(Px→Qx)' using '**P**(b/x).' Thus, the tree does not contain a completed open branch. Now consider what happens when '($\forall$x)(Px→Qx)' is decomposed using 'b' as a substitution instance.

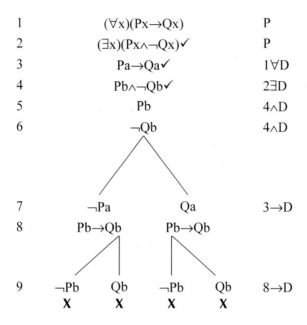

In decomposing '($\forall$x)(Px→Qx)' and using '**P**(b/x),' the tree turns out to close. Thus, it is important that clause (2) of ($\forall$D) be attended to because, as the example above shows, ignoring this feature will yield an open tree instead of a closed tree.

Consider a tree with the following stack of propositions:

($\forall$x)(¬Px→¬Rx), ($\forall$x)(Rx→Px)

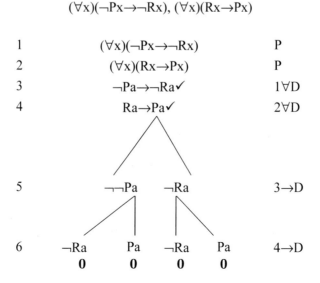

In the above example, there are two universally quantified expressions that are not checked off, yet the above tree is completed since for all universally quantified propositions '($\forall$x)**P**' occurring on the branch, there is a substitution instance '**P**(a/*x*)' for each constant that occurs on that branch. This constant is 'a.' The tree is completed, and since its branches do not close, the above tree is a completed open tree.

7.3.1 Semantic Analysis of Predicate Truth Trees

In PL, the basic unit of representation is the proposition, which is assigned a truth value (true or false). In RL, the truth or falsity of a predicate well-formed formula (wff, pronounced 'woof') is relative to an interpretation in a model (i.e., relative to a specification of the domain of discourse and an interpretation function). Using the notion of an interpretation, semantic properties can be defined for RL propositions, sets of propositions, and arguments as follows:

Tautology	A proposition '**P**' is a tautology in RL if and only if '**P**' is true on every interpretation.
Contradiction	A proposition '**P**' is a contradiction in RL if and only if '**P**' is false on every interpretation.
Contingency	A proposition '**P**' is a contingency in RL if and only if '**P**' is neither a contradiction nor a tautology.
Equivalence	Propositions '**P**' and '**Q**' are equivalent in RL if and only if there is no interpretation where the valuation of '**P**' is different from the valuation of '**Q**.'
Consistency	A set of propositions '{**A**, **B**, **C**, . . ., **Z**}' is consistent in RL if and only if there is at least one interpretation such that all of the propositions in the set are true.
Validity	An argument '**P**, **Q**, **R**, . . ., **Y**⊢**Z**' is valid in RL if and only if there is no interpretation such that all of the premises '**A**,' '**B**,' '**C**,' . . ., '**Y**' are true and the conclusion '**Z**' is false.

One way to show that a proposition, set of propositions, or argument has one of these properties is to construct an interpretation in a model. For example, to show that '{($\forall$x)Px, ($\exists$x)Rx}' is consistent in RL involves showing that there is at least one interpretation in a model where $v(\forall x)Px = T$ and $v(\exists x)Rx = T$. Here is an example of such a model:

$$D = \text{positive integers}$$
$$P = \{x \mid x \text{ is greater than } 0\}$$
$$R = \{x \mid x \text{ is even}\}$$

On this interpretation, $v(\forall x)Px = T$ since every positive integer is greater than zero. In addition, $v(\exists x)Px = T$ since there is at least one positive integer that is even. For example, four is a positive integer that is even. Since $v(\forall x)Px = T$ and $v(\exists x)Px = T$ in the interpretation above, there is at least one interpretation of the model such that all of the propositions from the set are true.

Similar procedures can be formulated for each of the above properties. The focus of this section, however, is to develop a clearer understanding of how these properties can be determined using truth trees. In PL, a completed open branch tells us that there is a valuation (truth-value assignment) that would make every proposition in the stack true. Similarly, in RL, a completed open branch tells us that there is an interpretation in a model for which every proposition in the stack is true. Thus, the presence of a completed open branch tells us that we can construct a model such that every proposition in the stack is true.

To illustrate, consider a very simple tree consisting of '$(\exists x)Px$' and 'Pa':

1	$(\exists x)Px\checkmark$	P
2	Pa	P
3	Pb	$1\exists D$
	0	

The above tree has a completed open branch, and so there is an interpretation for which all of the propositions in the branch are true. If we wanted, we could construct an interpretation in a model that would show '$(\exists x)Px$,''Pa,' and 'Pb' as being consistent. To do this, we would stipulate a domain of discourse involving two objects, letting 'a' stand for an object and 'b' stand for an object, and assign the one-place predicate 'P' an extension.

D:	{John, Fred}
Px:	x is a person {John, Fred}
a:	John
b:	Fred

In this interpretation of the model, 'Pa,' and 'Pb' are true, and thus '$(\exists x)Px$' is also true. '$(\exists x)Px$' is true because there is at least one object in the domain that is a person, while 'Pa' is true because 'a' refers to John, and John is in the extension of persons; likewise 'Pb' is true because 'b' refers to Fred, and Fred is in the extension of persons. Thus, using the tree, we can read off the propositions in the completed open branches and then give an interpretation in a model that shows the propositions in that set are true. And if this is the case, then truth trees offer us a method for determining certain properties of propositions, sets of propositions, and arguments. For instance, the above tree has a completed open branch, which shows that the stack of propositions is true under at least one interpretation and so is consistent.

Consider a slightly more complicated example:

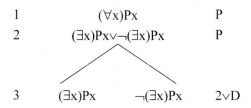

4	Pa			3∃D
5	Pa			1∀D
6	**0**		(∀x)¬Px	3¬∃D
7			¬Pa	6∀D
8			Pa	1∀D
			X	

In the above tree, there is a completed open branch and a closed branch. The closed branch on the right-hand side indicates that there is no interpretation for which '¬(∃x)Px' and '(∀x)Px' are true. However, the completed open branch (on the left-hand side) indicates that there is an interpretation for which all of the propositions in the branch are true. As such, we can construct an interpretation in a model such that the propositions in the branch are true.

D: {1}

Px: *x* is a number

In this model, '(∀x)Px' is true since every number in the domain is a number, and '(∃x)Px' is true since there is a number in the domain that is a number.

Finally, let's consider a tree involving a slightly more complicated proposition:

1	(∃x)(∃y)[(Ox∧Ey)∧Gxy]✓	P
2	(∃y)[(Oa∧Ey)∧Gay]✓	1∃D
3	(Oa∧Eb)∧Gab✓	1∃D
4	Oa∧Eb✓	3∧D
5	Gab	3∧D
6	Oa	4∧D
7	Eb	4∧D
	0	

The above tree has a completed open branch, and so there is an interpretation for which the propositions 'Gab,' 'Oa,' and 'Eb' are true, and thus '(∃x)(∃y)[(Ox∧Ey)∧Gxy]' is true. Again, a model can be constructed to reflect this fact:

D: {1, 2, 3}
Ox: *x* is an odd number
Ex: *x* is an even number
Gxy: *x* is greater than *y*
a: 3
b: 2

'Gab' is true since it is true that 3 is greater than 2. 'Oa' is true since three is an odd number. Lastly, 'Eb' is true since 2 is an even number. Thus, the predicate wff '(∃x)(∃y)[(Ox∧Ey)∧Gxy],' which says that there exists an odd number greater than some existent even number, is also true. In short, the truth tree, along with the model, demonstrates that the set '{(∃x)(∃y)[(Ox∧Ey)∧Gxy]}' is not a contradiction in RL.

It is important to note that the tree method can be analyzed semantically such that a completed open branch indicates that there is at least one interpretation that makes the propositions in the stack being decomposed true. For the remainder of this chapter, however, we avoid the discussion and construction of models for predicate truth trees and focus on how the truth-tree method can be used to determine whether a proposition, set of propositions, or argument has a particular logical property.

7.3.2 Consistency and Inconsistency

Using the truth-tree method, we can determine when a set of propositions '{**P, Q, R, . . ., Z**}' is consistent or inconsistent.

Consistency	A set of propositions '{**P, Q, R, . . ., Z**}' is shown by the truth-tree method to be consistent if and only if a tree of the stack '**P**,' '**Q**,' '**R**,' . . ., '**Z**' is an open tree; that is, there is at least one completed open branch.
Inconsistency	A set of propositions '{**P, Q, R, . . ., Z**}' is shown by the truth-tree method to be inconsistent if and only if a tree of the stack of '**P**,' '**Q**,' '**R**,' . . ., '**Z**' is a closed tree; that is, all branches close.

Below, we provide four examples of consistent and inconsistent trees. First, consider

$$(\forall x)(Px{\rightarrow}Rx), \neg(\forall x)(\neg Rx{\rightarrow}\neg Px)$$

1	$(\forall x)(Px{\rightarrow}Rx)$	P
2	$\neg(\forall x)(\neg Rx{\rightarrow}\neg Px)$✓	P
3	$(\exists x)\neg(\neg Rx{\rightarrow}\neg Px)$✓	2¬∀D
4	$\neg(\neg Ra{\rightarrow}\neg Pa)$✓	3∃D
5	$Pa{\rightarrow}Ra$✓	1∀D
6	$\neg Ra$	4¬→D
7	$\neg\neg Pa$	4¬→D
8	Pa	7¬¬D
9	$\neg Pa$ Ra	5→D
	X **X**	

The above tree is a closed tree and shows that '{(∀x)(Px→Rx), ¬(∀x)(¬Rx→¬Px)}' is inconsistent. Next, consider

$$(\forall x)(Px) \rightarrow (\forall y)(Ry), \neg(\forall y)(Ry), (\exists x)\neg(Px)$$

1	$(\forall x)(Px) \rightarrow (\forall y)(Ry)\checkmark$	P
2	$\neg(\forall y)Ry\checkmark$	P
3	$(\exists x)\neg Px\checkmark$	P
4	$(\exists y)\neg Ry\checkmark$	$2\neg\forall D$
5	$\neg Ra$	$4\exists D$
6	$\neg Pb$	$3\exists D$

7	$\neg(\forall x)Px\checkmark$	$(\forall y)Ry$	$1\rightarrow D$
8	$(\exists x)\neg Px\checkmark$		$7\neg\forall D$
9	$\neg Pc$		$8\exists D$
10	**0**	Ra	$7\forall D$
11		Rb	$7\forall D$
		X	

The above tree has a completed open branch and so shows that the stack composing the tree is consistent. Notice that in lines 10 and 11, two uses of ($\forall$D) are required to complete the tree since an 'a' and 'b' are found as object constants in the branch.

Consider a third tree involving the following propositions:

$$\neg\neg(\forall x)(Px)\vee(\forall y)(Ry), \neg(\forall y)(Ry), (Ra\wedge Rb)\wedge Pa$$

1	$\neg\neg(\forall x)(Px)\vee(\forall y)(Ry)\checkmark$	P
2	$\neg(\forall y)(Ry)\checkmark$	P
3	$(Ra\wedge Rb)\wedge Pa\checkmark$	P
4	$Ra\wedge Rb\checkmark$	$3\wedge D$
5	Pa	$3\wedge D$
6	Ra	$4\wedge D$
7	Rb	$4\wedge D$
8	$(\exists y)\neg(Ry)\checkmark$	$2\neg\forall D$
9	$\neg Rc$	$8\exists D$

10	$\neg\neg(\forall x)(Px)\checkmark$	$(\forall y)(Ry)$	$1\vee D$

11	$(\forall x)(Px)$			$10\neg\neg D$
12	Pa			$11\forall D$
13	Pb			$11\forall D$
14	Pc			$11\forall D$
15	**0**		Ra	$10\forall D$
16			Rb	$10\forall D$
17			Rc	$10\forall D$
			X	

The above tree has a completed open branch, which shows that the stack composing the tree is consistent. Notice again that lines 12 to 17 required multiple uses of $(\forall D)$ since each universally quantified proposition occurring on the branch requires a substitution instance '$\mathbf{P}(a/x)$' for each constant that occurs on that branch.

Consider one final example involving the following set of propositions:

$$\neg(\forall x)(\exists y)(Pxy)\wedge(\forall y)\neg(\exists x)(Rxy),\ \neg(\forall y)(\forall x)(Rxy),\ (Rab\wedge Rba)\wedge Pab$$

1	$\neg(\forall x)(\exists y)(Pxy)\wedge(\forall y)\neg(\exists x)(Rxy)\checkmark$	P
2	$\neg(\forall y)(\forall x)(Rxy)\checkmark$	P
3	$(Rab\wedge Rba)\wedge Pab\checkmark$	P
4	$Rab\wedge Rba\checkmark$	$3\wedge D$
5	Pab	$3\wedge D$
6	Rab	$4\wedge D$
7	Rba	$4\wedge D$
8	$\neg(\forall x)(\exists y)(Pxy)\checkmark$	$1\wedge D$
9	$(\forall y)\neg(\exists x)(Rxy)$	$1\wedge D$
10	$(\exists x)\neg(\exists y)(Pxy)\checkmark$	$8\neg\forall D$
11	$(\exists y)\neg(\forall x)(Rxy)\checkmark$	$2\neg\forall D$
12	$\neg(\exists y)(Pcy)\checkmark$	$10\ \exists D$
13	$(\forall y)\neg(Pcy)\checkmark$	$12\neg\exists D$
14	$\neg Pca$	$13\forall D$
15	$\neg Pcb$	$13\forall D$
16	$\neg Pcc$	$13\forall D$
17	$\neg(\forall x)(Rxe)\checkmark$	$11\exists D$
18	$\neg(\exists x)\neg(Rxe)\checkmark$	$17\neg\forall D$
19	$\neg Rfe$	$18\exists D$
20	$\neg(\exists x)(Rxa)\checkmark$	$9\forall D$
21	$\neg(\exists x)(Rxb)\checkmark$	$9\forall D$
22	$\neg(\exists x)(Rxe)\checkmark$	$9\forall D$
23	$\neg(\exists x)(Rxf)\checkmark$	$9\forall D$
24	$(\forall x)\neg(Rxa)$	$20\neg\exists D$
25	$(\forall x)\neg(Rxb)$	$21\neg\exists D$
26	$(\forall x)\neg(Rxe)$	$22\neg\exists D$

27	$(\forall x)\neg(Rxf)$	23¬∃D
28	¬Rab	23¬∃D
	X	

Yikes! the initial set of propositions is inconsistent because the tree is closed. In the above table, it is important to look for a proposition and its literal negation as soon as possible. Rather than starting by decomposing line 24 with multiple uses of (∀D), you can decompose line 25 into line 30 using one instance of (∀D) involving '**P**(b/x).' The reason for this is to generate the contradiction and close the tree.

Exercise Set #2

A. Using the truth-tree method, test the following sets of propositions for logical consistency and inconsistency.
   1. * $(\exists x)(Px\rightarrow Qx)$, $(\exists x)Px$
   2. $(\exists x)(Px\rightarrow Rx)$, ¬Pa, ¬Pb
   3. * $(\forall x)Px\vee(\exists y)Qy$, $(\exists x)(Px\wedge Qa)$
   4. $(\exists x)(Px\vee Gx)$, $\neg(\forall x)(Px\rightarrow\neg Gx)$
   5. * $(\forall x)(Px\rightarrow Mx)$, $(\exists x)Px$, $\neg(\exists x)Mx$
   6. $(\exists x)(\forall y)(Px\rightarrow Gy)$, $(\exists x)(\neg Gx\rightarrow\neg Px)$
   7. * $(\forall x)(Px\leftrightarrow Wx)$, $(\exists x)[Px\wedge(\exists y)(\neg Py\wedge Wy)]$
   8. $(\forall x)(Px\rightarrow Qx)$, $(\forall x)(Px\vee Qx)$
   9. * $(\forall x)(Px\wedge Mx)$, Pa, Pb, $(\exists x)Rx$
  10. $(\forall x)(\forall y)(Px\rightarrow Py)$, ¬Pb, ¬Pa
  11. * $(\forall x)(Px\rightarrow Rx)$, $(\exists x)(Mx\wedge\neg Rx)$, $(\exists x)\neg(Px\rightarrow Rx)$
  12. $\neg(\forall x)\neg(\forall y)(\forall z)[Px\rightarrow(My\rightarrow Tz)]$

Solutions to Starred Exercises in Exercise Set #2

A.
  1. * $(\exists x)(Px\rightarrow Qx)$, $(\exists x)Px$; consistent.

1	$(\exists x)(Px\rightarrow Qx)$✓	P
2	$(\exists x)Px$✓	P
3	Pa	2∃D
4	Pb→Qb✓	1∃D

| 5 | ¬Pb   Qb | 4→D |
| | **0**   **0** | |

3. * (∀x)Px∨(∃y)Qy, (∃x)(Px∧Qa); consistent.

1	(∀x)Px∨(∃y)Qy	P
2	(∃x)(Px∧Qa)✓	P
3	Pb∧Qa✓	2∃D
4	Pb	3∧D
5	Qa	3∧D

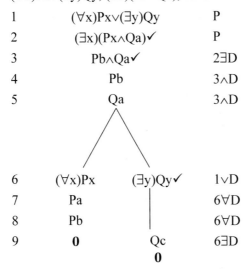

6	(∀x)Px	(∃y)Qy✓	1∨D
7	Pa		6∀D
8	Pb		6∀D
9	**0**	Qc	6∃D
		0	

5. * (∀x)(Px→Mx), (∃x)Px, ¬(∃x)Mx; inconsistent.

1	(∀x)(Px→Mx)	P
2	(∃x)Px✓	P
3	¬(∃x)Mx	P
4	Pa	2∃D
5	(∀x)¬Mx	3¬∃D
6	¬Ma	5∀D
7	Pa→Ma	1∀D

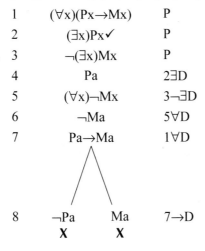

8	¬Pa	Ma	7→D
	X	**X**	

7. * (∀x)(Px↔Wx), (∃x)[Px∧(∃y)(¬Py∧Wy)]; inconsistent.

1	(∀x)(Px↔Wx)	P
2	(∃x)[Px∧(∃y)(¬Py∧Wy)]✓	P
3	Pa∧(∃y)(¬Py∧Wy)✓	2∃D
4	Pa	3∧D
5	(∃y)(¬Py∧Wy)✓	3∧D
6	¬Pb∧Wb✓	5∃D
7	¬Pb	6∧D
8	Wb	6∧D

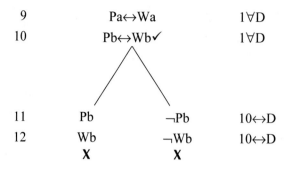

9	Pa↔Wa		1∀D
10	Pb↔Wb✓		1∀D

11	Pb	¬Pb	10↔D
12	Wb	¬Wb	10↔D
	X	X	

9. * (∀x)(Px∧Mx), Pa, Pb, (∃x)Rx; consistent.

1	(∀x)(Px∧Mx)	P
2	Pa	P
3	Pb	P
4	(∃x)Rx✓	P
5	Rc	4∃D
6	Pa∧Ma✓	1∀D
7	Pb∧Mb✓	1∀D
8	Pc∧Mc✓	1∀D
9	Pa	6∧D
10	Ma	6∧D
11	Pb	7∧D
12	Mb	7∧D
13	Pc	8∧D
14	Mc	8∧D
	0	

11. * (∀x)(Px→Rx), (∃x)(Mx∧¬Rx), (∃x)¬(Px→Rx); inconsistent.

1	(∀x)(Px→Rx)	P
2	(∃x)(Mx∧¬Rx)✓	P
3	(∃x)¬(Px→Rx)✓	P
4	Ma∧¬Ra✓	2∃D
5	¬(Pb→Rb)✓	3∃D
6	Ma	4∧D
7	¬Ra	4∧D
8	Pb	5¬→D
9	¬Rb	5¬→D
10	Pa→Ra✓	1∀D

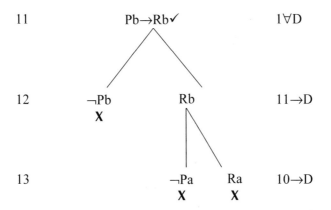

11	Pb→Rb✓	1∀D
12	¬Pb Rb	11→D
	X	
13	¬Pa Ra	10→D
	X **X**	

7.3.3 Tautology, Contradiction, and Contingency

In this section, the truth-tree method is devised to determine whether a predicate proposition 'P' is a tautology, contradiction, or contingency.

Tautology	A proposition 'P' is shown by the truth-tree method to be a tautology if and only if the tree '¬P' determines a closed tree; that is, all branches close.
Contradiction	A proposition 'P' is shown by the truth-tree method to be a contradiction if and only if the tree 'P' determines a closed tree; that is, all branches close.
Contingency	A proposition 'P' is shown by the truth-tree method to be a contingency if and only if 'P' is neither a tautology nor a contradiction; that is, the tree of 'P' does not determine a closed tree, and the tree of '¬P' does not determine a closed tree.

Consider the following proposition:

$$(\exists x)\neg(\forall y)[Px \rightarrow (Qx \vee \neg Ry)]$$

First, begin by testing to see whether or not it is a contradiction. This amounts to taking the whole proposition 'P' and testing whether or not all branches close.

1	$(\exists x)\neg(\forall y)[Px \rightarrow (Qx \vee \neg Ry)]$✓	P
2	$\neg(\forall y)[Pa \rightarrow (Qa \vee \neg Ry)]$✓	1∃D
3	$(\exists y)\neg[Pa \rightarrow (Qa \vee \neg Ry)]$✓	2¬∀D
4	$\neg[Pa \rightarrow (Qa \vee \neg Rb)]$✓	3∃D
5	Pa	4¬→D
6	$\neg(Qa \vee \neg Rb)$✓	4¬→D

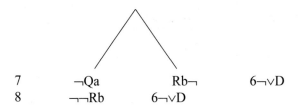

| 7 | ¬Qa | Rb¬ | 6¬∨D |
| 8 | ¬¬Rb | 6¬∨D | |

In the above example, '(∃x)¬(∀y)[Px→(Qx∨¬Ry)]' is not a contradiction since there is at least one completed open branch. Next, we test to see whether or not it is a tautology. This amounts to testing whether all of its branches close for '¬**P**.'

1	¬(∃x)¬(∀y)[Px→(Qx∨¬Ry)]✓	P
2	(∀x)¬¬(∀y)[Px→(Qx∨¬Ry)]	1¬∃D
3	¬¬(∀y)[Pa→(Qa∨¬Ry)]✓	2∀D
4	(∀y)[Pa→(Qa∨¬Ry)]	3¬¬D
5	Pa→(Qa∨¬Ra)✓	4∀D

| 6 | ¬Pa   Qa∨¬Ra✓ | 5→D |
| | **0** | |

| 7 | Qa   ¬Ra | 6∨D |
| | **0**   **0** | |

The above tree is shown not to be a tautology because all branches for '¬**P**' do not close. That is, there is at least one open and completed branch. Since it is not the case that trees for '**P**' and '¬**P**' close, the proposition '**P**' is neither a contradiction nor a tautology. And if '**P**' is neither a contradiction nor a tautology, it is a contingency.

Consider another proposition:

$$(\forall x)(Px \rightarrow Qx) \wedge (\exists x)(Px \wedge \neg Qx)$$

We begin by testing to see whether it is a contradiction. Namely, whether a tree for '**P**' is a closed tree.

1	(∀x)(Px→Qx)∧(∃x)(Px∧¬Qx)✓	P
2	(∀x)(Px→Qx)	1∧D
3	(∃x)(Px∧¬Qx)✓	1∧D
4	Pa∧¬Qa✓	3∃D
5	Pa	4∧D
6	¬Qa	4∧D

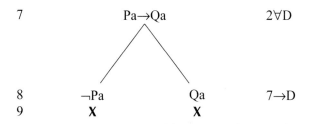

All of the branches of the above tree close. Thus, '(∀x)(Px→Qx)∧(∃x)(Px∧¬Qx)' is a contradiction.

Next, consider the following proposition:

$$(\forall x)(Px \rightarrow Px) \wedge (\forall y)(Qy \vee \neg Qy)$$

We test this proposition to see whether or not it is a tautology. A proposition '**P**' is a tautology if and only if the tree for '¬**P**' closes. Thus, we construct the tree below with '¬**P**.'

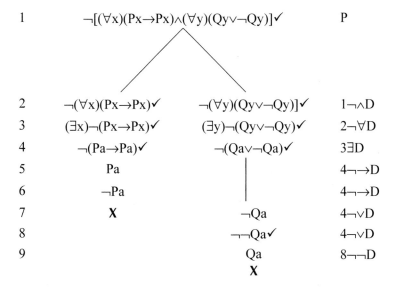

If all the branches for '¬P' close, then 'P' is a tautology. All of the branches for '¬[(∀x)(Px→Px)∧(∀y)(Qy∨¬Qy)]' close, thus '(∀x)(Px→Px)∧(∀y)(Qy∨¬Qy)' is a tautology.

Exercise Set #3

A. Using the truth-tree method, test the following propositions to determine whether each is a contradiction, tautology, or contingency.

1.   *   (∃x)Px∨¬(∃x)Px
2.   (∃x)Px∨(∃x)¬Px

3. * (∀x)(Px→Gx)
4.   (∀x)(Px∨¬Px)
5. * (∀x)(Px∧¬Mx)∨(∃x)(¬Px∨Mx)
6.   (∃x)(Fx∧Px)∨(∀y)(Py→Fy)
7. * (∀x)(∀y)Pxy∧(∃x)(∃y)¬Pxy
8.   (∃y)(∀x)(Pxy∧¬Pyx)
9. * (∀x)Pxx→Paa

Solutions to Starred Exercises in Exercise Set #3

A.
1. * (∃x)Px∨¬(∃x)Px; tautology.

    1   ¬[(∃x)Px∨¬(∃x)Px]✓   P
    2   ¬(∃x)Px   1¬∨D
    3   ¬¬(∃x)Px   1¬∨D
    4   (∃x)Px   3¬¬D
                    X

3. * (∀x)(Px→Gx); first tree, not a contradiction.

    1   (∀x)(Px→Gx)   P
    2   Pa→Ga✓   1∀D

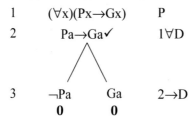

    3   ¬Pa   Ga   2→D
            0   **0**

    (∀x)(Px→Gx); second tree, not a tautology. Since it is neither a tautology
    nor a contradiction, '(∀x)(Px→Gx)' is a contingency.

    1   ¬(∀x)(Px→Gx)✓   P
    2   (∃x)¬(Px→Gx)✓   1¬∀D
    3   ¬(Pa→Ga)✓   2∃D
    4   Pa   3¬→D
    5   ¬Ga   3¬→D
                0

5. * (∀x)(Px∧¬Mx)∨(∃x)(¬Px∨Mx); first tree, not a contradiction.

    1   (∀x)(Px∧¬Mx)∨(∃x)(¬Px∨Mx)   P

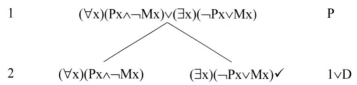

    2   (∀x)(Px∧¬Mx)   (∃x)(¬Px∨Mx)✓   1∨D

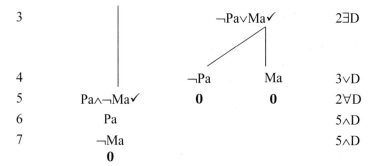

3		¬Pa∨Ma✓	2∃D
4		¬Pa Ma	3∨D
5	Pa∧¬Ma✓	**0** **0**	2∀D
6	Pa		5∧D
7	¬Ma		5∧D
	0		

(∀x)(Px∧¬Mx)∨(∃x)(¬Px∨Mx); second tree, a tautology. It is a tautology because all branches close for '¬[(∀x)(Px∧¬Mx)∨(∃x)(¬Px∨Mx)].'

1	¬[(∀x)(Px∧¬Mx)∨(∃x)(¬Px∨Mx)]✓	P
2	¬(∀x)(Px∧¬Mx)✓	1¬∨D
3	¬(∃x)(¬Px∨Mx)✓	1¬∨D
4	(∃x)¬(Px∧¬Mx)✓	2¬∀D
5	(∀x)¬(¬Px∨Mx)	3¬∃D
6	¬(Pa∧¬Ma)	4∃D
7	¬(¬Pa∨Ma)✓	5∀D
8	¬¬Pa✓	7¬∨D
9	¬Ma	7¬∨D
10	Pa	8¬¬D

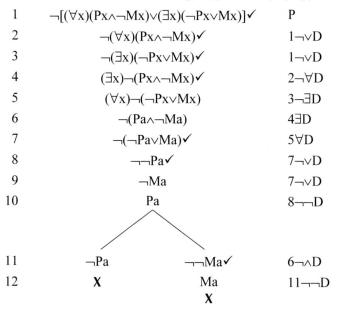

11	¬Pa ¬¬Ma✓	6¬∧D
12	**X** Ma	11¬¬D
	X	

7. * (∀x)(∀y)Pxy∧(∃x)(∃y)¬Pxy; contradiction.

1	(∀x)(∀y)Pxy∧(∃x)(∃y)¬Pxy	P
2	(∀x)(∀y)Pxy	1∧D
3	(∃x)(∃y)¬Pxy	1∧D
4	(∃y)¬Pay	3∃D
5	¬Pab	4∃D
6	(∀y)Pay	2∀D
7	Pab	6∀D
	X	

9. * (∀x)Pxx→Paa; tautology.

1	¬[(∀x)Pxx→Paa]	P
2	(∀x)Pxx	1¬→D
3	¬Paa	1¬→D
4	Paa	2∀D
	X	

7.3.4 Logical Equivalence

In this section, the truth-tree method is used to determine whether a pair of predicate propositions '**P**' and '**Q**' are equivalent in RL.

> Equivalence   A pair of propositions '**P**' and '**Q**' is shown by the truth-tree method to be equivalent if and only if the tree of the stack of '¬(**P**↔**Q**)' determines a closed tree; that is, all branches for '¬(**P**↔**Q**)' close.

First, consider the following two propositions: '(∀x)Px' and '¬(∃x)Px.' In order to test whether two propositions '**P**' and '**Q**' are logically equivalent, put them in negated biconditional form, '¬(**P**↔**Q**),' and use a truth tree to determine whether all branches close. If the tree closes, then '(∀x)Px' and '¬(∃x)Px' are equivalent. If there is a completed open branch, then '(∀x)Px' and '¬(∃x)Px' are not equivalent.

1	¬[(∀x)Px↔¬(∃x)Px]✔		P
2	(∀x)Px	¬(∀x)Px	1¬↔D
3	¬¬(∃x)Px✔	¬(∃x)Px	1¬↔D
4	(∃x)Px✔		3¬¬D
5	Pa		4∃D
6	Pa		2∀D
	0		

The above tree has a completed open branch, and so the truth-tree method shows that '(∀x)Px' and '¬(∃x)Px' are not equivalent.

Here is a more complicated example. Test the following two propositions for logical equivalence: '(∀x)¬(Px∨Gx)' and '(∀y)(¬Py∧¬Gy).'

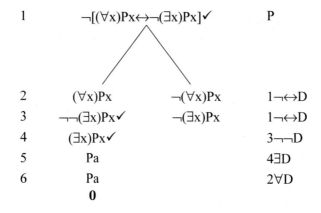

1	¬{[(∀x)¬(Px∨Gx)]↔[(∀y)(¬Py∧¬Gy)]} ✔		P
2	(∀x)¬(Px∨Gx)	¬(∀x)¬(Px∨Gx)✔	1¬↔D

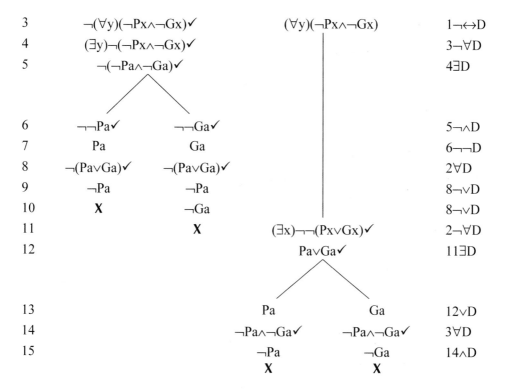

3	¬(∀y)(¬Px∧¬Gx)✓	(∀y)(¬Px∧¬Gx)	1¬↔D
4	(∃y)¬(¬Px∧¬Gx)✓		3¬∀D
5	¬(¬Pa∧¬Ga)✓		4∃D
6	¬¬Pa✓ ¬¬Ga✓		5¬∧D
7	Pa Ga		6¬¬D
8	¬(Pa∨Ga)✓ ¬(Pa∨Ga)✓		2∀D
9	¬Pa ¬Pa		8¬∨D
10	**X** ¬Ga		8¬∨D
11	**X**	(∃x)¬¬(Px∨Gx)✓	2¬∀D
12		Pa∨Ga✓	11∃D
13		Pa Ga	12∨D
14		¬Pa∧¬Ga✓ ¬Pa∧¬Ga✓	3∀D
15		¬Pa ¬Ga	14∧D
		X **X**	

In the above tree, we see that all branches of the negated biconditional close. Thus, '(∀x)¬(Px∨Gx)' and '(∀y)(¬Py∧¬Gy)' are logically equivalent.

Exercise Set #4

A. Using the truth-tree method, test the following sets of propositions for logical equivalence.
1. * (∀x)¬Px, ¬(∃x)Px
2.   ¬¬(∀x)Px, ¬(∃x)¬Px
3. * (∀y)Pyy, ¬(∃x)¬Pxx
4.   (∀y)Pyy∧(∀z)Pzz, ¬(∃x)¬Pxx
5. * (∀x)(Px→Qx), ¬(∃x)(Px∧¬Qx)
6.   (∃x)(Px∧¬Qx),¬(∀x)(¬Px∨Qx)
7. * (∃x)(∀y)(Px→Gy), (∃x)¬(∃y)¬(Px→Gy)
8.   (∃x)Px∧(∃y)Gy, (∀x)Px∧(∀y)Py
9.   (∀x)Mxx, (∃x)Mxx
10. (∀x)(∀y)Pxy, ¬(∃x)(∃y)¬Pxy

Solutions to Starred Exercises in Exercise Set #4

A.
1. * (∀x)Px, ¬(∃x)Px; not equivalent.

1	¬[(∀x)Px↔¬(∃x)Px]✓		P
2	(∀x)Px	¬(∀x)Px✓	1¬↔D
3	¬¬(∃x)Px	¬(∃x)Px✓	1¬↔D
4		(∃x)¬Px✓	2¬∀D
5		(∀x)¬Px	3¬∃D
6		¬Pa	4∃D
7		¬Pa	5∀D

0

3. * (∀y)Pyy, ¬(∃x)¬Pxx; equivalent.

1	¬[(∀y)Pyy↔¬(∃x)¬Pxx]✓		P
2	(∀y)Pyy	¬(∀y)Pyy	1¬↔D
3	¬¬(∃x)¬Pxx	¬(∃x)¬Pxx	1¬↔D
4	(∃x)¬Pxx		3¬¬D
5	¬Paa		4∃D
6	Paa		2∀D
7	**X**	(∃y)¬Pyy	2¬∀D
8		(∀x)¬¬Pxx	3¬∃D
9		¬Paa	7∃D
10		¬¬Paa	8∀D
11		Paa	10¬¬D

X

5. * (∀x)(Px→Qx), ¬(∃x)(Px∧¬Qx); equivalent.

1	¬[(∀x)(Px→Qx)↔¬(∃x)(Px∧¬Qx)]✓		P
2	(∀x)(Px→Qx)	¬(∀x)(Px→Qx)✓	1¬↔D
3	¬¬(∃x)(Px∧¬Qx)✓	¬(∃x)(Px∧¬Qx)✓	1¬↔D
4	(∃x)(Px∧¬Qx)✓		3¬¬D
5	Pa∧¬Qa✓		4∃D

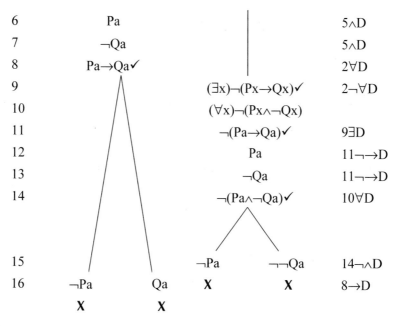

6	Pa			5∧D
7	¬Qa			5∧D
8	Pa→Qa✓			2∀D
9		(∃x)¬(Px→Qx)✓		2¬∀D
10		(∀x)¬(Px∧¬Qx)		
11		¬(Pa→Qa)✓		9∃D
12		Pa		11¬→D
13		¬Qa		11¬→D
14		¬(Pa∧¬Qa)✓		10∀D
15		¬Pa ¬¬Qa		14¬∧D
16	¬Pa Qa	X X		8→D
	X X			

7. * (∃x)(∀y)(Px→Gy), (∃x)¬(∃y)¬(Px→Gy); equivalent.

1	¬[(∃x)(∀y)(Px→Gy)↔(∃x)¬(∃y)¬(Px→Gy]✓		P
2	(∃x)(∀y)(Px→Gy)✓	¬(∃x)(∀y)(Px→Gy)✓	1¬↔D
3	¬(∃x)¬(∃y)¬(Px→Gy)✓	(∃x)¬(∃y)¬(Px→Gy)	1¬↔D
4	(∀x)¬¬(∃y)¬(Px→Gy)✓		3∃D
5	(∀x)(∃y)¬(Px→Gy)		4¬¬D
6	(∀y)(Pa→Gy)		2∃D
7	(∃y)¬(Pa→Gy)✓		5∀D
8	¬(Pa→Gb)		7∃D
9	Pa→Gb		6∀D
10	X	(∀x)¬(∀y)(Px→Gy)✓	2¬∃D
11		(∀x)(∃y)¬(Px→Gy)	10¬∀D
12		(∃x)(∀y)¬¬(Px→Gy)✓	3¬∃D
13		(∃x)(∀y)(Px→Gy)✓	12¬¬D
14		(∀y)(Pa→Gy)	13∃D
15		(∃y)¬(Pa→Gy)✓	11∀D
16		¬(Pa→Gb)	15∃D
17		Pa→Gb	14∀D
		X	

7.3.5 Validity

In this section, the truth-tree method is used to determine whether an argument '**P, Q, R, . . ., Y** ⊢ **Z**' is valid in RL.

Validity　　An argument '**P, Q, R, . . ., Y** ⊢ **Z**' is shown by the truth-tree method to be valid in RL if and only if the stack '**P**,' '**Q**,' '**R**,' . . ., '**Y**,' '¬**Z**' determines a closed tree.

Invalidity　An argument '**P, Q, R, . . ., Y** ⊢ **Z**' is shown by the truth-tree method to be invalid in RL if and only if the stack '**P**,' '**Q**,' '**R**,' . . ., '**Y**,' '¬**Z**' has at least one completed open branch.

First, we consider a very simple argument: '$(\forall x)Px$, Pa ⊢ Pa.' Remember that setting up the tree to test for validity requires listing the premises and the literal negation of the conclusion in the stack.

$$
\begin{array}{ccl}
1 & (\forall x)Px & P \\
2 & \neg Px & P \\
\end{array}
$$

Next, the tree is decomposed using the tree decomposition rules:

$$
\begin{array}{ccl}
1 & (\forall x)Pa & P \\
2 & \neg Pa & P \\
3 & Pa & 1\forall D \\
& \mathbf{X} &
\end{array}
$$

The above tree is closed. This shows that under no interpretation is it the case that all of the premises '**P**,''**Q**,''**R**,'. . ., '**Y**' and the negation of the conclusion '¬**Z**' are jointly true. In other words, under no interpretation is it the case that the premises are true and the conclusion is false. Thus, the above tree shows that the argument '$(\forall x)$Pa ⊢ Pa' is valid.

Moving on to a more complicated example, consider the following argument:

$$(\forall x)(Px \to Qx), (\exists y)(Py) \vdash (\exists x)(Px \to Qx)$$

In order to test this argument for validity, it is necessary to stack the premises and the negation of the conclusion. That is, if the argument is valid, then the following set of propositions should yield a closed tree:

$$\{(\forall x)(Px \to Qx), (\exists y)(Py), \neg(\exists x)(Px \to Qx)\}$$

$$
\begin{array}{ccl}
1 & (\forall x)(Px \to Qx) & P \\
2 & (\exists y)(Py)\checkmark & P \\
3 & \neg(\exists x)(Px \to Qx)\checkmark & P \\
4 & Pa & 2\exists D \\
5 & (\forall x)\neg(Px \to Qx) & 3\neg\exists D \\
\end{array}
$$

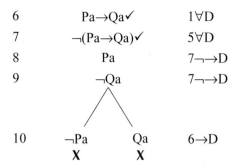

6	Pa→Qa✓	1∀D
7	¬(Pa→Qa)✓	5∀D
8	Pa	7¬→D
9	¬Qa	7¬→D

10   ¬Pa   Qa   6→D
       X   X

In the above tree, each of the branches is closed, so the tree is closed. Since the tree closes, the argument is valid.

Consider a final example:

$$Pa∧Qb, (∀x)(∀y)[Px→(Qy→Rx)] ⊢ Ra$$

Again, to determine whether or not this argument is valid, we test to see whether the following set of propositions is consistent:

$$\{Pa∧Qb, (∀x)(∀y)[Px→(Qy→Rx)], ¬Ra\}$$

If the set is inconsistent, the argument is valid. If it is consistent, then it is possible for the premises to be true and the conclusion to be false, and the argument is not valid.

1	Pa∧Qb✓	P
2	(∀x)(∀y)[Px→(Qy→Rx)]	P
3	¬Ra	P
4	Pa	1∧D
5	Qb	1∧D
6	(∀y)[Pa→(Qy→Ra)]	2∀D
7	(∀y)[Pb→(Qy→Rb)]	2∀D
8	Pa→(Qa→Ra)	6∀D
9	Pa→(Qb→Ra)✓	6∀D
10	Pb→(Qa→Rb)	7∀D
11	Pb→(Qb→Rb)	7∀D

12   ¬Pa   Qb→Ra✓   9→D
       X

13   ¬Qb   Ra   12→D
                 X   X

In the above tree, all the branches close; therefore the tree is closed. Thus, ' {Pa∧Qb, (∀x)(∀y)[Px→(Qy→Rx)], ¬Ra}' is inconsistent, and it is impossible for the premises to be true and the conclusion to be false. Therefore, the argument is deductively valid.

Exercise Set #5

A.   Using the truth-tree method, test the following arguments for validity.
   1. * (∀x)(Px→Gx), Pa ⊢ Ga
   2.   (∀x)(Px→Gx), Ga ⊢ Pa
   3. * (∀x)(∀y)(Pxy→Gxy), Pab ⊢ Gab
   4.   (∀x)(∀y)(Pyx→Gyx), Pab ⊢ (∃x)(∃y)Gyx
   5. * Pa, Pb, Pc, (∀x)(Px→Gx) ⊢ ¬(∃y)Gy

Solutions to Starred Exercises in Exercise Set #5

1. * (∀x)(Px→Gx), Pa ⊢ Ga; valid.

1	(∀x)(Px→Gx)	P
2	Pa	P
3	¬Ga	P
4	Pa→Ga✓	1∀D

| 5 | ¬Pa   Ga | 4→D |
| | X   X | |

3. * (∀x)(∀y)(Pxy→Gxy);Pab ⊢ Gab; valid.

1	(∀x)(∀y)(Pxy→Gxy)	P
2	Pab	P
3	¬Gab	P
4	(∀y)(Pay→Gay)	1∀D
5	(∀y)(Pby→Gby)	1∀D
6	Pab→Gab✓	4∀D

| 7 | ¬Pab   Gab | 6→D |
| | X   X | |

5. * Pa, Pb, Pc, $(\forall x)(Px \rightarrow Gx) \vdash \neg(\exists y)Gy$; invalid.

1	Pa	P
2	Pb	P
3	Pc	P
4	$(\forall x)(Px \rightarrow Gx)$	P
5	$\neg\neg(\exists y)Gy$✓	P
6	$(\exists y)Gy$✓	5¬¬D
7	Gd	6∃D
8	Pa→Ga✓	4∀D
9	Pb→Gb✓	4∀D
10	Pc→Gc✓	4∀D
11	Pd→Gd✓	4∀D

12	¬Pa Ga	8→D
	X	
13	¬Pb Gb	9→D
	X	
14	¬Pc Gc	10→D
	X	
15	¬Pd Gd	11→D
	0 0	

It is important to recognize that lines 8 to 11 are all necessary for the determination of a completed open branch at line 15. They are necessary because in order for there to be a completed open branch, for each universally quantified proposition '$(\forall x)\mathbf{P}$' occurring on a branch, there must be a substitution instance '$\mathbf{P}(a/x)$' for each constant already occurring on that branch. So, since object constants 'a,' 'b,' 'c,' and 'd' occur on lines 1, 2, 3, and 7, respectively, we need to make use of each of the following substitution instances for '$(\forall x)(Px \rightarrow Gx)$,' which occurs on line 4: '$\mathbf{P}(a/x)$,' '$\mathbf{P}(b/x)$,' '$\mathbf{P}(c/x)$,' '$\mathbf{P}(d/x)$.'

7.4 UNDECIDABILITY AND THE LIMITS OF THE PREDICATE TREE METHOD

Unlike PL, RL is undecidable. That is, there is no mechanical procedure that can always, in a finite number of steps, deliver a yes or no answer to questions about whether a given proposition, set of propositions, or argument has a property like consistency, tautology, validity, and the like. For some trees, the application of predicate decomposition rules will result in a process of decomposition that does not, in a finite number of steps, yield a closed tree or a completed open branch.

For example, consider the following tree for '$(\forall x)(\exists y)(Pxy)$':

1	$(\forall x)(\exists y)(Pxy)$	P
2	$(\exists y)Pay\checkmark$	$1\forall D$
3	Pab	$2\exists D$
4	$(\exists y)Pby\checkmark$	$1\forall D$
5	Pbc	$4\exists D$
6	$(\exists y)Pcy$	$1\forall D$
7	Pcd	$6\exists D$

.
.
.

Notice that in the above tree, the decomposition procedure will continue indefinitely since every time an $(\exists D)$ is used, another use of $(\forall D)$ will be required, followed by another $(\exists D)$, followed by another $(\forall D)$, and so on. This indefinite process of decomposition thus does not yield a completed open branch and so the truth-tree method does not show that '$(\forall x)(\exists y)(Pxy)$' is consistent. In order to avert this problem for predicate formulas that have finite models (i.e., interpretations of domains that are finite), there is a way to revise $(\exists D)$ to show that '$(\forall x)(\exists y)(Pxy)$' yields a completed open branch:[1]

New Existential Decomposition (N∃D)

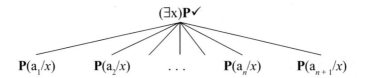

where $a_1, \ldots, a_n$ are constants already occurring on the branch
on which (N∃D) is being applied and a_{n+1} is a constant not occurring on that branch.

A use of (N∃D) requires that whenever we decompose an existentially quantified proposition '$(\exists x)\mathbf{P}$,' we create a separate branch for any substitution instance for any substitution instance $\mathbf{P}(a_1/x)$, $\mathbf{P}(a_2/x)$, $\ldots$, $\mathbf{P}(a_n/x)$, already occurring in the branch containing '$(\exists x)P$' and branch a substitution instance '$\mathbf{P}(a_{n+1}/x)$' that is not occurring in that branch.

With (N∃D) in hand, let's consider again the tree involving '$(\forall x)(\exists y)(Pxy)$,' in which $(\exists D)$ did not yield a completed open branch:

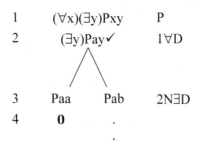

Notice that after (∀D) is applied to line 1, we have '(∃y)Pay.' When (N∃D) is applied at line 3, '**P**(a/*y*)' is used as a substitution instance on the left-hand side, and '**P**(b/*y*)' is used as a substitution instance on the right-hand side. Because of this, we can construct a finite model such that '(∀x)(∃y)Pxy' is true. However, notice that the right-hand side does not yield a closed branch or a completed open branch. Since the right-hand side involves an object constant 'b' and a universally quantified proposition, another round of (∀D) and (N∃D) will need to be undertaken.

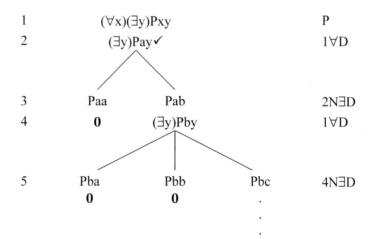

While another round of (∀D) and (N∃D) yields additional completed open branches, which will allow for the construction of finite models, the right-hand branch does not close. This indicates that the revision of (∃D) as (N∃D) will ensure that the truth-tree method can determine whether a set of propositions has a logical property provided that set has a finite model. In cases where a set of propositions has only an infinite model, the truth-tree method will neither yield a completed open branch nor a closed tree.

Although we have touched on the undecidability of predicate logic in this section, the proof of this feature of RL is a topic in meta theory and so is not discussed in this introduction to logic. Nevertheless, despite the undecidability of predicate logic, this undecidability only affects a small fraction of propositions in RL, and so the truth-tree method remains a useful method for determining logical properties

END-OF-CHAPTER EXERCISES

A. Using the truth-tree method, test the following sets of propositions to determine whether they are logically consistent.
1. * ¬(∀x)(Px→Qx), ¬(∃x)(Px∧¬Qx)
2. (∃x)Px∧(∃x)¬Px
3. * (∃x)(∃y)(Pxy∧Gxy), (∀y)(∀z)¬Gyz
4. ¬(∃x)¬(∀y)Pxy, (∃x)Px
5. * ¬(∀x)¬(Px→Mx), ¬(∃x)¬(Px∧¬Mx), (∀y)(Py→Zy)∨(∃x)(Mx∨Px)

B. Using the truth-tree method, test the following propositions to determine whether each is a contradiction, tautology, or contingency.
1. * (∃x)[(¬Fx∨¬Px)∨(∀y)(Py→Fy)]
2. (∃x)(∃y)(Pxy→Rxy)
3. (∃x)Px∧(∃x)¬Px
4. (∀x)(¬Px↔(Px∨Rx)]
5. (∃x)(∀y)(∀z)(Pxyz∨¬Rxyz)

C. Using the truth-tree method, test the following pairs of propositions to determine whether they are equivalent.
1. (∃x)Px, ¬(∀x)¬Px
2. (∃x)Px∧(∃x)Rx, (∃x)(Px∧Rx)
3. (∃x)(Px∧Rx), ¬(∀x)¬(¬Px∨¬Rx)
4. (∃x)Pxx, (∃x)(∃y)Pxy
5. (∃x)(∃y)Pxy, (∃x)(∃y)Pyx

D. Using the truth-tree method, test the following arguments for validity.
1. (∀x)(Px→Rx), (∃x)Px⊢(∃x)Rx
2. (∃x)(∃y)Pxy, (∃y)(∃x)Pyx→(∀x)Px⊢(∀y)Py
3. * (∃x)Px∧(∀y)¬Qy, (∃x)(Px∧Qx)⊢(∀x)(∀y)(∀z)Pxyz
4. (∃x)Px∨(∀y)Qy, ¬(∀y)Qy⊢¬(∃x)Px∨(∀x)(∀y)(∀z)Pxyz
5. * (∀x)[Px→¬(∀y)¬(Qy→My)], (∃x)Px, (∃x)Qx⊢(∃x)(Mx∨Px)

Solutions to Starred Exercises in End-of-Chapter Exercises

A.
1. * (∀x)(Px→Qx), ¬(∃x)(Px∧¬Qx); inconsistent.

1	¬(∀x)(Px→Qx)✓	P
2	¬(∃x)(Px∧¬Qx)✓	P
3	(∀x)¬(Px∧¬Qx)	2¬∃D
4	(∃x)¬(Px→Qx)✓	1¬∀D
5	¬(Pa→Qa)	4∃D

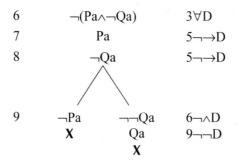

6	¬(Pa∧¬Qa)	3∀D
7	Pa	5¬→D
8	¬Qa	5¬→D

```
                    /\
                   /  \
9        ¬Pa       ¬¬Qa      6¬∧D
          X         Qa       9¬¬D
                    X
```

3. * (∃x)(∃y)(Pxy∧Gxy), (∀y)(∀z)¬Gyz, inconsistent.

1	(∃x)(∃y)(Pxy∧Gxy)	P
2	(∀y)(∀z)¬Gyz	P
3	(∃y)(Pay∧Gay)	1∃D
4	Pab∧Gab	3∃D
5	Pab	4∧D
6	Gab	4∧D
7	(∀z)¬Gaz	2∀D
8	¬Gab	7∀D
	X	

5. * ¬(∀x)¬(Px→Mx), ¬(∃x)¬(Px∧¬Mx), (∀y)(Py→Zy)∨(∃x)(Mx∨Px); inconsistent.

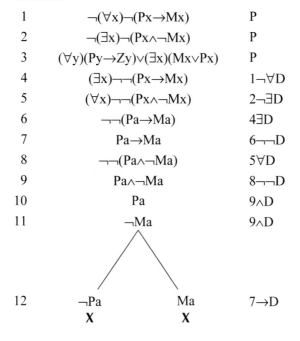

1	¬(∀x)¬(Px→Mx)	P
2	¬(∃x)¬(Px∧¬Mx)	P
3	(∀y)(Py→Zy)∨(∃x)(Mx∨Px)	P
4	(∃x)¬¬(Px→Mx)	1¬∀D
5	(∀x)¬¬(Px∧¬Mx)	2¬∃D
6	¬¬(Pa→Ma)	4∃D
7	Pa→Ma	6¬¬D
8	¬¬(Pa∧¬Ma)	5∀D
9	Pa∧¬Ma	8¬¬D
10	Pa	9∧D
11	¬Ma	9∧D

```
                   /\
                  /  \
12       ¬Pa          Ma        7→D
          X            X
```

B.

1. * $(\exists x)[(\neg Fx \vee \neg Px) \vee (\forall y)(Py \rightarrow Fy)]$; first tree, not a contradiction.

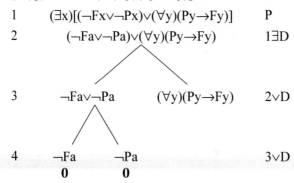

1	$(\exists x)[(\neg Fx \vee \neg Px) \vee (\forall y)(Py \rightarrow Fy)]$	P
2	$(\neg Fa \vee \neg Pa) \vee (\forall y)(Py \rightarrow Fy)$	1$\exists$D
3	$\neg Fa \vee \neg Pa$ $(\forall y)(Py \rightarrow Fy)$	2$\vee$D
4	$\neg Fa$ $\neg Pa$	3$\vee$D
	0 **0**	

$(\exists x)[(\neg Fx \vee \neg Px) \vee (\forall y)(Py \rightarrow Fy)]$; second tree, a tautology.

1	$\neg(\exists x)[(\neg Fx \vee \neg Px) \vee (\forall y)(Py \rightarrow Fy)]\checkmark$	P
2	$(\forall x)\neg[(\neg Fx \vee \neg Px) \vee (\forall y)(Py \rightarrow Fy)]$	1$\neg\exists$D
3	$\neg[(\neg Fa \vee \neg Pa) \vee (\forall y)(Py \rightarrow Fy)]\checkmark$	2$\forall$D
4	$\neg(\neg Fa \vee \neg Pa)\checkmark$	3$\neg\vee$D
5	$\neg(\forall y)(Py \rightarrow Fy)\checkmark$	3$\neg\vee$D
6	$\neg\neg Fa\checkmark$	4$\neg\vee$D
7	$\neg\neg Pa\checkmark$	4$\neg\vee$D
8	Fa	6$\neg\neg$D
9	Pa	6$\neg\neg$D
10	$(\exists y)\neg(Py \rightarrow Fy)\checkmark$	5$\neg\forall$D
11	$\neg(Pb \rightarrow Fb)\checkmark$	10$\exists$D
12	Pb	11$\neg\rightarrow$D
13	$\neg Fb$	11$\neg\rightarrow$D
14	$\neg[(\neg Fb \vee \neg Pb) \vee (\forall y)(Py \rightarrow Fy)]\checkmark$	2$\forall$D
15	$\neg(\neg Fb \vee \neg Pb)\checkmark$	14$\neg\vee$D
16	$\neg(\forall y)(Py \rightarrow Fy)$	14$\neg\vee$D
17	$\neg\neg Fb\checkmark$	15$\neg\vee$D
18	$\neg\neg Pb$	15$\neg\vee$D
19	Fb	17$\neg\neg$D
	X	

D.

3. * $(\exists x)Px \wedge (\forall y)\neg Qy,\ (\exists x)(Px \wedge Qx) \vdash (\forall x)(\forall y)(\forall z)Pxyz$; valid.

1	$(\exists x)Px \wedge (\forall y)\neg Qy\checkmark$	P
2	$(\exists x)(Px \wedge Qx)$	P
3	$\neg(\forall x)(\forall y)(\forall z)Pxyz$	P
4	$(\exists x)Px$	1$\wedge$D
5	$(\forall y)\neg Qy$	1$\wedge$D
6	$Pa \wedge Qa$	2$\exists$D
7	Pa	6$\wedge$D
8	Qa	6$\wedge$D
9	$\neg Qa$	5$\forall$D
	X	

5. * $(\forall x)[Px \rightarrow \neg(\forall y)\neg(Qy \rightarrow My)]$, $(\exists x)Px$, $(\exists x)Qx \vdash (\exists x)(Mx \lor Px)$; valid.

1	$(\forall x)[Px \rightarrow \neg(\forall y)\neg(Qy \rightarrow My)]$	P
2	$(\exists x)Px$	P
3	$(\exists x)Qx$	P
4	$\neg(\exists x)(Mx \lor Px)$	P
5	$(\forall x)\neg(Mx \lor Px)$	$4 \neg \exists D$
6	Pa	$2 \exists D$
7	Qb	$3 \exists D$
8	$\neg(Ma \lor Pa)$	$5 \forall D$
9	$\neg(Mb \lor Pb)$	$5 \forall D$
10	$\neg Ma$	$8 \neg \lor D$
11	$\neg Pa$	$8 \neg \lor D$
	X	

DEFINITIONS AND TREE RULES

Completed open branch	A branch is a completed open branch if and only if (1) all complex propositions that can be decomposed into atomic propositions or negated atomic propositions are decomposed; (2) for all universally quantified propositions '$(\forall x)$**P**' occurring in the branch, there is a substitution instance '**P**(a/x)' for each constant that occurs in that branch; and (3) the branch is not a closed branch.
Closed tree	A tree is a closed tree if and only if all branches close.
Closed branch	A branch is a closed branch if and only if there is a proposition and its literal negation (e.g., '**P**' and '¬**P**').
Consistency	A set of propositions '{**P, Q, R,** . . ., **Z**}' is shown by the truth-tree method to be consistent if and only if a complete tree of the stack of '**P**,' '**Q**,' '**R**,' . . ., '**Z**' is an open tree; that is, there is at least one completed open branch.
Inconsistency	A set of propositions '{**P, Q, R,** . . ., **Z**}' is shown by the truth-tree method to be inconsistent if and only if a completed tree of the stack of '**P**,' '**Q**,' '**R**,' . . ., '**Z**' is a closed tree; that is, all branches close.
Tautology	A proposition '**P**' is shown by the truth-tree method to be a tautology if and only if the tree '¬**P**' determines a closed tree; that is, all branches close.
Contradiction	A proposition '**P**' is shown by the truth-tree method to be a contradiction if and only if the tree '**P**' determines a closed tree; that is, all branches close.
Contingency	A proposition '**P**' is shown by the truth-tree method to be a contingency if and only if '**P**' is neither a tautology nor a contradiction; that is, the tree of '**P**' does not determine a closed tree, and the tree of '¬**P**' does not determine a closed tree.

Equivalence	A pair of propositions 'P' and 'Q' are shown by the truth-tree method to be equivalent if and only if the tree of the stack of '¬(P↔Q)' determines a closed tree; that is, all branches for '¬(P↔Q)' close.
Validity	An argument 'P, Q, R, ..., Y ⊢ Z' is shown by the truth-tree method to be valid in RL if and only if the stack 'P,' 'Q,' 'R,' ..., 'Y,' '¬Z' determines a closed tree.
Invalidity	An argument 'P, Q, R, ..., Y ⊢ Z' is shown by the truth-tree method to be invalid in RL if and only if the stack 'P,' 'Q,' 'R,' ..., 'Y,' '¬Z' has at least one completed open branch.

Negated Existential Decomposition (¬∃D)

¬(∃x)P✓
(∀x)¬P

Negated Universal Decomposition (¬∀D)

¬(∀x)P✓
(∃x)¬P

Existential Decomposition (∃D)

(∃x)P✓
P(a/x)

where 'a' is an individual constant (name) that does not previously occur in the branch.

Universal Decomposition (∀D)

(∀x)P
P(a . . . v/x)

Consistently replace every bound x with any individual constant (name) of your choosing (even if it already occurs in an open branch) under any (not necessarily both) open branch of your choosing.

Strategic Rules for Decomposing Predicate Truth Trees

1 Use no more rules than needed.
2 Decompose negated quantified expressions and existentially quantified expressions first.
3 Use rules that close branches.
4 Use stacking rules before branching rules.
5 When decomposing universally quantified propositions, it is a good idea to use constants that already occur in the branch.
6 Decompose more complex propositions before simpler propositions.

NOTE

1. See George Boolos, "Trees and Finite Satisfiability: Proof of a Conjecture of Burgess," *Notre Dame Journal of Formal Logic* 25, no. 3(1984): 193–97.

Chapter Eight

Predicate Logic Derivations

The entire set of derivation rules from propositional logic (PD) are imported into the natural deductive system of predicate logic (RD). In addition, new deductive derivation rules are formulated for derivations involving quantified expressions.

8.1 FOUR QUANTIFIER RULES

In this section, the natural deduction system of (RD) is articulated. (RD) consists of four quantifier derivation rules. Two derivation rules pertain to the elimination of quantifiers ('∀E' and '∃E') and the other two derivation rules pertain to the introduction of quantifiers ('∀I' and '∃I').

8.1.1 Universal Elimination (∀E)

Like propositional logic derivations, predicate logic derivations allow for adding a new line to the proof. Universal elimination (∀E) is a derivation rule that allows for the addition of a substitution instance '$P(a/x)$' of a universally quantified proposition. The substitution instance '$P(a/x)$' must be the result of a consistent (or uniform) replacement of all bound variables with any individual constant (name) of your choosing.

Universal Elimination (∀E)		
From any universally quantified proposition '$(\forall x)P$,' we can derive a substitution instance '$P(a/x)$' in which all bound variables are consistently replaced with any individual constant (name).	$(\forall x)P$ $P(a/x)$	∀E

Here is an example of the use of (∀E):

$$
\begin{array}{lll}
1 & (\forall x)Px & P \\
2 & Pa & 1\forall E
\end{array}
$$

Notice that x is bound in '$(\forall x)Px$,' and that the application of ($\forall$E) involves removing the quantifier and replacing x with an individual constant (name) of your choosing.

In the above example, ($\forall$E) is used on '$(\forall x)Px$' to derive 'Pa.' However, this is not the only proposition that can be inferred from '$(\forall x)Px$' since variables bound by the universal quantifier can be consistently replaced with any substitution instance '$\mathbf{P}(a/x)$,' '$\mathbf{P}(a_1/x)$,'. . .'$\mathbf{P}(a_n/x)$' of your choosing.

1	$(\forall x)Px$	P
2	Pa	1$\forall$E
3	Pb	1$\forall$E
4	Pc	1$\forall$E
5	Pd	1$\forall$E
6	Pe_{100}	1$\forall$E

Another feature of ($\forall$E) to keep in mind is that it demands that bound variables be *consistently* (or uniformly) replaced with any substitution instance '$\mathbf{P}(a/x)$' of your choosing.

1	$(\forall x)Pxx$	P
2	Paa	1$\forall$E
3	Pcc	1$\forall$E

Notice that each bound x has been uniformly replaced by an 'a' at line 2 and uniformly replaced by a 'c' at line 3. These are correct uses of ($\forall$E). It is important to note that in applying ($\forall$E) to '$(\forall x)Pxx$,' we cannot replace one x with 'a' and another with 'b':

1	$(\forall x)Pxx$	P
2	Pab	1$\forall$E—**NO!**
3	Pba	1$\forall$E—**NO!**

Notice that lines 2 and 3 are incorrect uses of ($\forall$E) because they do not uniformly replace bound variables with an individual constant. At line 2, one bound x is replaced by 'a' while another is replaced by 'b.' This is not a consistent replacement of bound variables with individual constants.

With a basic understanding of the correct formal use of ($\forall$E), consider the following English argument that involves ($\forall$E):

1	Everyone is a person.	P
2	If Alfred is a person, then Bob is a zombie.	P
3	Alfred is a person.	1$\forall$E
4	Therefore, Bob is a zombie.	2,3$\rightarrow$E

Notice that since line 1 states that everyone is a person, a use of ($\forall$E) allows for deriving the proposition that a particular person in the domain of discourse is a person. In the formal language of predicate logic the above argument is the following:

$$(\forall x)Px, Pa{\rightarrow}Zb \vdash Zb$$

1	$(\forall x)Px$	P
2	$Pa{\rightarrow}Zb$	P
3	Pa	$1\forall E$
4	Zb	$2,3{\rightarrow}E$

In the above example, since line 1 states that everything is a person, we are justified in inferring a substitution instance of that expression. In other words, from the quantified expression '$(\forall x)Px$' at line 1 we are justified in inferring a substitution instance 'Pa' at line 3. More concretely, since everything is a person, it follows in inferring that Alfred is a person. Universal elimination ($\forall E$) is valid for it would be impossible for *Everything is a person* to be true, yet *Alfred is a person* to be false.

Again, note that a different substitution instance '$\mathbf{P}(b/x)$' or '$\mathbf{P}(c/x)$' or '$\mathbf{P}(d/x)$' could have been chosen. To repeat, any substitution instance within the domain of discourse can be derived. That is, we could have inferred *Bob is a person*, or *Frank is a person*, or *Mary is a person* from line 1 since '$(\forall x)Px$' says that everything is a person. The reason '$\mathbf{P}(a/x)$' is chosen rather than any other proposition is that this selection will aid in the solution of the proof.

Here is another example of a proof involving ($\forall E$):

$$(\forall x)[Px{\rightarrow}(\forall y)(Qx{\rightarrow}Wy)], Pb{\wedge}Qb \vdash Wt$$

1	$(\forall x)[Px{\rightarrow}(\forall y)(Qx{\rightarrow}Wy)]$	P
2	$Pb{\wedge}Qb$	P
3	Pb	$2{\wedge}E$
4	$Pb{\rightarrow}(\forall y)(Qb{\rightarrow}Wy)$	$1\forall E$
5	$(\forall y)(Qb{\rightarrow}Wy)$	$3,4{\rightarrow}E$
6	$Qb{\rightarrow}Wt$	$5\forall E$
7	Qb	$2{\wedge}E$
8	Wt	$6,7{\rightarrow}E$

Notice three things about the above proof. First, when ($\forall E$) is applied to the left-most quantifier in line 1, every variable bound by *that* quantifier is replaced with the same substitution instance '$\mathbf{P}(b/x)$.' This is another way of saying that ($\forall E$) requires uniform replacement of bound variables with individual constants. Second, notice that at line 4 the universal quantifier ($\forall y$) is not the main operator, so ($\forall E$) cannot be applied to it. When using ($\forall E$), make sure that the proposition you are applying it to is a universally quantified proposition. Third, notice that at line 4, when ($\forall E$) is applied to line 1, '$\mathbf{P}(b/x)$' is chosen as a substitution instance. Also notice that at line 6, '$\mathbf{P}(t/y)$' is the substitution instance rather than '$\mathbf{P}(b/y)$.' Consider the following question:

When using ($\forall E$), I know that I can uniformly replace bound variables with any individual constant of my choosing. However, which individual constant should I choose?

The choice of individual constants is guided by two considerations:

(1) It is guided by individual constants already occurring in the proof.
(2) It is guided by individual constants in the conclusion.

To see the first consideration more clearly, take a look at the beginning portion of the above proof.

1	$(\forall x)[Px{\rightarrow}(\forall y)(Qx{\rightarrow}Wy)]$	P
2	$Pb{\land}Qb$	P
3	Pb	$2{\land}E$
4	**$Pb{\rightarrow}(\forall y)(Qb{\rightarrow}Wy)$**	**$1\forall E$**
5	$(\forall y)(Qb{\rightarrow}Wy)$	$3,4{\rightarrow}E$

Notice that when ($\forall$E) is used at line 4, 'b' is chosen because 'b' already occurs in the proof at lines 2 and 3, and because choosing 'b' will allow for inferring line 5. The general idea is that choosing an individual already found in the proof will facilitate the use of other derivation rules.

To see the second consideration more clearly, take a look at the latter portion of the above proof.

5	$(\forall y)(Qb{\rightarrow}Wy)$	$3,4{\rightarrow}E$
6	**$Qb{\rightarrow}Wt$**	**$5\forall E$**
7	Qb	$2{\land}E$
8	Wt	$6,7{\rightarrow}E$

Notice that when ($\forall$E) is used at line 6, 't' is chosen because 't' occurs in the conclusion.

These considerations allow for formulating a strategic rule for the use of ($\forall$E) and the rest of the quantifier rules more generally.

SQ#1($\forall$E)   When using ($\forall$E), the choice of substitution instances '**P**(a/x)' should be guided by the individual constants (names) already occurring in the proof and any individual constants (names) occurring in the conclusion.

Consider another example of a proof involving the use of ($\forall$E) for the following argument:

$$(\forall x)(Pxa{\land}Qxa) \vdash (Paa{\land}Pba){\land}Pma$$

1	$(\forall x)(Pxa{\land}Qxa)$	P
2	$Paa{\land}Qaa$	$1\forall E$
3	$Pba{\land}Qba$	$1\forall E$
4	$Pma{\land}Qma$	$1\forall E$

5	Paa	2∧E
6	Pba	3∧E
7	Pma	4∧E
8	Paa∧Pba	5,6∧I
9	(Paa∧Pba)∧Pma	7,8∧I

In the above proof, notice that the use of (∀E) at lines 2 to 4 is guided by the individual constants occurring in the conclusion.

8.1.2 Existential Introduction (∃I)

Existential introduction (∃I) is a derivation to an existentially quantified proposition '(∃x)**P**' from a possible substitution instance of it '**P**(a. . .v/x).' The procedure involves introducing an existentially quantified expression by consistently replacing at least one substitution instance with a variable that is bound by the existential quantifier.

Existential Introduction (∃I) From any possible substitution instance '**P**(a/x),' an existentially quantified proposition '(∃x)**P**' can be derived by consistently replacing at least one individual constant (name) with an existentially quantified variable.	**P**(a/x) (∃x)**P**	∃I

Here is a simple illustration:

$$Zr \vdash (\exists x)Zx$$

1	Zr	P
2	(∃x)Zx	1∃I

Notice that an existentially quantified proposition is derived by replacing at least one individual constant with an existentially quantified bound variable. In the above example, 'r' is replaced by the existentially quantified *x*.

Here is a slightly more complicated proof involving (∃I):

$$Pa \to Qa, Pa \vdash (\exists y)Qy \land (\exists x)Px$$

1	Pa→Qa	P
2	Pa	P
3	Qa	1,2→E
4	(∃x)Px	2∃I
5	(∃y)Qy	3∃I
6	(∃x)Px∧(∃y)Qy	4,5∧I

In the above example, from both 'Pa' and 'Qa,' two existentially quantified expressions are inferred. That is, '**P**(a/x)' and '**Q**(a/y)' are replaced by existentially quantified expressions.

It is important to note that it is only necessary to replace one individual constant, but if there is more than one replacement, the replacement must be uniform. Consider the following example:

1	Laab	P
2	(∃x)Lxab	1∃I
3	(∃x)Laxb	1∃I
4	(∃x)Lxxb	1∃I
5	(∃x)Lxxx	1∃I—**NO!**

Excluding line 5, each of the above uses of (∃I) is valid. Notice that at lines 2 and 3, only one individual constant from line 1 is replaced by the use of (∃I). This is accept-able since (∃I) says that an existentially quantified proposition '(∃x)Px' can be derived by replacing at least one individual constant with an existentially quantified variable. Also note that line 4 is correct since while more than one individual constant is being replaced, it is being replaced in a uniform manner. That is, more than one 'a' is being replaced by existentially quantified *x*'s. However, note that line 5 is not correct since the replacement of bound variables is not uniform. At line 5, not only is each 'a' being replaced by an *x* but so is every 'b.'

With a basic understanding of the correct formal use of (∃I), consider the following English argument that involves (∃I):

1	Rick is a zombie.	P
2	Therefore, someone is a zombie.	1∃I

The above example involves reasoning from an expression involving an individual constant (*Rick*), which is one of its possible substitution instances, to an existentially quantified expression.

Next, consider the following example, which illustrates that when using (∃I), an '(∃x)Px' can be inferred by consistently replacing at least one individual constant with an existentially quantified variable.

1	Alfred loves Alfred.	P
2	Someone loves Alfred.	1∃I
3	Alfred loves someone.	1∃I
4	Someone loves him- or herself.	1∃I

Each use of (∃I) is acceptable since if it is true that *Alfred loves himself*, then lines 2 to 4 must be true. The above argument formally corresponds to the following:

1	Laa	P
2	(∃x)Lxa	1∃I
3	(∃x)Lax	1∃I
4	(∃x)Lxx	1∃I

However, remember that replacement of individual constants with existentially quantified variables must be uniform. To see why this must be the case, consider the following:

1	Alice loves Bob.	P
2	Someone loves themselves.	1∃I—**NO!**

Formally, this corresponds to the following argument:

1	Lab	P
2	(∃x)Lxx	1∃I—**NO!**

The derivation from lines 1 to 2 does not follow for supposing that Alice is madly in love with Bob. She sees him, and her heart goes aflutter. There is nothing about Alice's love for another that entails that someone loves him- or herself.

However, suppose a slightly different conclusion to the above proof. That is, instead of 'Lab ⊢ (∃x)Lxx,' suppose the proof is 'Lab ⊢ (∃x)(∃y)Lxy.'

1	Lab	P/(∃x)(∃y)Lxy
2	(∃y)Lay	1∃I
3	(∃x)(∃y)Lxy	2∃I

Notice that at line 2 (∃I) can be applied to a proposition that is already existentially quantified. That is, (∃I) can be applied to '(∃y)Lay' by consistently replacing at least one individual constant with an existentially quantified variable.

Consider another example that illustrates mistaken uses of (∃I):

1	Lab	P
2	Lab→Rab	P
3	(∃x)Lxx	1∃I—**NO!**
4	(∃x)Lax→Rab	2∃I—**NO!**
5	(∃x)Lax→(∃x)Rax	4∃I—**NO!**

First, lines 4 and 5 are incorrect because (∃I), as a quantifier rule, can only be applied to the whole proposition 'Lab→Rab' at line 2 and not to subformulas within that proposition. Second, line 3 is invalid because there is not a consistent replacement of a possible substitution instance '**P**(a/x)' with an existentially quantified proposition '(∃x)**P**.' In the case of line 3, two different substitution instances are being replaced with a single bound variable, that is, '**P**(a/x)' and '**P**(b/x)' with '(∃x)**P**.'

With a clarified notion of how to use (∃I) and why it is valid, it is helpful both to develop a strategy for using (∃I) and to reinforce our understanding of (∃I) with a number of examples. Consider an example involving both (∀E) and (∃I):

$$(\forall x)(Px \land Rx), Pa \rightarrow Wb \vdash (\exists y)Wy$$

1	(∀x)(Px∧Rx)	P
2	Pa→Wb	P/(∃y)Wy
3	Pa∧Ra	1∀E
4	Pa	3∧E
5	Wb	2,4→E
6	(∃y)Wy	5∃I

The above example illustrates a strategy for using (∃I). Namely, if our aim is to obtain an existentially quantified proposition '(∃x)**P**,' try to formulate a substitution instance '**P**(a/x)' such that a use of (∃I) would result in '(∃x)**P**.' In the proof above, since the conclusion is '(∃y)Wy,' a subgoal of the proof is a possible substitution instance of '(∃y)Wy,' that is, 'W(a/x),' 'W(b/x),' 'W(c/x),' and so on.

> SQ#2(∃I)   When using (∃I), aim at deriving a substitution instance '**P**(a/x)' such that a use of (∃I) will result in the desired conclusion.

In other words, if the ultimate goal is to derive '(∃x)Px,' aim to derive a substitution instance of '(∃x)Px,' like 'Pa,' 'Pb,' 'Pr,' so that a use of (∃I) will result in '(∃x)Px.'

One way of thinking about this is to work backward in the proof. Consider the following again:

$$(∀x)(Px∧Rx), Pa→Wb ⊢ (∃y)Wy$$

1	(∀x)(Px∧Rx)	P
2	Pa→Wb	P/(∃y)Wy
.		
.		
k	W(a, . . ., v/x)	
k + 1	(∃y)Wy	k∃I

Notice that the above proof starts by moving one step backward from the conclusion and then making 'W(a. . .v/x)' the goal. Once, 'W(a. . .v/x)' is obtained, a simple use of (∃I) will yield the conclusion.

Here is another example. Prove the following:

$$(∀x)Px, (∀y)Zy ⊢ (∃x)(Px∧Zx)$$

Before beginning this proof, it can be helpful to mentally work backward from the conclusion.

1	(∀x)Px	P
2	(∀y)Zy	P/(∃x)(Px∧Zx)
.		
.		
k	P(a/x)∧Z(a/x)	
k + 1	(∃x)(Px∧Zx)	k∃I

If the goal of the conclusion is '(∃x)(Px∧Zx),' the subgoal will be something like 'Pa∧Za' since propositions like 'Pa∧Zb' or 'Pc∧Za' will not allow for a uniform replacement when (∃I) is used. With 'Pa∧Za' as a subgoal, consider the completed proof.

1	(∀x)Px	P
2	(∀y)Zy	P/(∃x)(Px∧Zx)
3	Pa	1∀E
4	**Za**	**2∀E**
5	Pa∧Za	3,4∧I
6	(∃x)(Px∧Zx)	5∃I

Notice that when using (∀E) at line 4, it is important to choose 'a' rather than some other individual constant (e.g., 'b') since using (∃I) on 'Pa∧Zb' would only allow for inferring '(∃x)(Px∧Zb)' and not '(∃x)(Px∧Zx).'

Exercise Set #1

A. Prove the following:
1. * Pa→Qb, Pa ⊢ (∃y)Qy
2.   Ma, (∃x)Mx→(Ra∧Tb) ⊢ (∃x)Rx
3. * (∀x)(Px→Rx), Pa∧Ma ⊢ (∃z)Rz
4.   Pa, [(∃x)Px∧(∃y)Py]→Qt ⊢ Qt
5. * (∀x)Px, (∀x)Px→Gb ⊢ (∃x)Gx
6.   (∀x)Pxxx, (Paaa∧Pbbb)→(∃x)Wx ⊢ (∃x)Wx
7. * (∀x)Px→(∃y)Gy, ¬(∃y)Gy ⊢ ¬(∀x)Px
8.   (∃x)Px→(∀x)Rx, (∀z)(Pz∧Fz) ⊢ (∃x)Rx∧(∃x)Fx
9. * (∀x)(∀y)(Pxy→Rxy), Paa ⊢ (∃x)Rxx
10.   ¬Pab→(∃x)Mx, ¬(∃x)Mx ⊢ (∃x)(∃y)Pxy
11. * Wab∧Qbc ⊢ [(∃y)(Way∧Qyc)∧(∃y)(Wyb)]∧(∃y)(Qyc)

Solutions to Starred Exercises in Exercise Set #1

A.

1. * Pa→Qb, Pa ⊢ (∃y)Qy

1	Pa→Qb	P
2	Pa	P/(∃y)Qy
3	Qb	1,2→E
4	(∃y)Qy	3∃I

3. * (∀x)(Px→Rx), Pa∧Ma ⊢ (∃z)Rz

1	(∀x)(Px→Rx)	P
2	Pa∧Ma	P/(∃z)Rz

3	Pa→Ra	1∀E
4	Pa	2∧E
5	Ra	3,4→E
6	(∃z)Rz	5∃I

5. * (∀x)Px, (∀x)Px→Gb ⊢ (∃x)Gx

1	(∀x)Px	P
2	(∀x)Px→Gb	P/(∃x)Gx
3	Gb	1,2→E
4	(∃x)Gx	3∃I

Notice that you cannot apply (∀E) on line 2 because it is not the main operator of the wff. Line 2 is a conditional, and since the antecedent of the conditional is in line 1, you can apply '→E.'

7. * (∀x)Px→(∃y)Gy, ¬(∃y)Gy ⊢ ¬(∀x)Px

1	(∀x)Px→(∃y)Gy	P
2	¬(∃y)Gy	P
3	¬(∀x)Px	1,2MT

Notice again that the main operator of line 1 is a conditional, and the main operator of line 2 is a negation. You cannot apply (∀E) or (∃I) to either of these rules since the quantifier is not the main operator of the wff. You can, however, apply MT since line 2 is the negation of the consequent in line 1.

9. * (∀x)(∀y)(Pxy→Rxy), ⊢ (∃x)Rxx

1	(∀x)(∀y)(Pxy→Rxy)	P
2	Paa	P
3	(∀y)(Pay→Ray)	1∀E
4	Paa→Raa	3∀E
5	Raa	2,4→E
6	(∃x)Rxx	5∃I

11. * Wab∧Qbc ⊢ [(∃y)(Way∧Qyc)∧(∃y)(Wyb)]∧(∃y)(Qyc)

1	Wab∧Qbc	P
2	(∃y)(Way∧Qyc)	1∃I
3	Wab	1∧E
4	Qbc	1∧E
5	(∃y)Wyb	3∃I
6	(∃y)Qyc	4∃I
7	(∃y)(Way∧Qyc)∧(∃y)(Wyb)	2,5∧I
8	[(∃y)(Way∧Qyc)∧(∃y)(Wyb)]∧(∃y)(Qyc)	6,7∧I

Notice that line 2 involves an application of (∃I) to the complex proposition 'Pab∧Qbc,' while lines 5 and 6 involve applications of (∃I) to atomic propositions. Note that (∃I) can be applied to a complex proposition, and this does not consist of a replacement of two substitution instances

'**W**(b/*y*)' and '**Q**(b/*y*)' with an existentially quantified proposition '(∃x)**P**.' It is a replacement of a single substitution instance '**P**(b/*y*)' with a quantified proposition '(∃x)**P**.'

8.1.3 Universal Introduction (∀I)

The remaining two quantifier rules—universal introduction (∀I) and existential elimination (∃E)—involve restrictions on their usage. Universal introduction (∀I) is a derivation to a universally quantified proposition '(∀x)**P**' from a substitution instance '**P**(a/*x*)' so long as (1) the instantiating constant 'a' does not occur in a premise or in an open assumption, and (2) the instantiating constant 'a' does not occur in the resulting '(∀x)**P**.'

	,	
Universal Introduction (∀I) A universally quantified proposition '(∀x)**P**' can be derived from a possible substitution instance '**P**(a/*x*)' provided (1) 'a' does not occur as a premise or as an assumption in an open subproof, and (2) 'a' does not occur in '(∀x)**P**.'	**P**(a/*x*) (∀x)**P**	∀I

Consider an example of a proof involving (∀I):

$$(∀x)Px ⊢ (∀y)Py$$

1	(∀x)Px	P
2	Pa	1∀E
3	(∀y)Py	2∀I

When using (∀I), always check to see if either of the restrictions has been violated. For instance,

(1) Does the substitution instance '**P**(a/*x*)' occur as a premise or as an open assumption?
(2) Does the substitution instance '**P**(a/*x*)' not occur in '(∀x)**P**?'

Notice that the use of (∀I) at line 3 does not violate the two restrictions placed on (∀I). Namely, (1) 'a' does not occur in any premise or open assumption in the proof, and (2) there is not an 'a' in the resulting quantified proposition '(∀y)Py.' In this case, it should be somewhat obvious why it is valid to infer '(∀y)Py' at line 3 from 'Pa' at line 2. This is because '(∀x)Px' and '(∀y)Py' are notational variants. That is, '(∀x) Px' says that everything in the domain of discourse has the property 'P,' and '(∀y) Py' says exactly the same thing. So, if '(∀x)Px' is true, then '(∀y)Py' cannot be false since '(∀y)Py' and '(∀x)Px' are logically equivalent.

Here is another example of a proof involving (∀I):

$$(∀x)(Px∧Rx) ⊢ (∀y)(Py)$$

1	$(\forall x)(Px \land Rx)$	P
2	$Pb \land Rb$	$1 \forall E$
3	Pb	$2 \land E$
4	$(\forall y)Py$	$3 \forall I$

Notice that the use of $(\forall I)$ at line 4 does not violate the two restrictions placed on $(\forall I)$. Namely, (1) 'b' does not occur in any premise or open assumption in the proof, and (2) there is not a 'b' in the resulting quantified proposition '$(\forall y)Py$.' Here, '$(\forall x)(Px \land Rx)$' and '$(\forall y)Py$' are not notational variants, but it should be obvious that it is impossible for '$(\forall x)(Px \land Rx)$' to be true and '$(\forall y)Py$' false since '$(\forall x)(Px \land Rx)$' says that everything is both 'P' and R, while '$(\forall y)Py$' says that everything is 'P.'

Consider a third example:

$$(\forall x)(Pxc \lor Qx), (\forall x)\neg Qx \vdash (\forall x)(Pxc)$$

1	$(\forall x)(Pxc \lor Qx)$	P
2	$(\forall x)\neg Qx$	P
3	$Pac \lor Qa$	$1 \forall E$
4	$\neg Qa$	$2 \forall E$
5	Pac	3,4DS
6	$(\forall x)Pxc$	$5 \forall I$

Notice that the use of $(\forall I)$ at line 6 does not violate the two restrictions placed on $(\forall I)$. Namely, (1) 'a' does not occur in any premise or open assumption in the proof, and (2) there is not an 'a' in the resulting quantified proposition '$(\forall x)Pxc$.' But in the above proof, it is no longer immediately obvious why $(\forall I)$ is valid. Thus, you might ask yourself the following question:

Why is it deductively valid to infer '$(\forall x)\mathbf{P}$' from '$\mathbf{P}(a/x)$'?

In fact, there seem to be at least two cases where a use of $(\forall I)$ is invalid. First, consider the statement *Johnny Walker is a criminal* (i.e., 'Cj'). This proposition involves an individual constant (name) that selects a single individual; that is, 'j' names Johnny Walker. It would be invalid to generalize from this proposition to *Everyone is a criminal*, or '$(\forall x)Cx$.' That is, it is possible for $v(Cj) = T$, yet $v(\forall x)Cx = F$. This is possible because there is nothing about the truth of Johnny Walker being a criminal that makes it such that everyone else is a criminal. Second, consider the statement *Someone is a criminal*, or '$(\exists x)Cx$.' Unlike in the first case where 'Cj' states that some specific individual is a criminal, '$(\exists x)Cx$' states that there exists a criminal in the domain of discourse but does not specifically identify that person. But, again, it would be invalid to generalize from '$(\exists x)Cx$' to *Everyone is a criminal* since it is possible for the former to be true while the latter is false. This is invalid for the same reason as in the first case: there is nothing about the truth of someone being a criminal that makes it such that everyone is a criminal. It is possible that the property of being a criminal is idiosyncratic to that existent person.

These invalid uses of (∀I) are sometimes known as *hasty generalizations*. That a single specific (or some nonspecific) object has a property does not entail that every object has that property. These two cases are what motivates the two restrictions placed on the use of (∀I). Namely, a use of (∀I) is not valid when it is used to generalize an instantiating constant that names either a single specific individual or some unknown individual. However, a use of (∀I) is valid when the instantiating constant it generalizes is an arbitrarily-selected individual, that is, an individual chosen at random from the domain of discourse.

Let's look at three concrete examples involving (∀I), each increasing in complexity. First, consider a domain of discourse consisting of living humans. Suppose that you wanted to prove the following claim: *All men are mortal*, or '(∀x)(Mx→Rx).' Before beginning, let's take some propositions as premises. Let's use *All organisms are mortal*, or '(∀x)(Ox→Rx),' as a premise and *All men are organisms*, or '(∀x)(Mx→Ox),' as a premise.

1	(∀x)(Ox→Rx)	P
2	(∀x)(Mx→Ox)	P

Your first step in trying to prove that *All men are mortal* would be to assume a randomly selected man from the domain of discourse.

1	(∀x)(Ox→Rx)	P
2	(∀x)(Mx→Ox)	P
3	Ma	A

In this case, the object constant 'a' refers not to a single man (e.g., Frank or some unknown man), but to a man randomly selected from the domain of discourse. So, in principle, it could be any man, although not every man. The next step would be to derive from these assumptions that the randomly selected man is mortal.

1	(∀x)(Ox→Rx)	P
2	(∀x)(Mx→Ox)	P
3	Ma	A
4	Oa→Ra	1∀E
5	Ma→Oa	2∀E
6	Oa	3,5→E
7	Ra	4,6→E
8	Ma→Ra	3–7→I

Line 8 reads, *If 'a' is a man, then 'a' is mortal.* The next step is to use ($\forall$I) on line 8.

1	$(\forall x)(Ox \rightarrow Rx)$	P
2	$(\forall x)(Mx \rightarrow Ox)$	P
3	Ma	A
4	Oa$\rightarrow$Ra	1$\forall$E
5	Ma$\rightarrow$Oa	2$\forall$E
6	Oa	3,5$\rightarrow$E
7	Ra	4,6$\rightarrow$E
8	Ma$\rightarrow$Ra	3–7$\rightarrow$I
9	$(\forall x)(Mx \rightarrow Rx)$	8$\forall$I

Line 9 reads, *All men are mortal.* Note two things. First, line 9 does not violate either of the two restrictions on the use of ($\forall$I), and it is not a hasty generalization because it generalizes not from some specific man and not from some unknown man but from a randomly selected man. We have shown then that if we randomly select a man from the universe of discourse and show that the man has the property of being mortal, we can generalize that every man has that property. This is valid because the generalization depends not on any idiosyncratic property that belongs to a specific man but on a property of all men.

As a second example, consider that the smallest and the only even prime number is 2. This follows from the fact that 1 is not a prime number (by definition), that every even number greater than 2 is divisible by 2, and that by definition, a prime number is divisible only by itself and 1. Suppose then that we want to claim *No positive, even integer greater than 2 is prime.* We know that this is true, but we want to prove that it follows from the considerations above. In other words, we want to prove something roughly translated as follows:

$$(\forall x)\{[(Ix \wedge Qx) \wedge (Ex \wedge Gx] \rightarrow \neg Px\}$$

The above formula reads, *No integer that is positive and even and greater than 2 is prime.* In order to derive this statement, start with the following assumption:

1 | $(Ia \wedge Qa) \wedge (Ea \wedge Ga)$ A

This says to assume that some randomly selected integer 'a' from a universe of discourse consisting of positive integers is an even integer greater than 2. From this assumption, the definition of a prime number, and the fact that every even number greater than 2 is divisible by 2, we can infer that this randomly selected integer is not prime.

1 | $(Ia \wedge Qa) \wedge (Ea \wedge Ga)$ A

$$
\begin{array}{c|l}
 & \quad\bullet \\
 & \quad\bullet \\
k & \neg Pa
\end{array}
$$

Next, we can make use of conditional introduction and exit the subproof.

$$
\begin{array}{cl}
1 & \begin{array}{|l} (Ia\wedge Qa)\wedge(Ea\wedge Ga) \qquad A \\ \quad\bullet \\ \quad\bullet \\ \quad\bullet \\ \end{array} \\
k & \;\;\neg Pa \\
k+1 & \;\;[(Ia\wedge Qa)\wedge(Ea\wedge Ga)]\to\neg Pa \quad 1-k,\to I
\end{array}
$$

Line $k+1$ states that this randomly selected integer is not prime. From here, we can make use of ($\forall$I) and conclude the proof.

$$
\begin{array}{cl}
1 & \begin{array}{|l} (Ia\wedge Qa)\wedge(Ea\wedge Ga) \qquad A \\ \quad\bullet \\ \quad\bullet \\ \quad\bullet \\ \end{array} \\
k & \;\;\neg Pa \\
k+1 & \;\;[(Ia\wedge Qa)\wedge(Ea\wedge Ga)]\to\neg Pa \qquad 1-k,\to I \\
k+2 & \;\;(\forall x)\{[(Ix\wedge Qx)\wedge(Ex\wedge Gx]\to\neg Px\} \qquad k+1,\forall I
\end{array}
$$

The above proof does not violate either of our two restrictions: 'a' occurs in neither a premise nor an open assumption. This is evident from the fact that the above proof makes use of no premises, and the subproof involving an assumption closes once '→I' is applied at line $k+1$. We can generalize at line $k+2$ because when we chose 'a' from the universe of discourse, we do not choose a specific positive integer that is even and greater than 2. That is, 'a' is not an abbreviation for 4, or 6, or 8. In the assumption, the selection of 'a' is random or arbitrary, for 'a' is any individual number that has the property of being an integer, positive, even, and greater than 2. So if we can say, *Take an integer that is positive, even, and greater than 2*, it follows that this integer will not be prime, and we can reason to *Any integer that is positive, even, and greater than 2 is not prime*. And from this we can generalize to *No integer that is positive, even, and greater than 2 is prime*.

As one final example, consider the general structure of Euclid's proof of the Pythagorean theorem involving a randomly selected right triangle. This is proposition I.47 in Euclid's *Elements*. The conclusion of the proof is the following:

For every right triangle, the square of the hypotenuse is equal to the sum of the squares of the legs. That is, $c^2 = a^2 + b^2$.

We can prove this by beginning with an assumption. That is, assume some randomly selected Euclidean triangle has an angle equal to 90 degrees. That is, assume $\triangle ABC$ where $\angle ACB$ is 90°. From this assumption, various assumptions about space and lines, and more general facts about triangles in general, we derive the consequence that the randomly selected triangle is such that the square of the hypotenuse is equal to the square of its two sides. After deriving this consequence, a use of '→I' is applied in order to exit the subproof, and then the conditional is generalized.

1	$\triangle ABC$ where $\angle ACB$ is 90°	A
.	.	.
.	.	.
.	.	.
k	$c^2 = a^2 + b^2$	.
$k+1$	ABC where $\angle ACB$ is 90° $\rightarrow c^2 = a^2 + b^2$	$1-k, \rightarrow$I
$k+2$	For all right-angled triangles, $c^2 = a^2 + b.$	$k+1, \forall$I

The proof of the Pythagorean theorem thus involves an inference from an arbitrarily selected individual constant to a universal statement about all right triangles.

Before considering a strategy for using ($\forall$I), it is helpful to examine a number of incorrect uses of ($\forall$I) so as to reinforce its correct usage.

1	Fbc∧Fbd	P
2	Fbc	1∧E
3	Fbd	1∧E
4	($\forall$x)(Fxc)	2$\forall$I—**NO!**
5	($\forall$x)(Fbx)	2$\forall$I—**NO!**
6	($\forall$x)(Fxd)	3$\forall$I—**NO!**
7	($\forall$x)(Fbx)	3$\forall$I—**NO!**

Lines 4 to 7 are all mistaken uses of ($\forall$I) because it violates the first restriction, namely, that the individuating constant does not occur as a premise or in an open assumption. In the above case, 'b', 'c,' and 'd' all occur in premise 1 and therefore do not refer to some arbitrarily selected individual that can be generalized. For example, the use of ($\forall$I) in line 4 on line 2 is incorrect since it involves replacing a substitution instance 'F(b/x)' with a universally quantified variable.

Consider another violation of this same restriction:

1	Fbb	P
2	Raa	A
3	($\forall$x)Rxx	2$\forall$I—**NO!**

Line 3 violates the first restriction, namely, that the individuating constant does not occur as a premise or in an open assumption. At line 2, 'a' occurs as an open assumption; therefore the use of (∀I) at line 3 is not acceptable. However, with this said, the following is an acceptable usage:

1	Fbb	P
2	Raa	A
3	Raa∨Pa	2∨I
4	Raa→(Raa∨Pa)	2–3→I
5	(∀x)[Rxx→(Rxx∨Px)]	4∀I

Line 5 is valid is because the instantiating constant 'a' no longer occurs in an open assumption since the subproof involving the assumption closes at line 4.

Consider an example of the second restriction on the use of (∀I):

1	(∀x)(Pxx→Lb)	P
2	Pcc→Lb	1∀E
3	(∀y)(Pyc→Lb)	2∀I—**NO!**
4	(∀y)(Pcy→Lb)	2∀I—**NO!**
5	(∀y)(Pyy→Lb)	2∀I

In the above example, lines 3 and 4 are incorrect insofar as they violate the second restriction. Namely, the substitution instance '**P**(c/x)' occurs in the quantified expression, that is, '(∀y)(Py**c**→Lb)' and '(∀y)(P**c**y→Lb).' However, line 5 is permissible, for when (∀I) is applied, the substitution instance is not in the quantified expression.

Finally, note a strategic rule for the use of (∀I).

SQ#3(∀I) When the goal proposition is a universally quantified proposition '(∀x)**P**,' derive a substitution instance '**P**(a/x)' such that a use of (∀I) will result in the desired conclusion.

Consider the following illustration of this strategic rule:

1	Rcc	A
2	Rcc	1R
3	Rcc→Rcc	1–2→I
4	(∀x)(Rxx→Rxx)	3∀I

Notice that 'Rcc→Rcc' is a proposition such that a use of (∀I) will result in '(∀x)(Rxx→Rxx).' It is important to see that 'Rcc' is assumed rather than '(∀x)Rxx.' Consider the above proof but this time not using SQ#3(∀I) as a strategy.

1	(∀x)Rxx	A
2	(∀x)Rxx	1R
3	(∀x)Rxx→(∀x)Rxx	1–2→I

Notice that line 3 is not the desired conclusion. This strategic rule suggests that if the goal is a universally quantified proposition '$(\forall x)\mathbf{P}$,' the first step should be a substitution instance '$\mathbf{P}(a/x)$' from which a universally quantified proposition '$(\forall x)\mathbf{P}$' can ultimately be derived using $(\forall I)$.

The strategic rule for $(\forall I)$ is similar to that for $(\exists I)$ insofar as one way of thinking about its use is to mentally work backward in the proof. Consider the following proof:

$$\vdash (\forall x)[Rxx \rightarrow (Sxx \vee Rxx)]$$

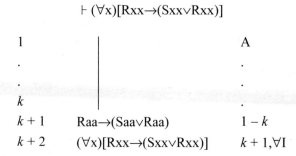

Notice that while the conclusion of the proof is '$(\forall x)[Rxx \rightarrow (Sxx \vee Rxx)]$,' the strategic rule for obtaining universally quantified propositions says that we ought to derive a substitution instance '$\mathbf{P}(a/x)$' such that a use of $(\forall I)$ will result in the desired conclusion. In the above proof, 'Raa$\rightarrow$(Saa$\vee$Raa)' or some other similar proposition is an example. With 'Raa$\rightarrow$(Saa$\vee$Raa)' as our subgoal, we can now use earlier strategic rules to solve the proof. For instance, since 'Raa$\rightarrow$(Saa$\vee$Raa)' is a conditional, the strategic rule for conditionals will help solve the proof.

1	Raa	A/Saa$\vee$Raa
2	Saa$\vee$Raa	1$\vee$I
3	Raa$\rightarrow$(Saa$\vee$Raa)	1–2$\rightarrow$I
4	$(\forall x)[Rxx \rightarrow (Sxx \vee Rxx)]$	3$\forall$I

8.1.4 Existential Elimination ($\exists$E)

Existential elimination ($\exists$E) is a derivation rule starting from an existentially quantified expression '$(\exists x)\mathbf{P}$' and an assumption '$\mathbf{P}(a/x)$' that is a possible substitution instance of '$(\exists x)\mathbf{P}$.' ($\exists$E) can be understood as the elimination (or replacement) of an existential quantified expression '$(\exists x)\mathbf{P}$' with a proposition that follows from it. There are two restrictions on its use:

Existential Elimination ($\exists$E)		
From an existentially quantified expression '$(\exists x)\mathbf{P}$,' an expression '$\mathbf{Q}$' can be derived from the derivation of an assumed substitution instance '$\mathbf{P}(a/x)$' of '$(\exists x)\mathbf{P}$' provided (1) the individuating constant 'a' does not occur in any premise or in an active proof (or subproof) prior to its arbitrary introduction in the assumption '$\mathbf{P}(a/x)$,' and (2) the individuating constant 'a' does not occur in proposition '$\mathbf{Q}$' discharged from the subproof.	$(\exists x)\mathbf{P}$ ⎪ $\mathbf{P}(a/x)$ ⎪ . ⎪ . ⎪ . ⎪ $\mathbf{Q}$ $\mathbf{Q}$	$\exists$E

To illustrate, consider a very simple use of (∃E):

1	(∃x)Px	P
2	Pa	A/∃E
3	(∃y)Py	2∃I
4	(∃y)Py	1,2–3∃E

Note three things about this use of (∃E). First, note that a use of (∃E) requires that an assumption be made at line 2. Second, note that '(∃y)Py' clearly follows from '(∃x)Px.' The two are notational variants. Third, and finally, consider whether the use of this derivation rule above has violated either of the two restrictions. Whenever using (∃E), always ask whether these two restrictions have been violated:

(1) Does the individuating constant 'a' occur in any premise or in an active proof (or subproof) prior to its arbitrary introduction in the assumption '**P**(a/x)'?
(2) Does the individuating constant 'a' occur in proposition '**Q**' discharged from the subproof?

If the answer to both of these questions is no, then you have not violated the restrictions placed on the use of this derivation rule. In the case of the above example, note that neither of the two restrictions is violated. That is, when 'Pa' is assumed at line 2, the individual constant 'a' does not occur in any premise or in an active proof (or subproof), and when '(∃y)Py' is discharged from the subproof, 'a' is not found in '(∃y)Py.'
Consider another slightly more complicated example involving (∃E):

$$(\exists x)Px \vdash (\exists x)(Px \lor \neg Mx)$$

1	(∃x)Px	P
2	Pa	A/∃E
3	Pa∨¬Ma	2∨I
4	(∃x)(Px∨¬Mx)	3∃I
5	(∃x)(Px∨¬Mx)	1,2–4∃E

In the above example, note that none of the restrictions have been violated. This is evident because (1) the individuating constant 'a' employed in the assumption at line 2 does not occur in the premise or any open assumptions. This includes the fact that the individuating constant 'a' employed in the assumption at line 2 is not in the existentially quantified expression '(∃x)Px' at line 1. Also, (2) the individuating constant 'a' does not occur in the proposition that is discharged from the subproof, that is, (∃x)(Px∨¬Mx).
Consider another example:

1	(∃z)(Wz∧Mz)	P
2	Wa∧Ma	A/∃E
3	Wa	2∧E

4	Ma	2∧E
5	(∃z)Wz	3∃I
6	(∃z)Mz	4∃I
7	(∃z)Wz∧(∃z)Mz	5,6∧I
8	(∃z)Wz∧(∃z)Mz	1,2–7∃E

Treating 'Wx' = *x is white* and 'Mx' = *x is a moose*, line 1 reads, *Something is a white moose.* From this proposition, *Something is white, and something is a moose* is inferred.

In order to get clearer on the correct use of (∃E), it can be helpful to examine some incorrect uses of the rule. It can be easy to overlook the two restrictions placed on the use of (∃E), so consider again the two restrictions on the use of (∃E):

(∃x)P

     P(a/x)   (1) The individuating constant 'a' should not occur in any premise or in an active proof (or subproof) prior to its arbitrary introduction in the assumption **P**(a/x).

     Q

Q   (2) The individuating constant 'a' should not occur in proposition **Q** discharged from the subproof.

Consider a violation of the first restriction, namely, that the individuating constant 'a' does not occur in any premise or in an active proof (or subproof) prior to its arbitrary introduction in the assumption '**P**(a/x).'

1	Ea→(∀x)Px	P
2	(∃x)Ex	P
3	Ea	A
4	(∀x)(Px)	1,3→E
5	(∀x)Px	2,3–4∃E—**NO!**

Line 5 is incorrect since the instantiating constant 'a' occurs in premise 1 as an individual constant. Assuming 'Ea' at line 3 allows us to use '→E' at line 4, which then allows us to incorrectly infer '(∀x)Px.' This is clearly invalid, for consider two English translations of lines 1 and 2.

1	If 3 is even, then every number is prime.	P
2	Some number is even.	P

3   .

4   .

5   Therefore, every number is prime.   2,3–4 ∃E—**NO!**

Propositions in lines 1 and 2 are both true, but line 5 is clearly false since 10 is not a prime number. The invalid use of (∃E) results from inferring more than what is asserted by the existential proposition at line 2. '(∃x)Px' says some number is prime, while the faulty use of (∃E) assumes that 3 is even. However, the following is not a violation of the first restriction:

1	(∃x)(Pxx∧Qxx)	P
2	Paa∧Qaa	A/∃E
3	Qaa	2∧E
4	(∃x)Qxx	3∃I
5	(∃x)Qxx	1,2–4∃E

Notice that the individuating constant 'a' at line 2 does not occur in any premise or assumption.

Consider another example where the first restriction is violated:

1	(∀x)(∃y)Txy	P
2	(∃y)Tay	1∀I
3	Taa	A/∃E
4	(∃x)Txx	3∃I
5	(∃x)Txx	2,3–4∃E—**NO!**

The first restriction states that the individuating constant 'a' does not occur in any premise or in an active proof (or subproof) prior to its arbitrary introduction in the assumption '**P**(a/x).' So, while no individuating constant occurs as a part of the premise of line 1, the individuating constant 'a' does occur as a part of an active proof at line 2.

Given the above restriction, the following strategic rule can be formulated for using (∃E):

SQ#4(∃E)   Generally, when deciding upon a substitution instance '**P**(a/x)' to assume for a use of (∃E), choose one that is foreign to the proof.

Next, consider an example that violates the second restriction on the use of (∃E). This restriction states that the instantiating constant 'a' in the assumed substitution instance '**P**(a/x)' does not occur in proposition '**Q**' discharged from the subproof.

1	(∃z)(Wzz∧Mz)	P
2	Wbb∧Mb	A/∃E

3	Wbb	2∧E
4	(∃x)(Wbx)	4∃I
5	(∃x)(Wbx)	1,2–5∃E—**NO!**

The above argument is invalid. The fallacious step of reasoning occurs at line 5 because the instantiating constant 'b' in the assumption at line 2 is in the proposition derived by '∃E' outside the subproof at line 5. To see that this is invalid, consider that line 1 only refers to there being at least one object *z* that is both 'Wzz' and 'Mz.' While we could validly infer '(∃x)(Wzz),' we could never validly infer a specific object constant.

Finally, note that the examples illustrating (∃E) all begin with an existentially quantified proposition '(∃x)**P**' and result in an existentially quantified proposition '(∃x)**P**.' However, note that (∃E) does not mandate this, and in a number of examples the derived proposition is not an existentially quantified proposition. Here are two, the first being somewhat trivial.

1	(∀x)Px	P
2	(∃x)Zx	P
3	Za	A/∃E
4	(∀x)Px	1R
5	(∀x)Px	2,3–4∃E

Notice that '(∀x)Px' does not violate either of the two restrictions placed on the use of (∃E).

Finally, consider the following derivation:

1	(∃y)Py→(∀y)Ry	P
2	(∃x)Px	P/(∀y)Ry
3	Pa	A/∃E
4	(∃y)Py	3∃I
5	(∀y)Ry	1,4→E
6	(∀y)Ry	2,3–5∃E

In the above example, notice that '(∀y)Ry' cannot be derived by using lines 1 and 2 and '→E' since '(∃y)Py' and '(∃x)Px' are not the same proposition.

Exercise Set #2

A. Prove the following:
1. * (∃x)(Gx) ⊢ (∃z)(Gz)
2. (∀x)(Fx∧Mx) ⊢ (∃z)Fz
3. * (∃x)[Rx∧(∃z)(Mz)] ⊢ (∃y)My
4. (∀x)(Zx∧Mx),(∃x)Zx→Ra ⊢ (∃y)Ry

5. * $(\forall x)(\forall y)(Zxa \wedge Mxy) \vdash (\forall z)(Zza \wedge Mzz)$

6. $\vdash (\forall x)(Px) \rightarrow (\forall x)(Px)$

7. * $\vdash (\forall x)(Px) \rightarrow (\forall x)(Px \vee Qx)$

8. $Zaaa \rightarrow Qb, (\exists x)Px, (\exists y)Py \rightarrow (\exists x)Lx, (\exists z)Lz \rightarrow (\forall x)Zxxx \vdash Qb$

Solutions to Starred Exercises in Exercise Set #2

1. * $(\exists x)(Gx) \vdash (\exists z)(Gz)$

1	$(\exists x)Gx$	P
2	Ga	A/$\exists$E
3	$(\exists z)Gz$	2$\exists$I
4	$(\exists z)Gz$	1,2–3$\exists$E

This proof is a straightforward example of ($\exists$E), although it illustrates that the choice of variable is not relevant. Rather than '$(\exists z)Gz$,' we could have inferred '$(\exists y)Gy$' or '$(\exists x)Gx$.'

3. * $(\exists x)[Rx \wedge (\exists z)(Mz)] \vdash (\exists y)My$

1	$(\exists x)[Rx \wedge (\exists z)(Mz)]$	P/$(\exists y)My$
2	$Ra \wedge (\exists z)Mz$	A/$\exists$E
3	$(\exists z)Mz$	2$\wedge$E
4	Mb	A/$\exists$E
5	$(\exists y)My$	4$\exists$I
6	$(\exists y)My$	3,4–5$\exists$E
7	$(\exists y)My$	1,2–6$\exists$E

The key to solving the above example is to make repeated use of ($\exists$E).

5. * $(\forall x)(\forall y)(Zxa \wedge Mxy) \vdash (\forall z)(Zza \wedge Mzz)$

1	$(\forall x)(\forall y)(Zxa \wedge Mxy)$	P/$(\forall z)(Zza \wedge Mzz)$
2	$(\forall y)(Zba \wedge Mby)$	1$\forall$E
3	$Zba \wedge Mbb$	2$\forall$E
4	$(\forall z)(Zza \wedge Mzz)$	3$\forall$I

7. * $\vdash (\forall x)(Px) \rightarrow (\forall x)(Px \vee Qx)$

1	$(\forall x)Px$	A/$(\forall x)(Px \vee Qx)$
2	$\neg(Pa \vee Qa)$	A/P$\wedge \neg$P
3	$\neg Pa \wedge \neg Qa$	2DEM
4	Pa	1$\forall$E
5	$\neg Pa$	3$\wedge$E
6	$Pa \vee Qa$	2–5$\neg$E
7	$(\forall x)(Px \vee Qx)$	6$\forall$I
8	$(\forall x)Px \rightarrow (\forall x)(Px \vee Qx)$	1–7$\rightarrow$I

The first step in solving the above theorem is recognizing that the theorem is a conditional. Recognizing that it is a conditional, we first assume the antecedent of the conditional. The second step is realizing that our goal formula is a universally quantified disjunction. In order to obtain this, our next assumption will not simply be a substitution instance of '$(\forall x)(Px \lor Qx)$' but will be guided by our strategic rule for a substitution instance that is a disjunction. Thus, we assume the negation of the disjunction, derive a contradiction, and use '$\neg E$' to exit with our desired conclusion.

8.2 QUANTIFIER NEGATION (QN)

In propositional logic, we devised a system of natural deduction (PD) and then added a number of additional rules to form (PD+), whose main purpose was to simplify or expedite proofs. Similarly, in predicate logic, $(\forall I)$, $(\forall E)$, $(\exists I)$, and $(\exists E)$ form a system of natural deduction (RD). In this section, we develop this system by adding an equivalence rule called *quantifier negation* (QN). The addition of QN to RD forms RD+. One of the central benefits of adding QN to the existing set of derivation rules is that it will allow us to readily deal with the negated universal '$\neg(\forall x)\mathbf{P}$' and negated '$\neg(\exists x)\mathbf{P}$' propositions.

Quantifier Negation (QN)		
From a negated universally quantified expression '$\neg(\forall x)\mathbf{P}$,' an existentially quantified expression '$(\exists x)\neg\mathbf{P}$' can be derived, and vice versa. Also, from a negated existentially quantified expression '$\neg(\exists x)\mathbf{P}$,' a universally quantified expression '$(\forall x)\neg\mathbf{P}$' can be inferred, and vice versa.	$\neg(\forall x)\mathbf{P} \dashv\vdash (\exists x)\neg\mathbf{P}$ $\neg(\exists x)\mathbf{P} \dashv\vdash (\forall x)\neg\mathbf{P}$	QN QN

Consider a very simple use of QN:

$$
\begin{array}{lll}
1 & \neg(\forall x)Px & P \\
2 & (\exists x)\neg Px & 1\,QN \\
3 & \neg(\forall x)Px & 2\,QN \\
\end{array}
$$

Notice that the application of QN on line 1 allows for replacing the negated universal '$\neg(\forall x)Px$' with an existential proposition that quantifies over a negated propositional form '$(\exists x)\neg Px$.' Also notice that QN allows for replacing the existential proposition that quantifies over a negated propositional form '$(\exists x)\neg Px$' with a negated universally quantified proposition '$\neg(\forall x)Px$.'

Consider another simple use of QN:

$$
\begin{array}{lll}
1 & \neg(\exists z)(Wzz \land Mz) & P \\
2 & (\forall z)\neg(Wzz \land Mz) & 1\,QN \\
3 & \neg(\exists z)(Wzz \land Mz) & 2\,QN \\
\end{array}
$$

In the example above, '¬(∃z)(Wzz∧Mz)' is an instance of '¬(∃x)**P**,' and QN tells us that we can derive '(∀x)¬**P**,' which is '(∀z)¬(Wzz∧Mz).'

Consider a more complex example:

1	¬(∀z)¬(Wzz→¬Mz)	P
2	(∃x)(Px∧Rx)	P
3	(∃z)¬¬(Wzz→¬Mz)	1QN
4	(∃z)(Wzz→¬Mz)	3DN
5	¬¬(∃x)(Px∧Rx)	2DN
6	¬(∀z)¬(Wzz→¬Mz)	3QN

Again, notice that at line 3, from a negated universally quantified proposition '¬(∀x)**P**,' we infer an existentially quantified negated proposition '(∃x)¬**P**.'

Since quantifier negation is an equivalence rule, it can be applied to subformulas within wffs and not merely to whole propositions whose main operator is the negation. For instance,

1	¬(∀z)¬(∃y)[Wzy→¬(∀x)¬My]	P
2	¬(∀z)¬(∃y)[Wzy→(∃x)¬¬My]	1QN
3	(∃z)¬¬(∃y)[Wzy→(∃x)¬¬My]	2QN
4	(∃z)(∃y)[Wzy→(∃x)¬¬My]	3DN
5	(∃z)(∃y)[Wzy→(∃x)My]	4DN

Notice that the first use of QN in line 2 is on the quantified antecedent of the conditional. That is, from a negated universally quantified subformula '¬(∀x)¬My,' we can infer the existentially quantified negated subformula '(∃x)¬¬My.'

QN is a derived rule and so can be proved using the underived quantifier rules ('∀E,' '∀I,' '∃I,' '∃E'). In order to prove this, the following needs to be shown without using QN:

$$¬(∀x)\mathbf{P} ⊣ ⊢ (∃x)¬\mathbf{P}$$
$$¬(∃x)\mathbf{P} ⊣ ⊢ (∀x)¬\mathbf{P}$$

The following proof shows '¬(∀x)Px ⊢ (∃x)¬Px.' The remainder of the proofs are left as exercises.

1	¬(∀x)Px	P/(∃x)¬Px
2	¬(∃x)¬Px	A/contra
3	¬Pa	A/contra
4	(∃x)¬Px	3∃I
5	¬(∃x)¬Px	2R
6	Pa	3–5¬E
7	(∀x)Px	6∀I
8	¬(∀x)Px	1R
9	(∃x)¬Px	2–8¬E

8.3 SAMPLE PROOFS

In this section, we illustrate the use of quantifier rules and various strategies for proving propositions in predicate logic.

Consider the following proof:

$$(\forall x)(Ax \rightarrow Bx), (\exists x)\neg Bx \vdash (\exists x)\neg Ax$$

1	$(\forall x)(Ax \rightarrow Bx)$	P
2	$(\exists x)\neg Bx$	P
3	$\neg Ba$	A/∃E
4	$Aa \rightarrow Ba$	1∀E
5	$\neg Aa$	3,4MT
6	$(\exists x)\neg Ax$	5∃I
7	$(\exists x)\neg Ax$	2,3–6∃E

There are two things of note about this proof. First, notice that since the goal of the proof is '$(\exists x)\neg Ax$,' a subgoal of the proof will be '$\neg Aa$' or some proposition such that we can use (∃I) to obtain '$(\exists x)\neg Ax$.' Second, notice that since '$(\exists x)\neg Bx$' will play a role in the proof, one should start the proof by assuming '$\neg Ba$' because starting the proof by using (∀E) on line 1 will not further the proof.

Consider another proof:

$$(\forall x)(\forall y)(Pxy \wedge Qxy), (\forall z)Pzz \rightarrow Sb \vdash (\exists x)Sx$$

Before moving forward in this proof, notice again that the conclusion is an existentially quantified proposition. Thus, the subgoal of the proof will be '$S(a. . .v/x)$.'

1	$(\forall x)(\forall y)(Pxy \wedge Qxy)$	P
2	$(\forall z)Pzz \rightarrow Sb$	P/$(\exists x)Sb$
3	$(\forall y)(Pay \wedge Qay)$	1∀E
4	$Paa \wedge Qaa$	3∀E
5	Paa	4∧E
6	$(\forall z)Pzz$	5∀I
7	Sb	2,6→E
8	$(\exists x)Sx$	7∃I

Finally, consider the following theorem of predicate logic:

$$\vdash (\forall x)[\neg(Qx \rightarrow Rx) \rightarrow \neg Px] \rightarrow (\forall x)[\neg(Px \rightarrow Rx) \rightarrow \neg(Px \rightarrow Qx)]$$

This is a zero-premise deduction, and while the size of the formula is intimidating, you should recognize that it is simply a complex instance of '$(\forall x)\mathbf{P} \rightarrow (\forall x)\mathbf{Q}$.' Since the proposition is a conditional, first assume the antecedent '$(\forall x)\mathbf{P}$.' Our goal is the consequent '$(\forall x)\mathbf{Q}$.'

1 | $(\forall x)[\neg(Qx \rightarrow Rx) \rightarrow \neg Px]$ A/$(\forall x)[\neg(Px \rightarrow Rx) \rightarrow \neg(Px \rightarrow Qx)]$

Next, we examine the main connective of '$(\forall x)\mathbf{Q}$,' which is the quantifier $(\forall x)$. This means that we will ultimately have to use $(\forall I)$ in order to obtain '$(\forall x)\mathbf{Q}$.' So, our assumption will be a possible substitution instance of '$(\forall x)\mathbf{Q}$.' Here we have two options. First, we could either assume '$\neg\mathbf{Q}(a/x)$' and derive a contradiction. Second, we could assume a substitution instance of the antecedent '$\mathbf{Q}(a/x)$,' derive the consequent, and then use '$\rightarrow$I.' We pursue this second option.

1 | $(\forall x)[\neg(Qx \rightarrow Rx) \rightarrow \neg Px]$ A/$(\forall x)[\neg(Px \rightarrow Rx) \rightarrow \neg(Px \rightarrow Qx)]$
2 | | $\neg(Pa \rightarrow Ra)$ A/$\neg(Pa \rightarrow Qa)$

Since our current goal is the negation of a conditional, we employ the strategic rule for negated propositions '$\neg\mathbf{R}$' and work toward a contradiction.

1 | $(\forall x)[\neg(Qx \rightarrow Rx) \rightarrow \neg Px]$ A/$(\forall x)[\neg(Px \rightarrow Rx) \rightarrow \neg(Px \rightarrow Qx)]$
2 | | $\neg(Pa \rightarrow Ra)$ A/$\neg(Pa \rightarrow Qa)$
3 | | | $Pa \rightarrow Qa$ A/$P \wedge \neg P$

At this point, no more assumptions are necessary, and what remains is to work toward the goals set on the right-hand side of the proof.

1	$(\forall x)[\neg(Qx \rightarrow Rx) \rightarrow \neg Px]$	A/$(\forall x)[\neg(Px \rightarrow Rx) \rightarrow \neg(Px \rightarrow Qx)]$
2	$\neg(Pa \rightarrow Ra)$	A/$\neg(Pa \rightarrow Qa)$
3	$Pa \rightarrow Qa$	A/$P \wedge \neg P$
4	$\neg(Qa \rightarrow Ra) \rightarrow \neg Pa$	1$\forall$E
5	$\neg(\neg Pa \vee Ra)$	2IMP
6	$Pa \wedge \neg Ra$	5DEM+DN
7	Pa	6$\wedge$E
8	Qa	3,7$\rightarrow$E
9	$\neg\neg Pa$	7DN
10	$Qa \rightarrow Ra$	4,9MT+DN
11	Ra	8,10$\rightarrow$E
12	$\neg Ra$	6$\wedge$E
13	$\neg(Pa \rightarrow Qa)$	3–12$\neg$I
14	$\neg(Pa \rightarrow Ra) \rightarrow \neg(Pa \rightarrow Qa)$	2–13$\rightarrow$I
15	$(\forall x)[\neg(Px \rightarrow Rx) \rightarrow \neg(Px \rightarrow Qx)]$	14$\forall$I
16	$(\forall x)[\neg(Qx \rightarrow Rx) \rightarrow \neg Px] \rightarrow (\forall x)$ $[\neg(Px \rightarrow Rx) \rightarrow \neg(Px \rightarrow Qx)]$	1–15$\rightarrow$I

END-OF-CHAPTER EXERCISES

A. Prove the following arguments:
1. * Fc∧Fb ⊢ (∃x)Fx∧Fb
2.   Fc∧Fb ⊢ (∃x)(Fx∧Fb)
3. * (∃x)Gx∧(∀y)My ⊢ (∃z)Gz
4.   (Pa∧Ra)∧Qc ⊢ (∃x)(Px∧Rx)∧Qc
5. * (∃x)(Px∧Wx),(∀x)(Px→Mx) ⊢ (∃x)(Mx)
6.   ¬(∃x)¬Rx∧¬(∀x)Px ⊢ (∀x)Rx∧(∃x)¬Px
7. * (∀x)(Rx∧Ma) ⊢ ¬(∃y)¬(Ry)
8.   (∀x)(Rx→Mx),(∀x)Rx ⊢ (∀x)Mx
9. * (∃x)(Px∧Mx),(∃y)(Py∨Ry)→(∀x)Sx ⊢ (∀x)Sx
10.   (∃x)(∀y)Pxy→(∀z)Rzz,(∀x)Pbx ⊢ Rcc
11. * (∀x)(Ax→Px),(∀x)Ax ⊢ (∀x)Px
12.   (∀x)(∃y)(∀z)(Pxy→Rz) ⊢ (∃y)(∃x)(Pyx→Rx)
13. * (∀z)(Fz→Gz),(∀w)(Gw→Mw) ⊢ (∀x)(Fx→Mx)
14.   (∀z)(∀y)(∀x)(Pxyz→Rxx),(∃y)Pyyy ⊢ (∃x)(∃y)Rxy
15. * ¬(∃x)¬Ax∨(∀x)¬Bx,¬(∀x)Ax,(∀x)(Mx→Bx) ⊢ (∃x)(¬Mx∨Rx)
16.   (∀w)(Aw→Bw),(∀x)(¬Ax→¬Cx),(∃z)Cz ⊢ (∃z)(Cz∧Bz)
17. * (∀x)(Ax→Bx),(∀x)[Bx→(Ax→¬Fx)],(∀x)[(¬Mx∧Dx)→Fx] ⊢ (∀x)[Ax→(Mx∨¬Dx)]
18.   (∃x)Px→(∀x)(Gx→Mx),(∃x)Zx→(∀x)(Mx→Fx),(∃x)(Px∧Zx) ⊢ (∀x)(Gx→Fx)
19. * (∃x)[Px∧(∀y)(Py→Ryxb)] ⊢ (∃x)(Px∧Rxxb)
20.   (∀x)(∃y)(Pxy→Zxx) ⊢ (∃z)(∃y)(Pyz→Zyy)

B. Prove the following zero-premise arguments:
1.   ⊢ (∀z)(Bz→Bz)
2.   ⊢ (∃x)(Bx→Bx)
3. * ⊢ (∀x)[Ax→(Bx→Ax)]
4.   ⊢ (∀x)[(Ax→Bx)∧(Ax→Dx)]→(∀x)[Ax→(Bx∧Dx)]
5. * ⊢ (∃x)Ax→(∃x)(Ax∧Ax)
6.   ⊢ (∀x)(Ax∨Bx)→(∀x)[(¬Ax∨Bx)→Bx]
7. * ⊢ ¬(∃x)¬[(Ax→Bx)∨(Bx→Dx)]
8.   ⊢ (∀x)Px→[(∃x)(Px→Lx)→(∃x)Lx]
9. * ⊢ (∀x)(Ax→Bx)→(∃x)[¬(Bx∧Dx)→¬(Dx∧Ax)]
10.   ⊢ (∀x)¬[(Ax→¬Ax)∧(¬Ax→Ax)]
11. * ⊢ (∀x)[Px→(Qx→Rx)]→(∀x)[(Px→Qx)→(Px→Rx)]

Solutions to End-of-Chapter Exercises

A.

1. * Fc∧Fb ⊢ (∃x)Fx∧Fb

1	Fc∧Fb	P/(∃x)Fx∧Fb
2	Fb	1∧E
3	(∃x)Fx	2∃I
4	(∃x)Fx∧Fb	2,3∧I

3. * (∃x)Gx∧(∀y)My ⊢ (∃z)Gz

1	(∃x)Gx∧(∀y)My	P/(∃z)Gz
2	(∃x)Gx	1∧E
3	Ga	A/∃E
4	(∃z)Gz	3∃I
5	(∃z)Gz	2,3–4∃E

5. * (∃x)(Px∧Wx), (∀x)(Px→Mx) ⊢ (∃x)(Mx)

1	(∃x)(Px∧Wx)	P
2	(∀x)(Px→Mx)	P/(∃x)(Mx)
3	Pa∧Wa	A/∃E
4	Pa	3∧E
5	Pa → Ma	2∀E
6	Ma	4,5→E
7	(∃x)Mx	6∃I
8	(∃x)Mx	1,3–7∃E

7. * (∀x)(Rx∧Ma) ⊢ ¬(∃y)¬(Ry)

1	(∀x)(Rx∧Ma)	P/¬(∃y)¬(Ry)
2	Rb∧Ma	1∀E
3	Rb	2∧E
4	(∀y)Ry	3∀I
5	¬¬(∀y)Ry	4DN
6	¬(∃y)¬Ry	5QN

Notice that we choose '**P**(b/x)' rather than '**P**(a/x)' as a substitution instance for '(∀x)**P**' at line 1. If you choose '**P**(a/x),' then you cannot make use of (∀I) at line 4 since one of the restrictions on (∀I) is that the instantiating constant does not occur as a premise or an open assumption.

9. * $(\exists x)(Px \wedge Mx), (\exists y)(Py \vee Ry) \rightarrow (\forall x)Sx \vdash (\forall x)Sx$

1	$(\exists x)(Px \wedge Mx)$	P
2	$(\exists y)(Py \vee Ry) \rightarrow (\forall x)Sx$	P/$(\forall x)Sx$
3	$\quad$ Pa$\wedge$Ma	A/$\exists$E
4	$\quad$ Pa	3$\wedge$E
5	$\quad$ Pa$\vee$Ra	4$\vee$I
6	$\quad (\exists y)(Py \vee Ry)$	5$\exists$I
7	$(\exists y)(Py \vee Ry)$	1,3–6$\exists$E
8	$(\forall x)Sx$	2,7$\rightarrow$E

11. * $(\forall x)(Ax \rightarrow Px), (\forall x)Ax \vdash (\forall x)Px$

1	$(\forall x)(Ax \rightarrow Px)$	P
2	$(\forall x)Ax$	P/$(\forall x)Px$
3	Aa$\rightarrow$Pa	1$\forall$E
4	Aa	2$\forall$E
5	Pa	3,4$\rightarrow$E
6	$(\forall x)Fx$	5$\forall$I

13. * $(\forall z)(Fz \rightarrow Gz), (\forall w)(Gw \rightarrow Mw) \vdash (\forall x)(Fx \rightarrow Mx)$

1	$(\forall z)(Fz \rightarrow Gz)$	P
2	$(\forall w)(Gw \rightarrow Mw)$	P/$(\forall x)(Fx \rightarrow Mx)$
3	Fa$\rightarrow$Ga	1$\forall$E
4	Ga$\rightarrow$Ma	2$\forall$E
5	$\quad$ Fa	A/Ma
6	$\quad$ Ga	3,5$\rightarrow$E
7	$\quad$ Ma	4,6$\rightarrow$E
8	Fa$\rightarrow$Ma	5–7$\rightarrow$I
9	$(\forall x)(Fx \rightarrow Mx)$	8$\forall$I

15. * $\neg(\exists x)\neg Ax \vee (\forall x)\neg Bx, \neg(\forall x)Ax, (\forall x)(Mx \rightarrow Bx) \vdash (\exists x)(\neg Mx \vee Rx)$

1	$\neg(\exists x)\neg Ax \vee (\forall x)\neg Bx$	P
2	$\neg(\forall x)Ax$	P
3	$(\forall x)(Mx \rightarrow Bx)$	P/$(\exists x)(\neg Mx \vee Rx)$
4	$(\exists x)\neg Ax$	2QN
5	$\neg\neg(\exists x)\neg Ax$	4DN
6	$(\forall x)\neg Bx$	1,5DS
7	Ma$\rightarrow$Ba	3$\forall$E
8	$\neg$Ba	6$\forall$E
9	$\neg$Ma	7,8$\neg\rightarrow$E
10	$\neg$Ma$\vee$Ra	9$\vee$I
11	$(\exists x)(\neg Mx \vee Rx)$	10$\exists$I

17. * $(\forall x)(Ax\rightarrow Bx)$, $(\forall x)[Bx\rightarrow(Ax\rightarrow\neg Fx)]$, $(\forall x)[(\neg Mx\wedge Dx)\rightarrow Fx]$ ⊢ $(\forall x)$ $[Ax\rightarrow(Mx\vee\neg Dx)]$

1	$(\forall x)(Ax\rightarrow Bx)$,	P
2	$(\forall x)[Bx\rightarrow(Ax\rightarrow\neg Fx)]$,	P
3	$(\forall x)[(\neg Mx\wedge Dx)\rightarrow Fx]$	P/$(\forall x)[Ax\rightarrow(Mx\vee\neg Dx)]$
4	Aa	A
5	$Aa\rightarrow Ba$	1$\forall$E
6	Ba	4,5$\rightarrow$E
7	$Ba\rightarrow(Aa\rightarrow\neg Fa)$	2$\forall$E
8	$Aa\rightarrow\neg Fa$	6,7$\rightarrow$E
9	$\neg Fa$	4,8$\rightarrow$E
10	$(\neg Ma\wedge Da)\rightarrow Fa$	3$\forall$E
11	$\neg(\neg Ma\wedge Da)$	9,10MT
12	$\neg\neg Ma\vee\neg Da$	11DeM
13	$Ma\vee\neg Da$	12DN
14	$Aa\rightarrow(Ma\vee\neg Da)$	4–13$\rightarrow$I
15	$(\forall x)[Ax\rightarrow(Mx\vee\neg Dx)]$	14$\forall$I

19. * $(\exists x)[Px\wedge(\forall y)(Py\rightarrow Ryxb)]$ ⊢ $(\exists x)(Px\wedge Rxxb)$

1	$(\exists x)[Px\wedge(\forall y)(Py\rightarrow Ryxb)]$	P
2	$Pa\wedge(\forall y)(Py\rightarrow Ryab)$	A/$\exists$E
3	Pa	2$\wedge$E
4	$(\forall y)(Py\rightarrow Ryab)$	2$\wedge$E
5	$Pa\rightarrow Raab$	4$\forall$E
6	Raab	3,5$\rightarrow$E
7	$Pa\wedge Raab$	3,6$\wedge$I
8	$(\exists x)(Px\wedge Rxxb)$	7$\exists$I
9	$(\exists x)(Px\wedge Rxxb)$	1,2–8$\exists$E

Notice that while the object constant 'b' occurs in lines 1 and 2, it is not a substitution instance '**P**(a/x)' used by ($\exists$E). Thus, it is valid to exit the subproof at line 9 with '$(\exists x)(Px\wedge Rxxb)$.'

B.

3. * ⊢ $(\forall x)[Ax\rightarrow(Bx\rightarrow Ax)]$

1	Aa		A/$Ba\rightarrow Aa$
2		Ba	A/Aa
3		Aa	1R
4	$Ba\rightarrow Aa$		2–3$\rightarrow$I
5	$Aa\rightarrow(Ba\rightarrow Aa)$		1–4$\rightarrow$I
6	$(\forall x)[Ax\rightarrow(Bx\rightarrow Ax)]$		5$\forall$I

5. * ⊢ (∃x)Ax→(∃x)(Ax∧Ax)

1	(∃x)Ax	A/(∃x)(Ax∧Ax)
2	Aa	A/∃E
3	Aa	1R
4	Aa∧Aa	2,3∧I
5	(∃x)(Ax∧Ax)	4∃I
6	(∃x)(Ax∧Ax)	1,2-5∃E
7	(∃x)Ax→(∃x)(Ax∧Ax)	1-6→I

7. * ⊢ ¬(∃x)¬[(Ax→Bx)∨(Bx→Dx)]

1	¬[(Aa→Ba)∨(Ba→Da)]	A
2	¬(Aa→Ba)∧¬(Ba→Da)	1DeM
3	¬(Aa→Ba)	2∧E
4	¬(Ba→Da)	2∧E
5	¬(¬Aa∨Ba)	3IMP
6	¬(¬Ba∨Da)	4IMP
7	¬¬Aa∧¬Ba	5DeM
8	¬¬Ba∧¬Da	6DeM
9	¬Ba	7∧E
10	¬¬Ba	8∧E
11	Ba	10DN
12	[(Aa→Ba)∨(Ba→Da)]	1–11¬E
13	(∀x)[(Ax→Bx)∨(Bx→Dx)]	12∀I
14	¬¬(∀x)[(Ax→Bx)∨(Bx→Dx)]	13DN
15	¬(∃x)¬[(Ax→Bx)∨(Bx→Dx)]	14QN

9. * ⊢ (∀x)(Ax→Bx)→(∃x)[¬(Bx∧Dx)→¬(Dx∧Ax)]

1	(∀x)(Ax→Bx)	A/(∃x)[¬(Bx∧Dx)→ ¬(Dx∧Ax)]
2	¬(Ba∧Da)	A/¬(Da∧Aa)
3	¬Ba∨¬Da	2DEM
4	Aa→Ba	1∀E
5	Ba→¬Da	3IMP
6	Aa→¬Da	4,5HS
7	¬Aa∨¬Da	6IMP

8	Da∧Aa	A/contra
9	Da	8∧E
10	Aa	8∧E
11	¬Aa	7,9DS
12	¬(Da∧Aa)	8-11¬I
13	¬(Ba∧Da)→¬(Da∧Aa)	2–12→I
14	(∃x)[¬(Bx∧Dx)→¬(Dx∧Ax)	13∃I
15	(∀x)(Ax→Bx)→(∃x)[¬(Bx∧Dx)→¬(Dx∧Ax)]	1–14→I

11. * ⊢ (∀x)[Px→(Qx→Rx)]→(∀x)[(Px→Qx)→(Px→Rx)]

1	(∀x)[Px→(Qx→Rx)]	A/(∀x)[(Px→Qx)→(Px→Rx)]
2	Pa→Qa	A/Pa
3	Pa	A/Ra
4	Qa	2,3→E
5	Pa→(Qa→Ra)	1∀E
6	Qa→Ra	3,5→E
7	Ra	4,6→E
8	Pa→Ra	3–7→I
9	(Pa→Qa)→(Pa→Ra)	2–8→I
10	(∀x)[(Px→Qx)→(Px→Rx)]	9∀I
11	(∀x)[Px→(Qx→Rx)]→(∀x)[(Px→Qx)→(Px→Rx)]	1–10→I

SUMMARY OF DERIVATION AND STRATEGIC RULES

Universal Elimination (∀E) From any universally quantified proposition '(∀x)**P**,' we can derive a substitution instance '**P**(a/x)' in which all bound variables are consistently replaced with any individual constant (name).	(∀x)**P** **P**(a/x)	∀E
Existential Introduction (∃I) From any possible substitution instance '**P**(a/x),' an existentially quantified proposition '(∃x)**P**' can be derived by consistently replacing at least one individual constant with an existentially quantified variable.	**P**(a/x) (∃x)**P**	∃I

Universal Introduction (∀I) A universally quantified proposition '(∀x)**P**' can be derived from a possible substitution instance '**P**(a/x)' provided (1) 'a' does not occur as a premise or as an assumption in an open subproof, and (2) 'a' does not occur in '(∀x)**P**.'	**P**(a/x) (∀x)**P**	∀I

Existential Elimination (∃E)
From an existentially quantified expression '(∃x)**P**,' an expression '**Q**' can be derived from the derivation of an assumed substitution instance '**P**(a/x)' of '(∃x)**P**' provided (1) the individuating constant 'a' does not occur in any premise or in an active proof (or subproof) prior to its arbitrary introduction in the assumption '**P**(a/x),' and (2) the individuating constant 'a' does not occur in proposition '**Q**' discharged from the subproof.
	(∃x)**P**
	   ⌐ **P**(a/x)
	   .
	   .
	   .
	   Q
	Q
		∃E

Quantifier Negation (QN) From a negated universally quantified expression '¬(∀x)**P**,' an existentially quantified expression '(∃x)¬**P**' can be derived, and vice versa. Also, from a negated existentially quantified expression '¬(∃x)**P**,' a universally quantified expression '(∀x)¬**P**' can be inferred, and vice versa.	¬(∀x)**P**⊣⊢(∃x)¬**P** ¬(∃x)**P**⊣⊢(∀x)¬**P**	QN QN

SQ#1(∀E) When using (∀E), the choice of substitution instances '**P**(a/x)' should be guided by the individual constants already occurring in the proof and any individual constants occurring in the conclusion.

SQ#2(∃I) When using (∃I), aim at deriving a substitution instance '**P**(a/x)' such that a use of (∃I) will result in the desired conclusion.

SQ#3(∀I) When the goal proposition is a universally quantified proposition '(∀x)**P**,' derive a substitution instance '**P**(a/x)' such that a use of (∀I) will result in the desired conclusion.

SQ#4(∃E) Generally, when deciding on a substitution instance '**P**(a/x)' to assume for a use of (∃E), choose one that is foreign to the proof.

Appendix

PROPOSITIONAL LOGIC

Propositional Operators

Sentence	Symbolization	Type
P and **Q**	**P**∧**Q**	Negation (¬)
not **P**	¬**P**	Conjunction (∧)
P or **Q**	**P**∨**Q**	Disjunction (∨)
if **P** then **Q**	**P**→**Q**	Conditional (→)
P if and only if **Q**	**P**↔**Q**	Biconditional (↔)
neither **P** nor **Q**	¬**P**∧¬**Q**	
not both **P** and **Q**	¬(**P**∧**Q**)	
P only if **Q**	**P**→**Q**	
P even if **Q**	**P** or **P**∧(**Q**∨¬**Q**)	
not-P unless **Q**	**P**∨**Q**	
P unless **Q**	(**P**∨**Q**)∧¬(**P**∧**Q**) or ¬(**P**↔**Q**)	

Truth Table Definitions for Propositional Operators

P	¬P
T	F
F	T

P	R	P∧R	P∨R	P→R	P↔R
T	T	T	T	T	T
T	F	F	T	F	F
F	T	F	T	T	F
F	F	F	F	T	T

Propositional Tree Decomposition Rules

Stacking	Branching	Stacking ∧ Branching
P∧Q **P** ∧D **Q** ∧D	**¬(P∧Q)** ¬P ¬Q ¬∧D	**P↔Q** P ¬P ↔D Q ¬Q ↔D
¬(P∨Q) **¬P** ¬∨D **¬Q** ¬∨D	**P∨Q** P Q ∨D	**¬(P↔Q)** P ¬P ¬↔D ¬Q Q ¬↔D
¬(P→Q) **P** ¬→D **¬Q** ¬→D	**P→Q** ¬P Q →D	
¬¬P **P** ¬¬D		

Nine Decomposable Proposition Types

Conjunction	**P∧R**
Disjunction	**P∨R**
Conditional	**P→R**
Biconditional	**P↔R**
Negated conjunction	**¬(P∧R)**
Negated disjunction	**¬(P∨R)**
Negated conditional	**¬(P→R)**
Negated biconditional	**¬(P↔R)**
Double negation	**¬¬P**

Tree Decomposition Strategies

1   Use no more rules than needed.
2   Use rules that close branches.
3   Use stacking rules before branching rules.
4   Decompose more complex propositions before simpler propositions.

Propositional Derivation (PD+) Rules

1	**Conjunction Introduction (∧I)** From 'P' and 'Q,' we can derive 'P∧Q.' Also, from 'P' and 'Q,' we can derive 'Q∧P.'		P Q P∧Q Q∧P	 ∧I ∧I
2	**Conjunction Elimination (∧E)** From 'P∧Q,' we can derive 'P.' Also, from 'P∧Q,' we can derive 'Q.'		P∧Q P Q	 ∧E ∧E
3	**Conditional Introduction (→I)** From a derivation of 'Q' within a subproof involving an assumption 'P,' we can derive 'P→Q' out of the subproof.		⎸ P ⎸ . ⎸ . ⎸ . ⎸ Q P→Q	A →I
4	**Conditional Elimination (→E)** From 'P→Q' and 'P,' we can derive 'Q.'		P→Q P Q	 →E
5	**Reiteration (R)** Any proposition 'P' that occurs in a proof or subproof may be rewritten at a level of the proof that is equal to 'P' or more deeply nested than 'P.'		P . . . P	 R
6	**Negation Introduction (¬I)** From a derivation of a proposition 'Q' and its literal negation '¬Q' within a subproof involving an assumption 'P,' we can derive '¬P' out of the subproof.		⎸ P ⎸ . ⎸ . ⎸ . ⎸ ¬Q ⎸ Q ¬P	A ¬I
7	**Negation Elimination (¬E)** From a derivation of a proposition 'Q' and its literal negation '¬Q' within a subproof involving an assumption '¬P,' we can derive 'P' out of the subproof.		⎸ ¬P ⎸ . ⎸ . ⎸ . ⎸ ¬Q ⎸ Q P	A ¬E

8	**Disjunction Introduction (∨I)** From 'P,' we can validly infer 'P∨Q' or 'Q∨P.'		P P∨Q Q∨P	∨I ∨I

9	**Disjunction Elimination (∨E)**
	From 'P∨Q' and two derivations of 'R'—one involving 'P' as an assumption in a subproof, the other involving 'Q' as an assumption in a subproof—we can derive 'R' out of the subproof.
			P∨Q
			
			  P
			  .
			  .
			  .
			  R
			
			  Q
			  .
			  .
			  .
			  R
			R
				A
				
				A
				
				∨E

10	**Biconditional Introduction (↔I)** From a derivation of 'Q' within a subproof involving an assumption 'P' and from a derivation of 'P' within a separate subproof involving an assumption 'Q,' we can derive 'P↔Q' out of the subproof.		P . . . Q Q . . . P P ↔ Q	A A ↔I
11	**Biconditional Elimination (↔E)** From 'P↔Q' and 'P,' we can derive 'Q.' And from 'P↔Q' and 'Q,' we can derive 'P.'		P↔Q P Q P↔Q Q P	 ↔E ↔E

12	**Modus Tollens (MT)** From 'P→Q' and '¬Q,' we can derive '¬P.'		P→Q ¬Q ¬P	MT
13	**Disjunctive Syllogism (DS)** From 'P∨Q' and '¬Q,' we can derive 'P.'		P∨Q ¬Q P P∨Q ¬P Q	DS DS
14	**Hypothetical Syllogism (HS)** From 'P→Q' and 'Q→R,' we can derive 'P→R.'		P→Q Q→R P→R	HS
15	**Double Negation (DN)** From 'P,' we can derive '¬¬P.' From '¬¬P,' we can derive 'P.'		P⊣⊢¬¬P	DN
16	**De Morgan's Laws (DeM)** From '¬(P∨Q),' we can derive '¬P∧¬Q.' From '¬P∧¬Q,' we can derive '¬(P∨Q).' From '¬(P∧Q),' we can derive '¬P∨¬Q.' From '¬P∨¬Q,' we can derive '¬(P∧Q).'	¬(P∨Q)⊣⊢¬P∧¬Q ¬(P∧Q)⊣⊢¬P∨¬Q	DeM DeM
17	**Implication (IMP)** From 'P→Q,' we can derive '¬P∨Q.' From '¬P∨Q,' we can derive 'P→Q.'	P→Q⊣⊢¬P∨Q	IMP

Propositional Derivation Strategies

SP#1(E+)   First, eliminate any conjunctions with '∧E,' disjunctions with DS or '∨E,' conditionals with '→E' or MT, and biconditionals with '↔E.' Then, if necessary, use any necessary introduction rules to reach the desired conclusion.

SP#2(B)   First, work backward from the conclusion using introduction rules (e.g., '∧I,' '∨I,' '→I,' '↔I'). Then, use SP#1(E).

SP#3(EQ+)   Use DeM on any negated disjunctions or negated conjunctions, and then use SP#1(E). Use IMP on negated conditionals, then use DeM, and then use SP#1(E).

SA#1(P,¬Q)   If the conclusion is an atomic proposition (or a negated proposition), assume the negation of the proposition (or the non-negated form of the negated proposition), derive a contradiction, and then use '¬I' or '¬E.'

SA#2(→)   If the conclusion is a conditional, assume the antecedent, derive the consequent, and use '→I.'

SA#3(∧)   If the conclusion is a conjunction, you will need two steps. First, assume the negation of one of the conjuncts, derive a contradiction, and then use '¬I' or '¬E.' Second, in a separate subproof, assume the negation of the other conjunct, derive a contradiction, and then use '¬I' or '¬E.' From this point, a use of '∧I' will solve the proof.

SA#4(∨)   If the conclusion is a disjunction, assume the negation of the whole disjunction, derive a contradiction, and then use '¬I' or '¬E.'

PREDICATE LOGIC

Predicate Operators

Sentence	Symbolization	Type
Everything is **P**.	$(\forall x)\mathbf{P}x$	Universal quantifier
Something is **P**.	$(\exists x)\mathbf{P}x$	Existential quantifier

Four Decomposable Proposition Types

Existential	$(\exists x)\mathbf{P}$
Universal	$(\forall x)\mathbf{P}$
Negated existential	$\neg(\exists x)\mathbf{P}$
Negated universal	$\neg(\forall x)\mathbf{P}$

Predicate Tree Decomposition Rules

Negated Existential Decomposition (¬∃D)	Negated Universal Decomposition (¬∀D)
$\neg(\exists x)\mathbf{P}$✓ $(\forall x)\neg\mathbf{P}$	$\neg(\forall x)\mathbf{P}$✓ $(\exists x)\neg\mathbf{P}$
Existential Decomposition (∃D)	**Universal Decomposition (∀D)**
$(\exists x)\mathbf{P}$✓ $\mathbf{P}(a/x)$ where 'a' is an individual constant (name) that does not previously occur in the branch.	$(\forall x)\mathbf{P}$ $\mathbf{P}(a \ldots v/x)$ Consistently replace every bound x with any individual constant (name) of your choosing (even if it already occurs in an open branch) under any (not necessarily both) open branch of your choosing.

Predicate Tree Decomposition Strategies

1   Use no more rules than needed.
2   Decompose negated quantified expressions and existentially quantified expressions first.
3   Use rules that close branches.
4   Use stacking rules before branching rules.
5   When decomposing universally quantified propositions, it is a good idea to use constants that already occur in the branch.
6   Decompose more complex propositions before simpler propositions.

Predicate Derivation Rules

1	**Universal Elimination ($\forall$E)** From any universally quantified proposition '($\forall$x) **P**,' we can derive a substitution instance '**P**(a/x)' in which all bound variables are consistently replaced with any individual constant (name).	($\forall$x)**P** **P**(a/x)	$\forall$E
2	**Existential Introduction ($\exists$I)** From any possible substitution instance '**P**(a/x),' an existentially quantified proposition '($\exists$x)**P**' can be derived by consistently replacing at least one individual constant with an existentially quantified variable.	**P**(a/x) ($\exists$x)**P**	$\exists$I
3	**Universal Introduction ($\forall$I)** A universally quantified proposition '($\forall$x)**P**' can be derived from a possible substitution instance '**P**(a/x)' provided (1) 'a' does not occur as a premise or as an assumption in an open subproof, and (2) 'a' does not occur in '($\forall$x)**P**.'	**P**(a/x) ($\forall$x)**P**	$\forall$I
4	**Existential Elimination ($\exists$E)** From an existentially quantified expression '($\exists$x) **P**,' an expression '**Q**' can be derived from the derivation of an assumed substitution instance '**P**(a/x)' of '($\exists$x)**P**' provided (1) the individuating constant 'a' does not occur in any premise or in an active proof (or subproof) prior to its arbitrary introduction in the assumption '**P**(a/x),' and (2) the individuating constant 'a' does not occur in proposition '**Q**' discharged from the subproof.	($\exists$x)**P** $\quad$ **P**(a/x) $\quad$. $\quad$. $\quad$. $\quad$ **Q** **Q**	A $\exists$E

| 5 | **Quantifier Negation (QN)**
From a negated universally quantified expression '¬(∀x)**P**,' an existentially quantified expression '(∃x)¬**P**' can be derived, and vice versa. Also, from a negated existentially quantified expression '¬(∃x)**P**,' a universally quantified expression '(∀x)¬**P**' can be inferred, and vice versa. | ¬(∀x)**P** ⊣ ⊢ (∃x)¬**P**
¬(∃x)**P** ⊣ ⊢ (∀x)¬**P** | QN
QN |

Predicate Derivation Strategies

SQ#1(∀E)   When using (∀E), the choice of substitution instances '**P**(a/x)' should be guided by the object constants already occurring in the proof and any object constants occurring in the conclusion.

SQ#2(∃I)   When using (∃I), aim at deriving a substitution instance '**P**(a/x)' such that a use of (∃I) will result in the desired conclusion.

SQ#3(∀I)   When the goal proposition is a universally quantified proposition '(∀x)**P**,' derive a substitution instance '**P**(a/x)' such that a use of (∀I) will result in the desired conclusion.

SQ#4(∃E)   Generally, when deciding on a substitution instance '**P**(a/x)' to assume for a use of (∃E), choose one that is foreign to the proof.

Further Reading

PHILOSOPHY OF LOGIC AND PHILOSOPHICAL LOGIC

Logic plays an important role in philosophical theorizing and is itself a topic of philosophical reflection. *Philosophical logic* is sometimes characterized as philosophy informed by and sensitive to the work done in logic. This makes philosophical logic an extremely broad discipline concerned with a variety of different linguistic, metaphysical, epistemological, or even ethical problems. In contrast, *philosophy of logic* is philosophy about logic; that is, it is an area of philosophy that takes logic as its primary object of inquiry.

Goldstein, Laurence. 2005. *Logic: Key Concepts in Philosophy*. New York: Continuum.
Grayling, A. C. 1997. *An Introduction to Philosophical Logic.* Malden, MA: Blackwell Press.
Haack, Susan. 1978. *Philosophy of Logics*. Cambridge: Cambridge University Press.
Jacquette, Dale, ed. 2002. *Philosophy of Logic: An Anthology.* Malden, MA: Blackwell Publishers.
———, ed. 2005. *A Companion to Philosophical Logic*. Malden, MA: Wiley-Blackwell.
Sider, Theodore. 2010. *Logic for Philosophy*. Oxford: Oxford University Press.
Wolfram, Sybil. 1989. *Philosophical Logic: An Introduction*. New York: Routledge.

MODAL LOGIC

Modal languages and syntax can be viewed as an extension of the language and syntax of propositional and/or predicate logic. This type of logic aims to provide a syntax, semantics, and deductive system for languages that involve modal expressions like *can, may, possible, must,* and *necessarily.* So, in continuing your study of logic, looking at an introductory modal logic textbook can be a great place to start.

Beall, J. C., and Bas C. van Fraassen. 2003. *Possibilities and Paradox*. Oxford: Oxford University Press.
Bell, John L., David DeVidi, and Graham Solomon. 2007. *Logical Options: An Introduction to Classical and Alternative Logics*. Orchard Park, NY: Broadview Press.

Haack, Susan. 1978. "Modal Logic." In *Philosophy of Logics*, 170–203. Cambridge: Cambridge University Press.

Hughes, G. E., and M. J. Cresswell. 1996. *A New Introduction to Modal Logic*. London and New York: Routledge.

Nolt, John. 1997. *Logics*. Belmont, CA: Wadsworth Publishing Company.

Priest, Graham. 2008. *An Introduction to Non-Classical Logic: From If to Is*. 2nd ed. Cambridge: Cambridge University Press.

NONCLASSICAL LOGICS, DEVIANT LOGICS, AND FREE LOGIC

The semantics articulated in this book is known as *classical semantics*. One crucial feature of classical semantics is that it operates under the assumption that every proposition is either true or false (not both and not neither), also known as the *principle of bivalence*. *Nonclassical logics* are logical systems that reject the principle of bivalence and develop a logical system built around a semantics involving more truth values than *true* and *false* (e.g., *indeterminate*). In addition to revising the semantics of propositional or predicate logic, logicians also take issue with certain derivation or inference rules (e.g., negation introduction or elimination) and suggest a more restrictive or alternative system of derivation rules. Such logics are sometimes known as *deviant logics*. Lastly, you may have noticed that the following derivation is valid in RD: Pa ⊢ (∃z)Pz. This is because the semantics of RL is such that every object constant (name) denotes a member of the domain of discourse D. What about names that do not have referents or refer to objects that do not exist? *Free logic* is a logic where individual constants may either fail to denote an existing object or denote objects outside D. This logic is useful for analyzing discourse where names do not refer to existing entities (e.g., works of fiction).

Bencivenga, Ermanno. 1986. "Free Logics." In *Handbook of Philosophical Logic*, edited by D. Gabbay and F. Guenthner, 373–42. Dordrecht: D. Reidel.

Bergmann, Merrie. 2008. *An Introduction to Many-Valued and Fuzzy Logic*. Cambridge: Cambridge University Press.

Haack, Susan. 1996. *Deviant Logic, Fuzzy Logic: Beyond the Formalism*. Chicago: University of Chicago Press.

Lambert, Karel. 2003. *Free Logic: Selected Essays*. Cambridge: Cambridge University Press.

Keefe, Rosanna. 2000. *Theories of Vagueness*. Cambridge: Cambridge University Press.

Priest, Graham. 2002. "Paraconsistent Logic." In *Handbook of Philosophical Logic*, edited by D. Gabbay and F. Guenthner, 287–393. Dordrecht: Kluwer Academic Publishers.

Williamson, Timothy. 1994. *Vagueness*. London and New York: Routledge.

NONDEDUCTIVE LOGICS

In this text, the focus has been on deductively valid arguments. But *symbolic logic*—a branch of logic that represents how we ought to reason by using a formal language consisting of abstract symbols—is not confined only to valid arguments. Logics have been developed for nondeductive arguments. That is, arguments where the truth of the premises does not guarantee the truth of the conclusion. These logics, known as *nondeductive* (or *inductive*) *logics*, range from those that deal with probability, how to make rational decisions under ignorance (decision theory), to those that deal with rational interaction between groups of individuals (game theory).

Hacking, Ian. 2001. *An Introduction to Probability and Inductive Logic*. Cambridge: Cambridge University Press.

Maher, Patrick. 1993. *Betting on Theories*. Cambridge: Cambridge University Press.

Resnik, Michael. *Choices: An Introduction to Decision Theory*. Minneapolis: University of Minnesota Press

Skyrms, Brian. 1966. *Choice and Chance: An Introduction to Inductive Logic*. Belmont, CA: Dickenson Publishing Company.

HIGHER-ORDER LOGIC

In this text, we have focused primary on *first-order logic*. In first-order logics, quantifiers range over individual objects. A *higher-order logic* is a system of logic in which quantifiers range over a variety of different types of objects (e.g., propositions by quantifying propositional variables or properties expressed by quantifying n-place predicate variables).

Bell, John L., David DeVidi, and Graham Solomon. 2007. *Logical Options: An Introduction to Classical and Alternative Logics*. Orchard Park, NY: Broadview Press.

Nolt, John. 1997. *Logics*. Belmont, CA: Wadsworth Publishing Company.

Van Benthem, Johan, and Kees Doets. 1983. "Higher-Order Logic." In *Handbook of Philosophical Logic*, edited by D. Gabbay and F. Guenthner, 275–329. Dordrecht: D. Reidel.

HISTORY OF LOGIC

Logic has a long history, from Aristotle's (384–322 BC) creation of the first logical system, through Leibniz's (AD 1646–1716) attempts at replacing all scientific thinking with a universal logical calculus, to modern-day research in logic by mathematicians and philosophers.

Adamson, Robert. 1911. *Short History of Logic*. Edinburgh: W. Blackwood and Sons.

Gabbay, Dov M., and John Woods, eds. 2004. *Handbook of the History of Logic*. Amsterdam: Elsevier North Holland.

Gensler, Harry J. 2006. *Historical Dictionary of Logic*. Lanham, MD: Scarecrow Press.

Haaparanta, Leila, ed. 2009. *The Development of Modern Logic*. Oxford: Oxford University Press.

King, Peter, and Stewart Shapiro. 1995. "The History of Logic." In *The Oxford Companion to Philosophy*, edited by Ted Honderich, 495–500. Ipswich, MA: Oxford University Press.

Kneale, William, and Martha Kneale. 1962. *The Development of Logic*. Oxford: Clarendon Press.

Prior, Arthur N. 1955. *Formal Logic*. Oxford: Clarendon Press.

LOGICS WITH ALTERNATIVE NOTATION

There are a number of alternative symbolizations of propositional and predicate logic. Two examples are Polish (or prenix) notation and Iconic (or diagrammatic) notation. In Polish notation (developed by Jan Łukasiewicz) truth-functional operators are placed before the propositions to which they apply. Instead of symbolizing 'p and q' as 'P∧Q,' in Polish notation 'p and q' is symbolized as 'Kpq,' where 'K' is the truth-functional operator for conjunction. Although not widely used in logic, prefixing and postfixing logical operators are often used in computer programming languages, specifically one of the oldest, LISP. Early efforts to represent expressions and conceptual relationships between propositions are found in Euler and Venn diagrams, but these notations lacked an accompanying proof system. One form of visual logic (developed by Charles S. Peirce) that overcame this feature is known as *existential graphs*. Instead of symbolizing 'not p and not q' using specific symbols for truth-functional operators, Peirce proposed various conventions, propositional letters, and the use of circles that circumscribe letters to represent the same proposition. That is, 'not not-p and not-q,' represented as '¬(¬P∧¬R)' in PL, is represented as follows in Peirce's existential graphs:

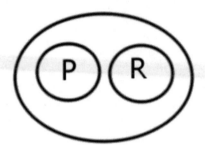

Roberts, Don D. 1973. *The Existential Graphs of Charles S. Peirce*. The Hague: Mouton.

Shin, Sun-Joo. 2002. *The Iconic Logic of Peirce's Graphs*. Cambridge, MA: MIT Press.

Sowa, John F. 1993. "Relating Diagrams to Logic." In *Lecture Notes in Artificial Intelligence*, edited by Guy W. Mineau, Bernard Moulin, and John F. Sowa, 1–35. Berlin: Springer-Verlag.

Woleński, Jan. 2003. "The Achievements of the Polish School of Logic." In *The Cambridge History of Philosophy, 1870–1945*, edited by Thomas Baldwin, 401–16. Cambridge: Cambridge University Press.

Index

About the Author

David W. Agler, Ph.D., is a lecturer in philosophy at The Pennsylvania State University. His research and publications focus primarily on topics in the philosophy of language and figures in classical American philosophy, especially the work of Charles S. Peirce.